Contribution of Vertical Farms to Increase the Overall Energy Efficiency of Cities

Mag.Arch.Daniel Podmirseg

„CONTRIBUTION OF VERTICAL FARMS TO INCREASE THE OVERALL ENERGY EFFICIENCY OF CITIES"

DOCTORAL THESIS

Presented in fulfillment of the doctoral thesis requirements for the degree of

Doctor of Technical Sciences

Academic Field
Architecture

Graz University of Technology

Academic advisor:
Univ. Prof. Brian Cody BSc(Eng) Hons CEng MCIBSE
Institute for Buildings and Energy

Graz, November 3rd 2015

Bibliografische Information der Deutschen Nationalbibliothek
Die Deutsche Nationalbibliothek verzeichnet diese Publikation in der
Deutschen Nationalbibliografie; detaillierte bibliografische Daten sind im Internet
über http://dnb.d-nb.de abrufbar.
1. Aufl. - Göttingen: Cuvillier, 2016
   Zugl.: Graz, Univ., Diss., 2015

© CUVILLIER VERLAG, Göttingen 2016
   Nonnenstieg 8, 37075 Göttingen
   Telefon: 0551-54724-0
   Telefax: 0551-54724-21
   www.cuvillier.de

Alle Rechte vorbehalten. Ohne ausdrückliche Genehmigung des Verlages ist
es nicht gestattet, das Buch oder Teile daraus auf fotomechanischem Weg
(Fotokopie, Mikrokopie) zu vervielfältigen.
1. Auflage, 2016
Gedruckt auf umweltfreundlichem, säurefreiem Papier aus nachhaltiger Forstwirtschaft.

   ISBN 978-3-7369-9305-1

up!

CONTRIBUTION OF VERTICAL FARMS TO INCREASE THE OVERALL ENERGY EFFICIENCY OF CITIES

Graz University of Technology

Mag. Arch. Daniel Podmirseg

Academic advisor:
Univ. Prof. Brian Cody BSc(Eng) Hons CEng MCIBSE
Institute for Buildings and Energy
Graz University of Technology

External Expert:
Univ. Prof. Dr. Nirmal Kishnani
Department of Architecture
National University of Singapore

External Experts and consultants:
Univ.Prof. Dipl.-Ing.sc.agr. Dr.sc.agr.Prof. Anna Keutgen
Ass.Prof. Dipl.-Ing. Dr.nat.tech. Univ.Ass. Johannes Balas
Institute for Plant Sciences
University of Natural Resources and Life Sciences, Vienna

Lectured by **y'plus**

© 2015 Copyright: Daniel Podmirseg

Beschluss der Curricula-Kommussion für Bachelor-, Master- und Diplomstudien vom 10.11.2008
Genehmigung des Senates am 1.12.2008

Deutsche Fassung:

Ich erkläre Eides statt, dass ich die vorliegende Arbeit selbstständig verfasst, andere als die angegebenen Quellen / Hilfsmittel nicht benutzt, und die den benutzten Quellen wörtlich und inhaltlich entnommenen Stellen als solche kenntlich gemacht habe.

Graz, am 03.11.2015

English Version:

I declare that I have authored this thesis independently, that I have not used other than the declared sources / resources, and that I have explicitly marked all material which has been quoted either literally or by content from the used sources.

Graz, November 3rd 2015

food for the next generations

For you, Sophia.

"How inappropriate to call this planet ‚Earth', when it is clearly ‚Ocean'."

Arthur C. Clarke

"The thesis, "UP"!, presented by Dr. Daniel Podmirseg, as partial fulfillment for his doctoral thesis, compares and contrasts in detail the global input (land use, energy, water, etc.) needed to feed 9.4 billion people using traditional and non-traditional farming technologies.

It further examines the impact of these two agricultural approaches from a climate, social, and political perspective. The analysis is thorough and decisive, coming down strongly in favor of alternative agricultural strategies, namely vertical farming, to help solve the impending crisis of population growth, ultra-urbanization, climate change, crop failures, and meeting basic human needs.

Conclusions are based on a detailed examination of what it takes to operate traditional farms compared with existing vertical farms and the outcome of those two efforts with regards to yields and energetics (indoor farming using LED grow lights vs. outdoor farming's use of fossil fuels). In each section, an extensive literature review is presented that provides data on the issues of land use, energy needed for a given crop (mostly focusing on tomatoes), and the social benefits of vertical farming in urban centers.
Basic metrics for the daily needs of an average male and female adult human being (calories consumed, energy needs for each organ system, etc.) are listed, as well as global overall needs for the entire human population. The author concludes that each human needs around 2,500 m2 on a yearly basis.
This places a huge amount of responsibility on the global food "system" to provide enough calories per year for each person. He further examines the metrics involved in food production from an energy use perspective and concludes that if things do not change, there will be food crises and energy shortages, as demand goes up over the next 20-50 years.

In the final section, the author examines the realities of establishing a robust vertical farming industry worldwide, the metrics needed to do so, and the potential positive impact vertical farming could have on a variety of pressing issues; climate change, urbanization, energy use, health.

In conclusion, this exhaustive body of work presents the best view so far published on what the world would be like if alternative agricultural strategies were employed to feed some 9.6 billion people.

In the opinion of this reviewer, it is a tour de force blueprint for how to proceed into the next millennium, enabling humans to finally achieve a peaceful co-existence with the rest of nature."

Prof. Dickson Despommier
Emeritus professor of microbiology and Public Health at Columbia University, New York

„Dr. Podmirseg makes a timely and compelling case for the integration of food within the city limits, with the vertical farm as a means to that end.

A vertical farm, seen in various guises across the world, consists of vertically stacked surfaces for indoor food production. The technical challenge is how to grow anything under glass so that energy use - for lighting, heating or cooling - is kept low.

The study begins with an iterative and systematic study of envelope and growing conditions, seeking the sweet spot between energy demand, indoor light and temperature. Not surprisingly, soil-based farming fares better on the energy score card. The vertical farm however does better on the question of land use, reducing demand for agricultural land in rural and peri-urban areas.

The findings, seemingly inconclusive, are highly significant. The author's ability to bridge disciplinary knowledge – architecture and agriculture – is exemplary and an important step in the right direction. He leaves us with new questions on how a single intervention must bridge multiple systems and performance criteria.

With urbanisation at its current pace, the vertical farm seems inevitable. With the help of Dr. Podmirseg, we are one step closer to understanding how to get it right."

Prof. Nirmal Kishnani
Department of Architecture, National University of Singapore

Table of Contents

    Book review by Prof. Dickson Despommier 11

    Book review by Prof. Nirmal Kishnani 13

I. Preface 15

II. Acknowledgements 19

III. Synopsis 21

IV. Abstract 22

1. **Introduction** **35**

2. **Landuse, Biocapacity and Energy Consumption** **41**

2.1. Compiling a status quo model of traditional agriculture 41

2.2. Land requirement for daily coverage of essential nutritional value for the human body 43

    2.2.1 Food Balance Sheets 49

    2.2.2 World land masses and the ratio of land used for agricultural production . 57

    2.2.3 Biocapacity of the Earth 61

2.3. Energy Consumption 71

    2.3.1 Looking back a century 71

    2.3.2 Energy and the Food Supply Chain 75

    2.3.3 Sketching the Big Picture of World Energy Consumption by the Food System 93

    2.3.4 Landuse, Energy Consumption and Biocapacity - Resumée 95

3. **The Vertical Farm Reference Models** **99**

3.1. Goals and Process 99

3.2. Evaluation of components - Modelling a Vertical Farm - Prototype 101

    3.2.1 Cultivation Methods - Overview and Evaluation of the advantages and challenges 101

    3.2.2 Production Methods - Building footprint, cultivation area and soil-based-area-equivalent 111

3.3 Typological comparison of existing Vertical Farms 126

4. Vertical Farm - Substitution of Natural Growth Factors 163

4.1. Goals and methods 163

4.2. Introduction and basics 165
    4.2.1 Photometry and Radiometry................................ 165
    4.2.2 Light for Plants - Light for humans 170

4.3. The Tomato - *Lycopersicon esculentum* (Mill.)................ 179
    4.3.1 General Data - Origin and Distribution 179
    4.3.2 *L. esculentum* and Light 184
    4.3.2 *L. esculentum* and Photosynthesis 186
    4.3.4 Factors affecting the rate of Photosynthesis 187
    4.3.5 Temperature and other Growing Conditions 191
    4.3.5 Growing media and plant density 195

4.4. Setting up Vertical *L. esculentum* - Farm 199
    4.4.1 Light availability in Vienna and Greenhouse Practises 199
    4.4.2 Artificial Lighting - General Data 202
    4.4.3 Artificial Lighting - LED LumiGrow 325 and LumiGrow 650 .. 205
    4.4.4 *L. esculentum* - Organization of the vertical cultivation area 209

5. The Vertical Farm - Simulationmodel 215

5.1. Introduction 215

5.2. Goals 217

5.3. Method 219
    5.3.1 Simulation Software 219
    5.3.2 Location 221
    5.3.3 Parametric Simulation Building 225

5.4. Lighting Parameters 235
    5.4.1 Lighting demand and Lighting schedule 235

5.5. Heating Parameters 239
    5.5.1 EnergyPlus Material Parameters 239
    5.5.2 EnergyPlus Simulation Parameters 247
    5.5.3 Limitations and Restrictions 248

5.6. Lighting Demand - Simulation Results 249
    5.6.1 Annual Results 249
    5.6.2 Lighting Schedule 249
    5.6.3 VF 32 - Lighting Demand - Daysim Results 255
    5.6.4 VF 14 - Lighting Demand - Daysim Results 259
    5.6.5 VF 7 - Lighting Demand - Daysim Results 263

5.7. Heating Demand - Simulation Results 267
    5.7.1 VF 32 - Heating Demand 267
    5.7.2 VF 14 - Heating Demand 277
    5.7.3 VF 7 - Heating Demand 290

5.8. Annual Simulation Results - Discussion 297
    5.8.1 Total Primary Energy Supply for the Vertical Farms 303
    5.8.2 Vertical Farming and Land use 307
    5.8.3 Energy Land - Resumée 309

6. Conclusio 315

6.1. From Modernism to Sustainism 315

6.2. Vertical Farming and Energy Consumption 318

6.3. Vertical Farming and Land use 321

6.4. Final Considerations 323

V.   **List of Figures** 331

VI.   **List of Tables** 339

VII.   **List of Abbreviations** 343

VIII.   **Bibliography** 349

IX.   **Nomenclatura** 363

X.   **Appendix** 379

Impressum 391

I. Preface

I first came into contact with the Hubbert's curve and the Peak-Oil-theory in 2006. From then on, I proceeded to deepen my knowledge on this topic, initially by reading the publication of ASPO[1], the Peak-Oil-Protocol[2] and several books written by Colin Campbell, founder of ASPO and author of the protocol. I had the privilege of meeting him twice in Ireland. The awareness of the finite resource of hydrocarbon energy and the significance of the consumption of this source of energy, which has led to the total dependency of an entire culture, developed through modernism. This situation in turn prompted my decision to postulate "The Last Skyscraper of the 21st century" as a diploma at the Academy of Fine Arts in Vienna, with the skyscraper monumentally ignoring every reflection of scale and meaning, keeping 80% of its volume free from every use and re-establishing nature within the void.
It need not be said that finding arguments in support of this huge vertical green structure proved to be an intellectual hike to nowhere. Continuing discussions with Prof. Wolfgang Tschapeller on this topic, however, and his persistence in wishing me to find a plausible argument to keep the concept afloat, led me as an incidental, side-effect to the website of Prof. Dickson Despommier of Columbia University, where he published his first experimental works developed with his students all around the issue of Vertical Farming.

This idea exerted an immediate attraction on me. The sculptural dystopian monumentality of the developed skyscraper project turned overnight into an interest in what the actual potential of a vertical greenhouse could be and what it might offer both in an architectonic and an urban context. Furthermore, intensive research was called for to examine the extent to which Vertical Farming might actually make sense in practical terms. Both the city and the site for the diploma project were found immediately - London, Canary Wharf, on the banks of the River Thames. A DEFRA-study[3] came to the conclusion that over 80% of all food consumed in London is imported from abroad and the food footprint for this one city was equal to that for the whole of Sweden. Based on the statistics published on Prof. Dickson Despommier's website, a building was developed which would have the capability of feeding 1% of London's population by 2050. I am very grateful for the support and for the discussions I had with Prof. Dickson Despommier during the diploma work and during a meeting in Sweden.

---

1   Association for the Studies of Peak Oil and gas. http://www.aspo2012.at/
2   CAMPBELL, C. J. 1996. The Oil Depletion Protocol. Available: http://richardheinberg.com/odp/theprotocol.
3   WATKISS, P., SMITH, A., TWEEDLE, G., MCKINNON, A., BROWNE, M., HUNT, A., TRELEVEN, C., NASH, C. & CROSS, S. 2005. The Validity of Food Miles as an Indicator of Sustainable Development: Final report produced for DEFRA. AEA Technology.

The interest in Vertical Farms[1] and their challenges and potentials has been kept alive in the meantime. However, the necessity grew for deepening the approach to the question and critically seeking a raison d'être for this structural typology, which on a superficial analysis might turn out to consume a vast amount of energy by substituting artificial light for most of the sunlight needed for plant growth and also for providing ideal climate conditions.

> This controversy came to a head in the question of to what extent Vertical Farming could actually increase the overall energy efficiency of cities, or in other words: if the energy consumption of this typology could balance out all side effects of our actual world food system in terms of energy consumption.

I was conscious of the need here for in-depth knowledge inputs from many fields of expertise, ranging from detailed understanding of plant physiology to quantum physics, together with the need to acquire advanced knowledge in simulation processes in order to establish the value approximations for energy consumption assessments.

To cut a long story short, this ambitious scheme provided more than enough work for five years of work and to fill a book with additional contributions for this fascinating and highly controversial discussion.

---

1 DESPOMMIER, D. 2010. The Vertical Farm, New York, St. Martin's Press.

II. Acknowledgements

I would like to thank Prof. Brian Cody and Prof. Nirmal Kishnani for their support and contributions throughout the last few years.

Thanks to "mi moglie" Maria Huber, probably the only art historian (involuntarily) able to explain what PPFD is. Thank you for your support, your contribution on every imaginable level, and for your patience, as well as for your encouragement during the summer "we didn't have".

Markus Godez, Kathi Schegula, it must be said that you not only made this period possible in the first place, but very much easier, too. Your extraordinary hospitality must not only be gratefully acknowledged, but praised in its every detail as such a valuable contribution to my activity at the institute.

Special thanks to my external experts and consultants Prof. Anna Keutgen, Ass. Prof. Johannes Balas and Rita Kappert for your interest in my work and for introducing me to the fascination of plant physiology.

Prof. Anselm Wagner, Prof. Erwin Fiala, Prof. Hans Gangoly, Prof. Jörg Schütze and Ass. Prof. Peter Slepcevic-Zach from the Doctoral School at the TU Graz must be mentioned with gratitude for their remarkable contributions to the quality of this doctoral thesis. Thanks to all my colleagues at the Institute for Buildings and Energy, especially Sebastian Sautter, Wolfgang Löschnig, Eduard Petriu and Daniel Korwan. Thank you Doris Damm for having brought structure into my organization, time and again.

Special thanks to Ass. Prof. Abraham Yezioro from Israel Institute of Technology, Research Associate and Adjunct Professor at the University of Pennsylvania Mostapha Sadeghipour Roudsari and MIT-graduate Building Scientist Chris Mackey for supporting me so frequently during the period I was setting up the simulation process with Honeybee and EnergyPlus.

Ashley Veach and Jimmy Byrtus from Lumigrow must be mentioned for their generous support.

Thanks, Mathilde Podmirseg for your constant moral support and for financing the copies of the first edition. Thanks to my friends directly involved in the process of this work, Lucas Kulnig, Peter Franzelin (my host at 2,000 m at "Nuihitt"), the Zrim Clan, above all Philipp Zrim, Federico Rovetta, Tajda Potrc, Markus Godez, Helmut Holleis and Doris Steinacher for sharing VF-ideas with me through the years.

Lastly, I thank my voluntary lecturers, Maria Huber, Georg Holländer and Lucas Kulnig.

III. Synopsis

Vertical Farming has been an issue of controversial discussion since the publication of the manifesto by Dickson Despommier[1] . This doctoral thesis with the search for a raison d'être for Vertical Farming by sketching the current situation of world agriculture in terms of energy consumption, land use, potential and the consequences in increasing productivity on the actual agricultural land in use and also the potential increase of natural land conversion into agricultural land, exploiting the total biocapacity of the world.

Typologies and the cultivation- and production methods currently in use on existing Vertical Farms are compared, before proceeding to the development and analysis of the lighting- and heating demand for three specific Vertical Farm building types.

World total primary energy supply (TPES) in 2014 was around 550 Exajoule (EJ)[2]. A third of this is consumed by the food sector.[3] For every calorie we need to cover our daily energy requirement, we consume nearly six calories of total primary energy. One percent of the global landmass is defined as built-up land, where with the exception of a small percentage of indigenous populations, more than 7 billion people live. The area required to supply the world population with food is ten times higher. A food production network has been required for emerging and developed countries over the past few, which is completely dependent on hydrocarbon energy on a global scale.

The world population will continue to grow over the next decades, reaching a plateau in 2075 at 9.22 billion people before it starts to decline.[4] This work intends to contribute to the discussion on urban and Vertical Farming, aiming to find indicators for answering the question of to what extent Vertical Farming could actually increase the overall energy efficiency of cities.

---

1 DESPOMMIER, D. 2010. The Vertical Farm, New York, St. Martin's Press.
2 http://www.iea.org/publications/freepublications/publication/KeyWorld2013.pdf, retrieved 05.04.2014
3 FOOD AND AGRICULTURE ORGANIZATION OF THE UNITED NATIONS 2011.Energy-Smart Food for People and Climate, Issue Paper, Rome:FAO, p.10
4 http://www.un.org/esa/population/publications/longrange2/WorldPop2300final.pdf, p.1, retrieved 10.09.2015

IV. Abstract

Mag.Arch. Daniel Podmirseg
University of Technology, Institute for Buildings and Energy, Graz
Rechbauerstraße 12/III
8010 Graz, Austria

Academic Advisor: Prof. Brian Cody, Institute for Buildings and Energy, Graz University of Technology
External Expert: Prof. Nirmal Kishnani, Department of Architecture, National University of Singapore

Keywords: Vertical Farming, world agriculture, world population growth, land use, biocapacity, world energy consumption, fossil fuels, food production, *Lycopersicon Esculentum* (Mill.)

While currently still being constructed as prototypes and for research purposes, Vertical Farming facilities are nevertheless providing food for thought for architects everywhere. The purpose of this work is to answer the question as to what extent Vertical Farming can contribute to disburdening the current alarming situation in conventional soil based agriculture in terms of land use, and to sketch in the potentials of whether Vertical Farms have the capacity to increase the overall energy efficiency of cities.

As world population is expected to peak in 2075 with an estimated 9.22 billion people[1] and changes in diet[2] are most likely to be expected, especially in emerging countries, additional food production is needed to cover the total nutritional energy requirements of both humans and livestock. Potential exists on various levels here, e.g. by increasing productivity, or expanding the area for soil based agriculture. Biocapacity of the earth is adequate[3] for feeding future generations.

Does this mean that the raison d'être for Vertical Farming is shrinking and it is thus a lost cause? By no means. When the potentials are turned into practice, dramatic side effects are entailed on the energy and climatic levels. This work defines the potential of land use reduction and frames the impetus to what extent Vertical Farming actually can contribute to making cities more energy efficient.

---

1 http://www.un.org/esa/population/publications/longrange2/WorldPop2300final.pdf, p.1, retrieved 10.09.2015
2 KASTNER, T., IBARROLA RIVAS, M. J., KOCH, W. & NONHEBEL, S. 2012. Global changes in diets and the consequences for land requirements for food. Available: http://www.pnas.org/content/early/2012/04/10/1117054109.full.pdf+html.
3 FISCHER, G., VELTHUIZEN, H. V. & NACHTERGAELE, F. O. 2000. Global Agro-Ecological Zones Assessment: Methodology and Results. International Institute for Applied Systems Analysis, FAO. Executive Summary

FOOD AND ENERGY

World total primary energy supply (TPES) was around 550 Exajoule (EJ) in 2014.[1] A third of this energy is required by the food sector.[2] For every calorie we need to cover our daily nutritional energy requirement, we consume nearly six calories of primary energy. One percent of the global landmass is defined as built-up land, where, except for a small percentage of indigenous populations, more than 7 billion people live. The area required for cropland to supply the world population with food is ten times higher. A food production network which is completely dependent on hydrocarbon energy on a global scale has been required for emerging and developed countries over the past few decades.

This work is structured primarily in three parts: Chapter 2 and Chapter 3 investigate whether there is a raison d'être for Vertical Farming, or to put it in other words, if the necessity exists for developing additional production and cultivation methods within cities. Statistical analysis of different research results by the Food and Agriculture Organization, IIASA[3] and PNAS[4] are compared quantitatively for the purpose of sketching the consequences of changing current actions in the traditional world agriculture and attempting to define the limits for the biocapacity of the earth capable of use for food production.

Existing Vertical Farms were examined qualitatively in terms of food cultivation methods and compared by means of ratio assumption as to their potential to reduce the footprint of agricultural land, related to annual crop yield.

Part two correlates to Chapter 4 where parameters needed to substitute primary growth factors are defined, primarily concentrating on light and temperature demand for *Lycopersicon Esculentum* (Mill.). Based on these factors, lighting- and heating Schedules will be developed that serve the simulation model.

The third part, Chapter 5, includes three parametrically generated Vertical Farms which are compared on the basis of their energy consumption and capacity for reducing agricultural land use.

LAND USE, BIOCAPACITY AND ENERGY CONSUMPTION

Covering the total energy requirement of a sedentary male requires 11.3 MJ, or 8.82 MJ for a sedentary female.[5] Since human beings are heterotrophs, energy must first be captured from sunlight by plants either for direct human consumption or indirectly through use as feed for livestock. Considering cultural, and thus dietary differences, the size of the food footprint is different in every region of the world. We can claim that the higher the vegetal ratio in everyday diet, the lower the footprint tends to be.

In addition, Kastner et al.[6] identified three main drivers which influence the food footprint: population numbers, diet and the level of technological development. The main findings of this study are that the biggest driver in cropland expansion is not population growth, but socioeconomic development. What has been observed is that by increasing GDP population growth slows down, but the effects on dietary change still make increases essential in the area of food pro-

---

1 http://www.iea.org/publications/freepublications/publication/KeyWorld2013.pdf, retrieved 05.04.2014
2 FOOD AND AGRICULTURE ORGANIZATION OF THE UNITED NATIONS 2011.Energy-Smart Food for People and Climate, Issue Paper, Rome:FAO
3 International Institute for Applied Systems Analysis, Vienna
4 Proceedings of the National Academy of Sciences of the United States of America
5 http://ajcn.nutrition.org/content/51/2/241.abstract, retrieved 12.06.2014
6 KASTNER, T., IBARROLA RIVAS, M. J., KOCH, W. & NONHEBEL, S. 2012. Global changes in diets and the consequences for land requirements for food. Available: http://www.pnas.org/content/early/2012/04/10/1117054109.full.pdf+html.

duction. This means that by area and on a global scale, the average cropland needed for feeding a single person (food supply) is 1,732 m2/a.[1]

The agricultural land of 15,529,767 km2 (10% of the earth's land mass) produced more than 9.5 bn metric tons of primary products in 2011.[2] By assuming a per capita food supply of 900 kg/a which corresponds to FBS[3] of a European high GDP country and dividing it by the total primary production every person could be supplied with 1,399.80 kg of food annually. Enough food for all? Not so. Roughly 795,000,000 people are undernourished or suffer from hunger.[4] From the total primary production of animal feed (for a world livestock of 57,064,502,778 animals)[5], seeds, wastes, other (non-food/feed) uses and food manufacture has to be subtracted - with remaining 688,03 kg per person/a. In terms of calories only 55% of the global crops produced are consumed directly by humans. By theoretically eliminating every calorie which is lost from the food sector (both feed and biofuel production), an additional four billion people could be fed, enough to feed the expected world population by 2075.[6] Additional potential also exists in changing diet, although it is very unlikely that policies in this area will be supported by social acceptance. The trends clearly go in a quite different direction.

Research findings on the global agro-ecological zones (GAEZ) and the biocapacity of the world estimate that agricultural land could be more than doubled to exploit all land that is very suitable, or at least suitable for agricultural production.[7] Natural land, of which over 40% is currently covered by forests, would thus need to be converted into arable land. This is a scenario that is not desirable for two reasons: the vast CO_2 release from slash-and-burn-practices and the loss of natural habitats.

Productivity could be intensified on the land we already cultivate. By learning from history in this and by looking back at the 20th century we see that to increase yield by 600% between 1900 to 2000 energy subsidies had to increase 8,500%.[8] The energy dependency of conventional soil based agriculture would most likely continue to increase further along the same projection track by continuing to follow this policy.

Some 32% of global energy demand is currently used by the food sector, whereas 24% is consumed until the farm gate. 14% is used for transportation and distribution and twice this value for food processing.[9] The difference of the total 176 EJ TPES is consumed by retail, for preparation and cooking. Agriculture is dependent on hydrocarbon energy, from production of macronutrients to the global transportation network, which is almost entirely petroleum driven. Food prices are thus strongly related to oil prices. Their fluctuations, primarily in the developing countries, have a negative impact and endanger global food security and threaten inequality of distribution.

---

1. KASTNER, T., IBARROLA RIVAS, M. J., KOCH, W. & NONHEBEL, S. 2012. Global changes in diets and the consequences for land requirements for food. Available: http://www.pnas.org/content/early/2012/04/10/1117054109.full.pdf+html. p.2
2. http://faostat3.fao.org/download/Q/*/E
3. Food Balance Sheets, http://faostat.fao.org/site/354/default.aspx, retrieved 14.09.2015
4. http://www.fao.org/docrep/018/i3434e/i3434e.pdf, p.4 retrieved 13.08.2015
5. faostat3.fao.org/ live animals, 2011, retrieved 28.08.2015
6. CASSIDY S. EMILY, WEST C. PAUL, GERBER S JAMES and JOLEY A JONATHAN, 2013. Redefining agricultural yields: from tonnes to people nourished per hectare. Environmental Research Letters, IOP PUblishing, p.1, p.4
7. FISCHER, G., VELTHUIZEN, H. V. & NACHTERGAELE, F. O. 2000. Global Agro-Ecological Zones Assessment: Methodology and Results. International Institute for Applied Systems Analysis, FAO
8. SMIL, V. 2008. Energy in Nature and Society, Cambridge, Mass., MIT Press., p.304
9. FOOD AND AGRICULTURE ORGANIZATION OF THE UNITED NATIONS 2011.Energy-Smart Food for People and Climate, Issue Paper, Rome:FAO

If Vertical Farming has the capacity to disburden the current situation of the world agricultural system, primarily through reduction in land use and energy consumption, this structural typology could well be worth considerable further investigation. Before setting up a Vertical Farm simulation model, an investigation of the indicators for vertical greenhouses that are already built and operational is recommended.

THE VERTICAL FARM REFERENCE MODELS

Four verticalized cultivation methods are compared for estimating the actual potential in land reduction for agricultural production. Four prototypical Vertical Farms are selected with different production methods and the same cultivation methods (hydroponics).

A unit established 120 m2 at „Paignton Zoo" Devon, UK, in 2009, has horizontally rotating elements and produces leafy vegetables for the zoo animals. The building footprint is 144.45 m2, the cultivation area 388.32m2. Comparing the annual yield with soil based agriculture only 9.09% of the required soil based area is used. The soil based equivalent would be some 1,580 m2.

The horizontal conveyor system enables equal light exposure to the stacked vegetables.

A Vertical Farm with a climatically induced short period for plant growth and a combination of horizontally static layers and vertically rotating elements for fresh vegetable and herbs production has recently been established in Jackson, Wyoming, USA. An annual yield is expected within the greenhouse volume of the building (roughly 2,000 m3), which corresponds to more than 1.5 ha. The building „Vertical Harvest" footprint is 488.44 m2, a reduction of the food footprint compared to conventional agriculture of nearly 97%.

SkyGreens in Singapore implemented vertically rotating elements. On the principle of a classical greenhouse, combined with this technique the building height can be expanded, the rotation enables equal light distribution to the plants throughout the day. The salad production on a building footprint of 196.16 m2 achieves an annual yield where 2,369.15m2 would be needed, a reduction of nearly 92%.

Lastly, the most promising Vertical Farm both in terms of production method used and the ambition of developing a new typology, is Plantagon's Vertical Farm for Linköping in Sweden, which had its ground breaking ceremony by the year 2012, and referring to information provided during the „Urban Agriculture Summit" in Linköping two years ago, should in all probability be built within the next years.

The production is done on a 3D-conveyor belt where seedlings are planted at the top of the spiral and move down to the ground floor level throughout the crop rotation when ready to harvest. From an architectural perspective it should be mentioned that an office building is situated on the north side of the productive greenhouse, enabling synergy potentials in terms of energy flows, oxygen- and CO_2-cycles.

Pak choi is produced in a vertical greenhouse volume of 15,003 m2 on a building footprint roughly of 1,000 m2. The annual yield of this Vertical Farm reaches an estimated quantity for which over 8 ha would be needed if it were to be produced conventionally. A yield on an area corresponding to only 1.18% of that require for soil based agriculture.

The right choice of crop type combined with the appropriate production method can drastically reduce land use for food production. But the question still to answer is at what energy cost? All these listed reference models were mostly transparent on the top level, with some reduction taking place „Vertical Harvest". What potential do Vertical Farms currently have for crop production in a stacked greenhouse? And to increase the challenge, what if we produce crops with a high light demand? To answer these question a clearer picture needs to be drawn on what growth factors must be established within a building to establish ideal conditions.

SUBSTITUTION OF NATURAL GROWTH FACTORS

Greenhouses have been established, largely in temperate zones, since the 17th century. They are used for growing more sensitive plants and also to produce crops. Crops in greenhouses were planted mainly to enlarge the crop rotation scope and to make fresh food available over a longer period of time.

Greenhouses are now established everywhere around the world and now cover an area of some 4,000 km2 worldwide[1], although this area is very likely a significant underestimation by the FAO of the real greenhouse area now in use. High-tech-greenhouses not only boost the crop rotation, but also offer a means to benefit from the greenhouse effect, since the photoperiod throughout the day has now been extended around the world and the conversion of light into sugar is the key for food production.

Plants need a specific part of the electromagnetic spectrum for photosynthesis. From 400-700 nm light affects photosynthesis. The ratio of the total spectrum is thus termed PAR, or photosynthetic active radiation. This photosynthetically active radiation is the waveband 400 to 700 nm, this being the wavelength limitations that are of primary importance for plant photosynthesis. The PPFD, photosynthetic photon flux density is the number of photons in the PAR waveband that are incidental on a surface in a given time period ($\mu mol/m-2/s-1$). The quantum sensor will measure this value."[2] To set up the simulation parameters, requirements for greenhouse tomatoes, Lycoper sicon esculentum (Mill.), was chosen. This cultivar has a high requirement in terms of light and a relatively high requirement in terms of temperature.

THE VERTICAL FARM - SIMULATION MODEL

The location for the simulation model is Vienna, Austria with 4,401 daylight hours, 43% of them are sunshine hours. The annual total solar horizontal radiation is 1,119.32 kWh/m2 which corresponds to 559.66 kWh/m2 PAR. 263.62 kWh/m2/a PAR is the lighting demand for L. esculentum through the sigmoidal growing curve.

Three different building types are parametrically generated and compared. The volume, is oriented to the volume of the Vertical Farm planned in Linköping, Sweden.
The volume of each VF is oriented to 15,000 m3. The dimensions: VF7 (36m x 7.2m x 61m), VF14 (36m x 14.4m x 33m) and VF32 (36m x 32m x 12m). All farms get simulated with three different building envelopes: Single glazing (U-value= 5.88 W/m2/K, VT=0.85, SHGC=0.8), Double-ETFE (U-value= 2.90W/m2/K, VT=0.85, SHGC=0.65) and double-glazing (U-value= 1.70 W/m2/K, VT=0.91, SHGC=0.7). L. esculentum will obtain daylight through the facade, the difference to the DLI needed will be supplied by LED -lighting (Lumigrow 325PRO).

On top of the building, if DLI exceeds the needed value, LED will be turned off the whole day. At all other level, without excess light, LEDs will be turned on to cover 57,600 seconds or 16 hours of photoperiod. Ventilation and infiltration is not considered. Key findings are that Vertical Farms, developed with intermediate levels as stacked greenhouses, connected to a conventional energy grid and producing crops with high lighting- and heating demand in temperate climate zones can't compete with nowaday's practise of soil based agriculture.

---

1 FAO Good agriclutural practices for greenhouse vegetable crops.pdf, p.9
2 GIACOMELLI, G. 1998. Components of Radiation Defined: Definition of Units, Measuring Radiation Transmission, Sensors. CCEA, Center for Controlled Environment Agriculture, rutgers university, Cook College.

Low lighting demand show VF32 with its compactness which led to the biggest rooftop surface, where a third of the cultivars gain daylight throughout the whole dayhours. Maximizing all facades to all cardinal directions (nearly) equally, also positively influences the relatively low lighting demand. The fact that 1,144 m2 (0C= 572 m2 and 1C = 572m2), which is 33.10% of the cultivation area, are offset by 5m from the facade and therefor has the maximum lighting demand, still makes results comparable to VF7.

Compactness, activation of the top level for cultivation and optimizing the building orientation towards the sunpath seems to be the recommended way for following studies to optimize the building shape for Vertical Farming.

The difference of the results to VF7 only are around 2.4% (SG) to 1.5% (DG). VF7 with its minimized building depth might also be worth to be investigated more deeply for future Vertical Farm building typology studies. The building depth of 7.2m and south orientation has the lowest light requirement of all three Vertical Farm building types analyzed. Although, through its highest A/V ratio of 0.36 heating demand is the highest, this picture doesn't add up in the moment when the values, shown on this pages, get changed from end energy use to total primary energy supply (TPES), visualized on the next pages.

TOTAL PRIMARY ENERGY DEMAND AND LAND USE OF VERTICAL FARMS

In terms of building types we see a strong difference in energy consumption. Whereas lighting demand is strongly dependent from the building type, heating demand is more influenced by the building envelope. Theoretical crops with lower lighting and heating demand in ratio, though, have a stronger impact in reducing TPES[1] than an optimized building envelope or the building type.

> The simulation results of the different building types show that a careful followed design strategy for Vertical Farms can reduce the energy consumption up to 800%.

In numerical terms encapsulating the simulation results, a Vertical Farm with some 15,000m3 within a temperate zone, must envisage a TPES of 376.56 kWh/m2/a, with three quarters of this related to lighting demand (353.65 kWh/m2/a), 22.91 kWh/m2/a for heating for crops with high light requirement and relatively high temperature needs. This leads to CO_2-emissions of 311.17 t/a or 0.51 kg CO_2/kg L. esculentum.

By considering these values we see that vertical production is more energy intense than the actual practise in world agriculture. Around 1.50 GWh/a (400 kWh/m2/a TPES) per square meter are needed for annual production of L. esculentum. The actual world average of energy supply for the food sector per square meter agricultural land is 11.73 MJ/m2/a or 3.25 kWh/m2/a. Subtracting the energy for retail, preparation and cooking, this number is reduced to 7.80 MJ/m2 or 2.16 kWh/m2/a.[2]

---

[1] We assume the ratio that 24% of TPES of the food sector is related to the energy consumption until the farmgate. See Chapter 2.
[2] FOOD AND AGRICULTURE ORGANIZATION OF THE UNITED NATIONS 2011.Energy-Smart Food for People and Climate, Issue Paper, Rome:FAO

The effect on reducing land use for agricultural production based on the upmentioned simulation models and considering the assumptions of other Vertical Farms draws a clear picture: Land use can be reduced up to 50 times comparing the cultivation area of the production entity to the alternatively needed area for traditional soil based agriculture. More precisely, depending on the building types VF32 uses 1/10th of SBA-area, VF14 1/25th and VF7 uses a ratio of 1/53 compared to SBA. Compared to traditional greenhouse practises, VF 32's ratio is 1/6.5, for VF14 1/16 and VF7 1/33.

The advantage of land set free by optimized cultivation practises and stacking principle, though, with high-energy requiring crops, can be canceled out by adding to calculation the area needed to cover the energy demand with renewable energy.

The thesis, though, also reveals potentials which could make Vertical Farming competitive with nowaday's agriculture practise: Beyond adapting an intelligent energy concept though the following decisions (meant as future fields of research) can reduce the energy demand for Vertical Farms:

- Optimization of the building shape [1]

- Sunlight analysis, daylight availability and solar heat gain within the Vertical Farm zones are the decision making factors which crop type will be cultivated throughout the year or shorter crop rotations will be defined to adopt products to the specific seasonal conditions.[2]

- Requirement of light vary strongly from crop to crop. (*L. esculentum* has been chosen within this dissertation because it has the highest light requirement of all our food items.)[1] Results clearly picture expectable TPES on the top of the scale.

The aim therefor for future Form Follows Energy¹-studies for Vertical Farms is to optimize the building shape to reduce TPES, secondly reading urban food production as a structural entity of the urban system, and furthermore the city as an ecosystem. "The one characteristic they all [Ecosystems, Ed.] share is that primary productivity (the total mass of plants produced over a year in a given geographically defined region) is limited by the total amount of energy received and processed."²

The strategic design decisions have shown that energy consumption can be reduced significantly. Furthermore, seeing the Vertical Farm as a structural element of a system, within an urban context, enables the potential to reduce the overall energy consumption by activating synergy potentials with other programmes and functions of the city.

---

1 CODY, B. 2012. „Form follows Energy - Beziehungen zwischen Form und Energie in der Architektur und Urban Design, DBZ Deutsche BauZeitschrift, Bauverlag BV GmbH, Gütersloh. p.211 ff.

2 By adapting Harald von Witzke's postulate that each region (in our case ‚zone' „produces the food most appropriate to that region at a relatively low, affordable cost, and these products are subsequently made available to the market (...). WITZKE, H. V. 2011. Bananas from Bavaria?, Augsburg, Ölbaum-Verlag., p.9

1. Introduction

Vertical Farming is defined as a highly industrialized year round cultivation method for food production, adaptable for multiple crop types, where the verticalized building typology, its programme and functions primarily focus on optimum plant growth. The building is seen as a structural element of the urban ecosystem. In addition to food production, the Vertical Farm must incorporate elements of the food sector which, at present, are spatially detached from each other on a global scale, something which has a severe impact on energy consumption and the environment. Form Follows Energy[1] for Vertical Farms in three ways: to grant optimum growing conditions for crops, optimized to follow the position of the sun all year round and guaranteeing energy flows, meaning phenomenologically, on the ground level of the city, or preferably, also vertically for public use. Primarily the development of the building itself must follow two main goals: Increasing the overall energy efficiency of a city and also attempting to bring about a considerable reduction in land use, as a result of the favorable comparison between vertically achieved yield and traditional agricultural practices.

Vertical Farming is a subject of controversial discussion. Throughout my last exhibitions, lectures and public presentations, the boundless fascination this theme unleashes among some audiences is as notable as the emphatic refusal it provokes from others. The typology of the Vertical Farm has deepened considerably since my diploma at the Academy of Fine Arts in Vienna, supported by Prof. Markus Schäfer. This progress is primarily due to the potential of the Vertical Farm to re-establish local social and economic interdependencies within the city on the one hand, while on the other, even if it does not seem to be the full solution at first sight, it at least presents a partial opportunity to relieve the burden on the current situation of conventional world agriculture practices and the dependency of the urban population on it.

> Energy consumption, soil erosion, the conversion of natural land for farming use, especially using slash-and-burn methods in rain forests to make additional arable land available, i.e. the Neolithic Revolution, probably the biggest revolution of humankind in which hunters and gatherers became farmers, is now turning into a dystopia.

---

1   CODY, B. 2012. „Form follows Energy - Beziehungen zwischen Form und Energie in der Architektur und Urban Design, DBZ Deutsche BauZeitschrift, Bauverlag BV GmbH, Gütersloh, p.48 ff.

INTRODUCTION

World total primary energy supply (TPES) in 2014 was around 550 Exajoule (EJ).[1] A third of it is used by the food sector.[2] On a global scale, for every calorie we need to cover our daily energy requirement, we consume nearly six calories of total primary energy. One percent of the global landmass is defined as built-up land, where with the exception of a small percentage of indigenous populations, more than 7 billion people live. The area required to supply world population with food is ten times higher. Countries with emerging economies and above all the developed countries require and have established a food production network over the past few decades, which has reached a global scale and is completely dependent on hydrocarbon energy.

The world population will continue to grow within the next decades, reaching a plateau in 2075 of 9.22 billion people before it starts to decline.[3] This work aims to contribute to the discussion on whether Vertical Farming entails the potential to increase the overall energy efficiency of cities.

The architectural interest in how this typology could be interwoven into the city fabric first needed to be reset before fundamental questions could be answered, at least in part. There is no doubt, even without simulating the energy demand of a verticalized food production entity that this must be higher than it is on the field. In addition, the building is not planned as a principle for humans who have completely different requirements regarding the indoor climate. Light is perceived differently, humidity and temperature must be in a different relationship. What does a crop plant actually need to turn light into sugar to be a relevant deliverer of calories and nutrients for human consumption? These questions led to the decision to start an excursion through plant physiology and quantum physics for the development of parametric Vertical Farm models and to develop an aligned simulation tool especially for this calculation.

Throughout this research work, however, a number of limitations must be made, for reasons of time and complexity. The parametric Vertical Farm primarily attempts to find answers to the influence of different orientations or, more precisely, to find guidelines for future typological developments, especially in the context of the zoning of the building and the building depth. Water evaporation from the plants was not considered, although this clearly has an impact on heating demand. Plant growth, especially growth in height has a strong impact on the lighting demand. Although techniques are being developed by the author of this thesis to simulate the autoshading of the plants themselves, this emerging research did not find a place in this work by the time the dissertation was completed.

---

1 http://www.iea.org/publications/freepublications/publication/KeyWorld2013.pdf, retrieved 05.04.2014
2 FOOD AND AGRICULTURE ORGANIZATION OF THE UNITED NATIONS 2011.Energy-Smart Food for People and Climate, Issue Paper, Rome:FAO, p.10
3 http://www.un.org/esa/population/publications/longrange2/WorldPop2300final.pdf, p.1, retrieved 10.09.2015

INTRODUCTION

STATE OF DESIGN

There are already plenty of design proposals for Vertical Farming with most of them unfortunately stopping at the design level. Over the past few years, since my diploma in 2008, some prototypes and research entities of Vertical Farms have now been built or are about to be built ranging from Suwon in South Korea to Paignton Zoo in Devon, or the exciting "Vertical Harvest" project in Wyoming, USA, and above all the strong architectural statement that has been made in Linköping, Sweden, where Plantagon had its ground breaking ceremony in 2012.

STATE OF RESEARCH

In most cases, academic research papers, dissertations or master theses dealing with Vertical Farming are an attempt to frame the state of (research)design, touching as raw assumptions the widely discussed topics on (Vertical) farming, namely water, land use and energy consumption. Vertical Farming is complex and current speed of growth of companies, industries, plant physiologists, horticulturists, urbanists and architects dealing with this topic makes it necessary to accept that the practice of stacking the cultivation area is still in a state of infancy.

On a qualitative level, Gordon Graff's thesis has to be mentioned here.[1] His work highlights the necessity of reading the Vertical Farm-building as an integrative part of the city's metabolism. A work which delivers quantitative values was written by Chirantan Banerjee.[2] The market analysis of a Vertical Farm elaborates predictions in energy and investment costs.

Basic research data for crop production in controlled environments delivers, above all, the National Aeronautics and Space Administration.[3] Abundance of activity in research on high-tech greenhouses is noticeable especially in the Netherlands[4] and Germany[5] as well as in the US[6]. Recommendations for literature can be retrieved from the bibliography of this work. The quickly growing interest on agriculture within controlled environments within the last years makes it understandable that the list on this page must be considered as incomplete.

The doctoral thesis at hand is enlarging the research on energy consumption of Vertical Farming. By concentrating on tomatoes, *Lycopersicon Esculentum* (Mill.), it was possible to consider specific plant needs, to highlight parameters influencing pho-

---

1 GRAFF, G. 2011. Skyfarming. Master of Architecture, University of Waterloo, Ontario, Canada.
2 BANERNJEE, C. 2012. Market Analysis for Terrestrial Application of Advanced Bio-Regenerative Modules: Prospects for Vertical Farming. Masterarbeit, Rheinische Friederichs-Wilhems-Universität, Hohe Landwirtschaftliche Fakultät.
3 https://www.nasa.gov/image-feature/space-farming-yields-a-crop-of-benefits-for-earth, retrieved 31.10.2015
4 http://www.wageningenur.nl/en/Expertise-Services/Research-Institutes/Wageningen-UR-Greenhouse-Horticulture.htm, retrieved 31.10.2015
5 http://www.zineg.net/ZINEG_E/, retrieved 31.10.2015
6 http://ag.arizona.edu/ceac/, retrieved 31.10.2015

INTRODUCTION

tosynthesis in more detail and to integrate them into a parametric simulation model. The simulation method, especially developed for the thesis, unlike simulation software widely used for building simulations, evaluates year round solar radiation in WPAR within a Vertical Farm by using specific climate data. The simulation model was built up in a way so that parameters such as plant needs, climate data and building geometry can easily be substituted and therefore help to optimize future studies on an architectural level and will facilitate predictions relating to energy consumption of specific crops cultivated in Vertical Farms in specific climate zones.[1]

The integration of agriculture into discussions about architecture and urbanism is actually experiencing a revival. Concepts on (horizontal) urban farming from Ebenezer Howard to Frank Lloyd Wright and Le Corbusier are well known and documented. Vertical Farming as a substitution of traditional soil based agriculture or a supplement in food production is increasingly becoming an integral part of research works, theses, design projects and competitions dealing with urbanism in general[2][3][4], smart cities[5], „productive cities"[6][7] or „Hyperbuilding cities"[8].

Vertical Farms as buildings or elaborated design proposals can be retrieved from the world map on page 128.

---

1 CODY, B. 2012. „Form follows Energy - Beziehungen zwischen Form und Energie in der Architektur und Urban Design, DBZ Deutsche BauZeitschrift, Bauverlag BV GmbH, Gütersloh, p.48 ff.
2 VIE:BRA - Vienna-Bratislava-City, urban strategies: http://www.dieangewandte.at/jart/prj3/angewandte/main.jart?rel=en&reserve-mode=active&content-id=1234966513566&AktId=4493, retrieved 31.10.2015
3 http://www.braincitylab.org/, die angewandte, Coop Himmelb(l)au, retrieved 31.10.2015
4 http://milliardenstadt.at. University of Technology, Vienna. Project initiator: Lukas Zeilbauer
5 LIM CJ, ED LIU. 2010. Smartcities + Eco-warriors. Oxfordshire (first published), New York. Routledge.
6 NELSON, N. 2009. Planning the productive city. Available: http://www.nelsonelson.com/DSA-Nelson-renewable-city-report.pdf. Delft Technical University, Wageningen University and Research, NL
7 http://www.futurarc.com/index.cfm/competitions/2013-fap/. Addressing „adaptation of existing building typologies for agriculture (...) urban networks for production [and] distributions (...)
8 CODY, B. 2014. Form Follows Energy - Die Zukunft der Energie-Performance, energy2121, Bilder zur Energiezukunft, Klima- und Energiefonds, Vienna, omninum, p. 121 ff.

2. Landuse, Biocapacity and Energy Consumption

2.1. Compiling a status quo model of traditional agriculture

Ever since agriculture became more and more structurally coupled with industry, especially the oil- and armament's industries[1] [2], agricultural production has not only completely changed in practice and scale, but also in its energy consumption patterns. From the Neolithic Revolution to the Green Revolution the only energy source for food production was direct solar radiation and human labor which was then supplemented increasingly by the use of electricity and, above all, by fossil fuels. Agricultural production is becoming ever more energy intensive, if not altogether dependent on cheap and abundant oil and gas.[3] It is becoming a widespread concern that the reliance of the global food system on fossil fuel increases drastically."[4] In fact there is an intrinsic factor of energy consumption in conventional food production that lies behind the structural coupling of the oil- and the food industries. Regarding future food supply, it is necessary to understand if Vertical Farming can make cities more energy independent, especially from hydrocarbon energy. At the present time one third of world energy consumption is accounted for the "nutrition" system (food sector), 25 % of this within the farm gate.

Beyond the production entity of the Vertical Farm although a significant reduction of hydrocarbon energy and CO_2-emissions with urban Vertical Farming can be assumed. Substitution of natural sunlight with electrical power and heating demand for year round crop production however, might well increase energy demand in urban agglomerations. The question is if the reduction of energy consumption beyond the Vertical Farm gate (Vertical Farm-gate) can balance out the surplus in energy consumption for indoor crop production.

Before this question can be investigated with appropriate depth, a brief digression on the issue of energy consumption in the global food sector is appropriate at this

---

[1] FOOD AND AGRICULTURE ORGANIZATION OF THE UNITED NATIONS 2011.Energy-Smart Food for People and Climate, Issue Paper, Rome:FAO
[2] FOOD AND AGRICULTURE ORGANIZATION OF THE UNITED NATIONS 2011.Energy-Smart Food for People and Climate, Issue Paper, Rome:FAO. p.6
[3] ibid. p.6
[4] "It is this increased reliance of global food systems on fossil fuels that is now becoming cause for concern." Energy-Smart" Food for People and Climate Issue paper, FAO, 2011, http://www.bigpictureagriculture.com/2011/12/fao-report-warns-about-fossil-fuel.html (19.05.2014)

point. The objective of this chapter is to investigate the current situation of world agriculture in terms of land use efficiency and to what extent it can be increased, while additionally presenting an all-round view of the limits to the current biocapacity of the earth for meeting future food demand. Investigations of landuse, energy consumption and biocapacity have the purpose of establishing the degree of pertinence Vertical Farming has gained in the face of this situation.

2.2. Land requirement for daily coverage of essential nutritional value for the human body

The human body, like that of every living being needs energy to sustain its biological functions and life. There are multiple calculation models to define the specific energy need per person. In addition to several prediction equations one of the most notable of these for calculating the Basal Metabolic Rate (BMR)[1] is the Harris-Benedict equation, created in 1919. This equation was revised in 1984 using new insights in biology. This was widely regarded and used as the best prediction equation until 1990, when Mifflin et al. introduced the Mifflin St. Jeor- Equation[2]. A simplification based on this equation will be used in this work to define the basal metabolic rate of the human body - which is „relatively constant among population groups of a given age and gender. (...)"[3]

- male: 1 kg of body mass consumes 24 kcal/day
- female: 1 kg of body mass consumes 24 * 0.9 kcal/day

BMR, broken down on organs and muscles of the human body, are divided as follows:

- liver   26 %
- muscles   26 %
- brain   18 %
- heart   9 %
- kidney   7 %
- other organs 14 %[4]

In addition to BMR, the Physical Activity Level (PAL) is of importance to calculate the total energy requirement. The Food and Agriculture Organization defines three ranges of values:

- sedentary or light activity lifestyle   1.40 - 1.69
- active or moderately active lifestyle   1.70 - 1.99
- vigorous or vigorously active lifestyle   2.00 - 2.40[5]

---

1   Basal metabolic rate (bmr), index of the general level of activity of an individual's body metabolism, determined by measuring its oxygen intake in the basal state—i.e. during absolute rest, but not sleep, 14 to 18 hours after eating. The higher the amount of oxygen consumed in a certain time interval, the more active is the oxidative process of the body and the higher is the rate of body metabolism. (...) http://www.britannica.com/topic/basal-metabolic-rate, retrieved 25.08.2015
2   http://ajcn.nutrition.org/content/51/2/241.abstract, retrieved 12.06.2014
3   http://www.fao.org/docrep/007/y5686e/y5686e07.htm, retrieved 25.08.2015
4   http://www.fao.org/3/contents/3079f916-ceb8-591d-90da-02738d5b0739/M2845E00.HTM, retrieved 25.08.2015
5   „PAL values higher than 2.40 are difficult to maintain over a long period of time.", ibid.

This PAL-value is the factor multiplied by, BMR to obtain the needed daily energy requirement for the human body.[1]

With these figures we can now make an assumption about the daily energy requirement of an adult person, irrespective nationality or culture:

- adult male, 75 kg, sedentary: 75*24*1.5 = 2,700 kcal/day
- adult female, 65 kg, sedentary: 65*24*0.9*1.5 = 2,106 kcal/day

The total energy requirement of a sedentary male can therefore be assumed as

- 2,700   kcal or
- 11.30   MJ or
- 3.14   kWh or
- 0.30   OE[2].

The total energy requirement of a sedentary female therefor can be assumed as

- 2,160   kcal or
- 8.82   MJ or
- 2.45   kWh or
- 0.24   OE.

world population		FOOD REQUIREMENT / CAPUT						FOOD REQUIREMENT / CAPUT / YEAR			
7,325,965,000		kCAL	kJ	Wh	MJ	kWh	OE	kCAL	GJ	kWh	OE
3,787,523,905	male	3,200	13,397.89	3,721.64	13.40	3.72	0.36	1,168,000	4.89	1,358.40	131.7
3,538,441,095	female	2,300	9,629.73	2,674.93	9.63	2.67	0.26	839,500	3.51	976.35	94.7
7,325,965,000	fao average	2,750	11,513.81	3,198.28	11.51	3.20	0.31	1,003,750	4.20	1,167.37	113.2

Tab.1.   World food energy content requirement estimation

---

1   It is explanatory, that there are additional variables throughout a human lifetime which cannot be considered here, such as pregnancy-periods, lactating women, the length of adolescence, etc. Additional information about the principles followed by the 1985 FAO/WHO/UNU expert consultations can be found on http://www.fao.org/docrep/007/y5686e/y5686e07.htm, retrieved 19.08.2015

2   http://www.aie.org.au/AIE/Energy_Info/Energy_Value.aspx, retrieved 09.09.2015: 1 l oil = 42 MJ, 1 kg of oil equals 37 MJ

This total energy requirement[1] (Tab.1) for the human body must be get provided through a continuously operating food supply system, at the beginning of which agriculture is to be found, with the exception of some very few aboriginal populations the provision is thus based on the cultivation care of fertile land.

By the end of the Paleolithic period, human societies made a sweeping change in their habits. „(...) [A]fter hundreds of thousands of years of biological and cultural evolution, human societies were able to make increasingly varied, sophisticated, and specialized tools, thanks to which they developed differentiated modes of predation (hunting, fishing, gathering), adapted to the most diverse environments."[2]

Back then, with the emergence of a radical change in food provision, a single person would have needed between 40 and 150 ha to cover the total energy requirement of an estimated 3,960 kcal (75*24*2.2) per day, by hunting, fishing and gathering, depending on the fertility and topography of the land. That means a family of five would have required approx. 200 ha. „This estimate is based on an ideal ecosystem, one containing those wild plants and animals that are most suitable for human consumption."[3]

We are entered the final period of prehistory, around 10,000 years ago. Several societies, among the most advanced ones of the time, enabled one of the most radical and influential changes in human history - the Neolithic Revolution.
„At the beginning of this change, the very first practices of cultivation and animal raising, which we will call protocultivation and proto-animal raising, were applied to populations of plants and animals which had not yet lost their wild characteristics.

WORLD FOOD REQUIREMENT / YEAR					
kCAL	EJ	TWh	barrel oil eq.	% of world oil prod.	% of world TPES
4,423,827,921,040,000	18.52	5,144.96	3,138,749,808.20	8.92	3.38
2,970,521,299,252,500	12.44	3,454.75	2,107,614,338.69	5.99	2.27
7,394,349,220,292,500	30.79	8,599.71	5,246,364,146.89	14.91	5.62

But, as a result of such practices, these populations acquired new characteristics, typical of domestic species, which are the origin of most of the species that are still cultivated or bred today."[4]

---

1 Data retrieved for the extrapolation of Tab. 1: https://www.cia.gov/library/publications/the-world-factbook/fields/2018.html, retrieved 01.06.2015
   http://www.fao.org/docrep/meeting/009/ae906e/ae906e35.htm, retrieved 01.06.2015
   http://www.eia.gov/forecasts/steo/report/global_oil.cfm, retrieved 01.06.2015
   http://www.eia.gov/cfapps/ipdbproject/iedindex3.cfm?tid=5&pid=53&aid=1&cid=ww,&syid=2010&eyid=2014&unit=TBPD, retrieved 01.06.2015
   „The reference man and woman", FAO, http://www.fao.org/docrep/meeting/009/ae906e/ae906e35.htm, retrieved 31.10.2015
2 MAZOYER, M. & ROUDART, L. 2006. A History of World Agriculture, London, Earthscan. p. 71
3 PIMENTEL, D. et al. 2008. Food, Energy and Society, third edition, CRC Press Boca Raton. p. 45-46.
4 MAZOYER, M. & ROUDART, L. 2006. A History of World Agriculture, London, Earthscan. p. 71,

Within this change of habits, the creation of new social organizations were possible, or necessary. To plant grains in an already prepared fertile ground, or to capture and raise wild animals is not the challenge that was faced here. The difficulties at this stage of evolution were the following:

- „To arrange a social organization and rules that make it possible for units (or groups) of producer-consumers to subtract from immediate consumption an important part of the annual harvest in order to save it as seed stocks"(...)[1]
- „To exempt from slaughter enough reproductive and young animals to make it possible for the herd to reproduce itself" (...)[2]
- „To protect the fields planted by one group from the previously recognized right of other groups to ‚gather' in those areas and to protect the animals being raised from the right of those groups to ‚hunt' them."[3]
- „Lastly, what is difficult is to ensure the distribution of the fruits of agricultural work among the producer-consumer of each group, not only every day, but above all (...) when the eldest die and when the group becomes too large and must be subdivided into several smaller groups."[4]

Still it took four thousand years until the first state-governmental structures in Egypt arose, but by settling down permanently for the first time in human history it was possible to build the first fortified villages and towns. The surplus of density of people around agriculture land directly led „to the rise of cities and civilization because it allowed people to develop and concentrate on manufacturing, trading and other specializations (...) like advances in technology, art and other innovations."[5]

The natural landscape became divided in two cultured landscapes - land for agriculture and land for cities, spatially united - as the nucleus for further civilizations.

Approximately 50 million people lived on earth by the beginning of the Neolithic Revolution.[6] This revolution now started a steady and continuous growth of the world population. Several factors have been attributed to this:

---

1 MAZOYER, M. & ROUDART, L. 2006. A History of World Agriculture, London, Earthscan. p.71
2 ibid. p.71
3 ibid. p.71-72
4 ibid. p.72
5 MC.KINNEY, M. et al. 2012 „Environmental Science: Systems and Solutions", Burlington, Logan Yonavjak Jones & Bartlett Publishers. p. 36
6 ibid. p. 35

- „Settlement on farms may have allowed women to bear and raise more children; freed from the nomadic lifestyle, women no longer had to carry young offspring for great distances (...)"
- Labor capacities of children can more easily used in agriculture than in gathering and hunting
- „Agriculture and domestication may have made softer foods available, which allowed mothers to wean their children earlier." More children a mother therefor could bear.
- Higher densities of people were possible, as a consequence of agriculture and domestication. „(...)[W]ith farming, one family or group of persons could raise more food than they personally needed."

The upcoming developments of tools, achieved knowledge in plant culture, seed production and husbandry and especially the capacity in storing sun energy through feed and food storage for periods when nothing can be harvested because weather or seasonal conditions radically reduced the area needed to supply human beings with their daily energy requirements to guarantee a personal healthy life for the individual and on a communal level - maintaining social cohesion.

Before we come up with a concluding ratio in land use between hunters/gatherers and sedentary people, an important concept for food supply needs to be explained, the food balance sheets, as defined by the Food and Agriculture Organization.

2.2.1. Food Balance Sheets

Food balance sheets create a picture of the pattern of food supply of a specific country in a defined period. This information sketches the daily consumption of food items of a country both in terms of the amount (g/day) and the nutritional value (kcal). In addition FBS provide information about the quantity of imports and exports, items used for feed (livestock), used for seeds and for food losses or food waste. Apart from several weaknesses, e.g. they do not provide any information on differences in food supply within a country or seasonal differences, the FBS are (...) the only source of standardized data that permit international comparison over time.[1] FBS „provide an approximate picture of the overall food situation in a country and can be useful for (...) nutritional studies. In addition FBS provide data to estimate future changes in food supply or, more correctly, food consumption, especially in countries with emerging markets."[2]

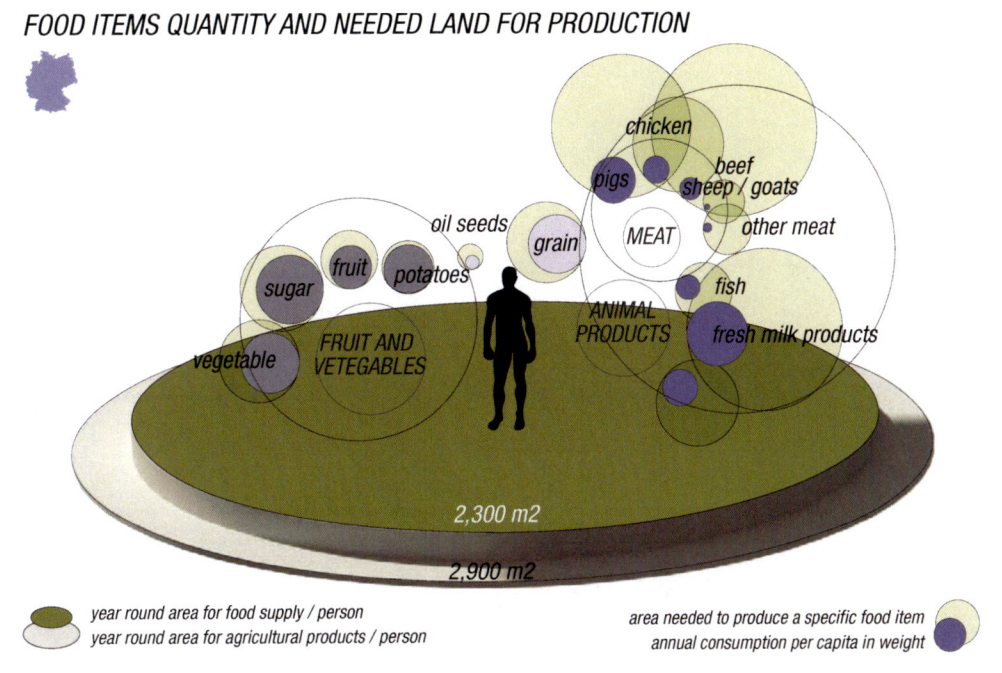

Fig.1. Arable land per person and food items consumption ratio (blue) and the related area needed (green) exemplary for Germany

As we have seen before in our two assumed examples, an average person needs around 2,500 kcal/day. By contrast with the data provided by FBS, this is data for food consumption, specifically the amount of energy a human being needs per day to sustain biological functions and life. Before food can be consumed, it must first be provided - this is defined as food supply. The values for this are always are high-

---

[1] FOOD AND AGRICULTURE ORGANISATION OF THE UNITED NATIONS 2001. Food balance sheets. A handbook. Rome: FAO. p.4
[2] ibid.

er than those for consumption, since supplied food will not be consumed completely because of food losses, food waste and various other differences. What area does single human being need to cover his daily nutritional value requirement? In January 2012 Steffen Noleppa and Harald von Witzke[1] published a detailed account of the situation in Germany with the aim of providing recommendations for the German Society for Nutrition (Deutsche Gesellschaft für Ernährung, DGE). Single food items of daily consumption were analyzed by their land requirements and related food losses and food wastes. For more detailed information I recommend the above mentioned study. For this work it is of interest to approximate the effective use of land to cover the food supply. The following diagram shows the area a German needs for food production, 2,300 m2. The total agricultural land per person is 2,900 m2. This area also includes land considered for agriculture products, which are not intended for direct consumption but are used for other purposes such as those agricultural products used for industry and agricultural products for clothings such as cotton or for rubber production and also plants cultivated for biofuel production. A factor that is also visible is the enormous difference between the different food items, compared with the actual amount of food consumed and the land area required to produce it, e.g. potato provision for a year needs only 15 m2, whereas grain requires 115 m2 and pork production 498 m2.[2] In brief of the 2,900 m2 needed per person for agriculture products annually 1,099 m2 are for animal products, including meat, milk and dairy products and eggs.[3]

Coming back to the initial steps of the Neolithic Revolution we can say that the land needed for people to cover the total energy requirement for the human body has shrunk within the past 11,000 years from 40 ha (highly fertile land) per person to 1,3 ha. In terms of the ecological food footprint[4] (Fig. 2), this is a reduction of more than 30 times, taking into account that the ecological food footprint of hunters and gatherers was close to 0.

These data now are focused on the one single country Germany, which is highly technologized, with one of the highest GDPs in the world and, as in most other countries, with its own unique and specific culturally characterized diet. How can we obtain an image of world food data in order to establish firm statistics for the effective land use of the global population and simultaneously to venture a picture of future land use?

Tab. 2 on page 46 shows the difference between the consumption of food items of different countries (USA, Italy, Germany, Austria and China). We know that the area we need for food supply is dependent on the items consumed. The more the diet of

---

1 NOLEPPA, S. & WITZKE, H. V. 2012. Tonnen für die Tonne. Berlin: WWF
2 ibid. p. 42.
3 ibid. p.42-43
4 VALE, R. & VALE, B. 2009. Time To Eat The Dog?, London, Thames & Hudson. p.36

culture is oriented to food of vegetable origins, the less area is required per person. The question here is whether this is the only parameter for defining the area needed.

AGRICULTURE LAND AND ECOLOGICAL FOOD FOOTPRINT

2,300 m2

13,000 m2

year round area for food supply / person
ecological food footprint

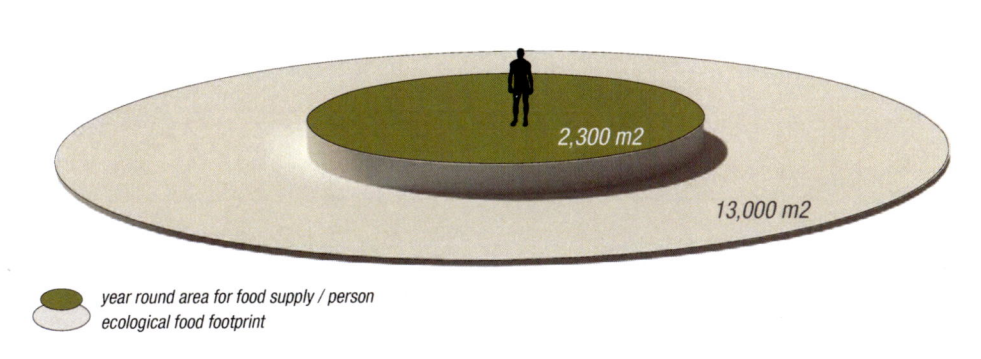

Fig.2.   Land use per person for food production and related ecological food footprint

FOOD BALANCE SHEETS COMPARISON

	USA	Italy	Germany	Austria	China
Cereals - Excluding Beer	302	432	315	325	423
Starchy Roots	162	108	196	168	187
Sugar & Sweeteners	188	85	140	131	21
Pulses	12	15	2	2	3
Treenuts	9	21	15	17	4
Oilcrops	15	8	8	13	18
Vegetable Oils	81	76	47	58	23
Vegetables	336	448	248	259	757
Fruits - Excluding Wine	301	411	249	405	168
Stimulants	21	21	24	26	2
Spices	2	1	2	2	1
Alcoholic Beverages	272	206	375	399	96
Meat	338	245	234	286	149
Offals	3	9	3	4	8
Animal Fats	16	29	57	51	6
Milk - Excluding Butter	692	718	673	627	72
Eggs	40	31	33	37	47
Fish, Seafood	66	67	41	37	71
Aquatic products, other					22

Tab.2. FBS comparison of different countries (gramms supplied per person/day) for USA, Italy, Germany, Austria and China (left to right)

In an effort to investigate this question more deeply, Kastner et al.[1] was looking for the main drivers in changes in land use and future requirements of croplands by comparing FBS of different subcontinents. Major findings are that the size of the population, the average food consumption pattern and the output per unit land define the cropland needed. In other words, it is population change, diet (FBS) and agriculture technology in use. „The amount of cropland needed depends on population numbers, average food consumption patterns, and output per unit of land."[2] (...) „Population, diets, and production techniques change over time and show large spatial variation. With socioeconomic development, population growth rates decrease and diets change: typically, consumption of animal protein, vegetable oils, fruits and vegetables increases, while starchy staples become less important. The change from these staples toward richer diets implies that cropland demand for average diets will in general increase. By contrast, the introduction of new technologies leads to improvements in agricultural area productivity through time."[3] [4]

---

[1] KASTNER, T., IBARROLA RIVAS, M. J., KOCH, W. & NONHEBEL, S. 2012. Global changes in diets and the consequences for land requirements for food. Available: http://www.pnas.org/content/early/2012/04/10/1117054109.full.pdf+html.
[2] ibid. p.1
[3] ibid. p.1
[4] VALE, R. & VALE, B. 2009. Time To Eat The Dog?, London, Thames & Hudson.

In order to form a clear picture of the actual food consumption and the required land to cover daily food requirements Kastner et al. used FAO FBS and grouped them on subcontinental levels. For the task of evaluating which are the main drivers related to land use for agricultural production, FAO data from 1961 to 2007 was analyzed. Major findings, relevant for this work are that on a global level, the „average land area needed to feed a person in 2005 was two thirds of the corresponding value in 1963, decreasing from approximately 2,650 to just over 1,732 m2/person/y. (...) Across the regions, per capita cropland requirements for food in 2005 were lowest in much of Asia, with approximately 1,300 m2/person/y in Southeast Asia (...). The highest values, with more than 3,000 m2/person/y, were found in Oceania and Southern Europe, two dry regions with a large annual variability. Western Africa and Northern Europe, two regions at very different ends of the global spectrum in terms of per capita food supply, show the same per capita values in 2005, at approximately 2,350 m2/person/a."[1]

While the cropland needed per capita has decreased since 1963 the effective area increased from 8,400,000 km2 to 11,000,000 km2, an increase of 32%, while world population increased from 3,201,000,000 to 6,540,000,000 in the decades from 1963 to 2005 (=103%). „This was mostly driven by growing land demand for animal products, which accounted for almost 50% of the total increase."[2]

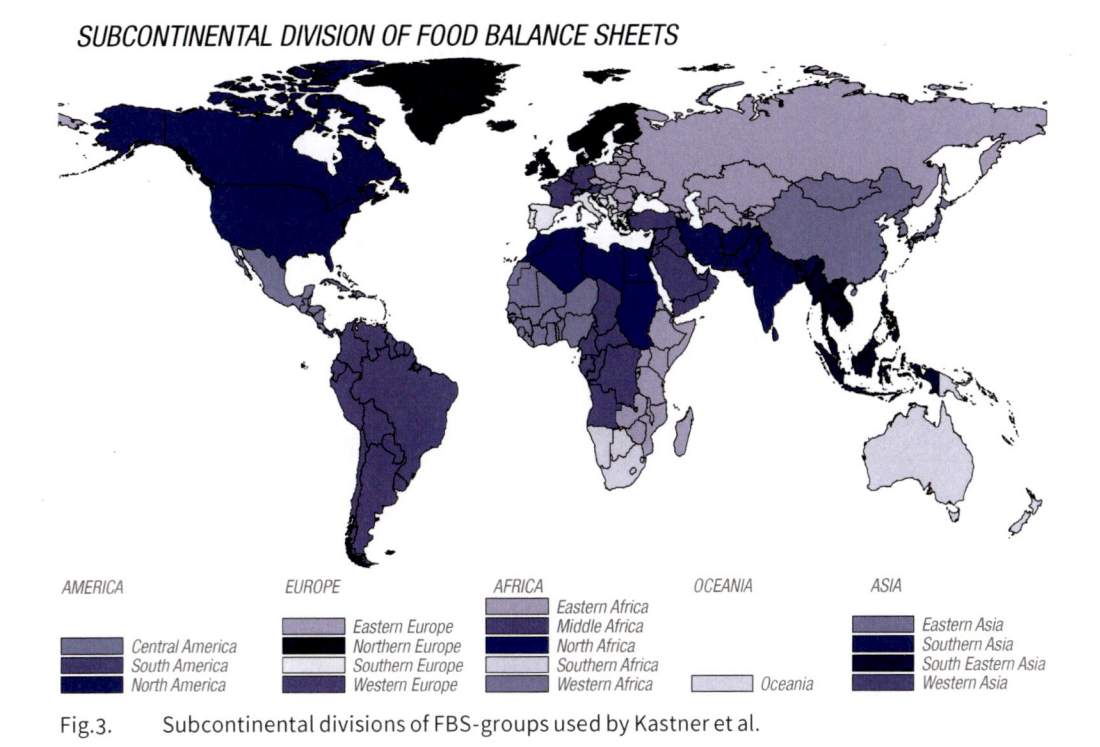

Fig.3.   Subcontinental divisions of FBS-groups used by Kastner et al.

---

1   KASTNER, T., IBARROLA RIVAS, M. J., KOCH, W. & NONHEBEL, S. 2012. Global changes in diets and the consequences for land requirements for food. Available: http://www.pnas.org/content/early/2012/04/10/1117054109.full.pdf+html. p.2
2   ibid. p.2

	Domestic Supply					Domestic Utilization			
	1000 Metric tons								
FBS World 2011 Population: 6,847,859,000	Prod.	Import	Stock Variation	Export	Total	Food	Food Manufacture	Feed	Seed
Cereals - Excluding Beer	2,341,981	371,750	-21,877	384,712	2,307,110	1,007,765	89,441	816,104	66,441
Starchy Roots	796,696	58,437	-5,858	56,229	793,044	434,756	14,765	176,134	35,520
Sugar Crops	2,083,717	1,104	-219	1,383	2,083,221	31,309	1,518,351	50,930	28,251
Sugar & Sweeteners	204,530	70,936	-8,327	73,326	193,798	165,142	7,349	429	0
Pulses	68,111	12,091	1,471	13,471	68,200	47,156	0	12,930	3,965
Treenuts	15,322	6,916	180	7,232	15,186	15,060	0		0
Oilcrops	549,152	126,432	-8,673	127,577	539,307	47,973	421,868	34,661	11,068
Vegetable Oils	158,746	79,718	-1,455	82,933	154,062	79,507	438	773	
Vegetables	1,082,283	65,496	244	72,122	1,075,893	930,001	605	52,604	113
Fruits - Excluding Wine	631,787	123,182	1,836	121,264	635,525	513,761	55,251	5,497	
Stimulants	18,412	20,138	561	20,641	18,456	17,843	45	12	
Spices	9,317	2,259	-25	2,296	9,250	8,718		0	
Alcoholic Beverages	290,351	35,748	4,050	39,324	290,815	257,967	3,239		
Meat	295,984	40,760	163	43,816	293,098	290,078	454	74	
Offals	18,078	3,893	40	4,770	17,241	15,250	0	1,041	
Animal fats	36,529	8,940	-243	9,277	35,947	22,622	217	2,063	
Eggs	70,399	2,164	5	2,464	70,106	61,304	0	73	4,709
Milk - Excluding Butter	738,107	102,994	363	107,240	734,233	621,333	26	78,637	
Fish, Seafood	153,001	60,562	-623	55,802	157,128	129,609		23,463	466
Aquatic Products, Other	23,144	700	0	602	23,240	14,278		159	0

Tab.3. World FBS 2011

Before we take a look at the total agriculture land in use, the main highlights of this study must be pointed out:

The biggest drivers in cropland expansion is not population growth, but socioeconomic development (Tab. 3 on page 48). What has been observed is that by increasing GDP population growth slows down, but the effects on dietary changes still have the result of an increase in the area needed for food production. „It suggests that pressures on land resources linked to the provision of food are likely to remain high in the coming decades, as these dietary changes affect a large share of global population."[1]
In addition, an increase in land use efficiency, something that mostly occured in the industrialized countries within the last fifty years, and led to a decrease in land use per person on a global scale, can only be possible with an increase in external inputs, mostly hydrocarbon energy used for fertilizers, pesticides etc. but also for machinery and irrigation infrastructure with all the environmental impacts entailed which are to expect.

---

[1] KASTNER, T., IBARROLA RIVAS, M. J., KOCH, W. & NONHEBEL, S. 2012. Global changes in diets and the consequences for land requirements for food. Available: http://www.pnas.org/content/early/2012/04/10/1117054109.full.pdf+html. p.4

		Per Capita Supply			
		Total		Prot.	Fat
Waste	Other Uses	Kg / Yr	KCal / Day	Gr / Day	Gr / Day
	Grand Total	**688.03**	**2,870**	**80.49**	**82.56**
	Vegetal Products	**521.53**	**2,362**	**48.67**	**45.53**
	Animal Products	**166.50**	**508**	**31.82**	**37.03**
101,061	226,914	147.17	1,296	31.92	5.88
78,436	53,585	63.49	141	2.27	0.26
148,052	306,360	4.57	4	0.02	0.02
259	20,788	24.12	230	0.04	0.01
3,460	709	6.89	65	4.05	0.42
457	68	2.20	15	0.39	1.14
12,964	13,458	7.01	57	2.72	4.18
567	74,075	11.61	278	0.03	31.42
92,399	536	135.81	93	4.77	0.79
60,083	1,881	75.03	94	1.10	0.60
336	390	2.61	7	0.52	0.43
232	328	1.27	11	0.39	0.35
1,699	28,070	37.67	69	0.37	0.00
878	1,791	42.36	231	14.26	18.80
70	905	2.23	7	1.09	0.21
142	11,146	3.30	61	0.08	6.78
3,252	777	8.95	35	2.72	2.46
18,221	16,379	90.73	139	8.26	7.59
	3,588	18.93	34	5.23	1.19
	8,800	2.08	2	0.16	0.01

We now have a clearer picture in what area is needed to provide people with food, what main drivers are responsible for the change of cropland requirements and the uneven distribution in total land use of different subcontinents.

The study of Kastner et al. focused on 11 mostly cultivated food categories produced on croplands. „(…) [M]ore than 90% of all food calories and approximately 80% of all food protein and fats available in the world were derived from croplands."[1] But the output of the agricultural land per person is not all used for direct human consumption (diagram on the next page). Before coming back to yield or calories as output of cropland directly available for human consumption, we will investigate to what extent this area currently can be extended by answering the question of how much is in use at the present time and what is the expectable biocapacity of the world.

---

1   KASTNER, T., IBARROLA RIVAS, M. J., KOCH, W. & NONHEBEL, S. 2012. Global changes in diets and the consequences for land requirements for food. Available: http://www.pnas.org/content/early/2012/04/10/1117054109.full.pdf+html. p.2 and http://faostat.fao.org/, retrieved 13.05.2013

	m²/P/a	kcal/P/d	MJ/P/d	MJ/P/a
World	**1,732**	**2,780**	**11.639**	**4,248.35**

	m²/P/a	kcal/P/d	MJ/P/d	MJ/P/a
Eastern Africa	1,774	1,018	4.262	1,555.69
Middle Africa	1,545	1,242	5.200	1,898.00
Nothern Africa	2,289	1,239	5.187	1,893.42
Souther Africa	1,673	1,349	5.648	2,061.52
Western Africa	2,434	1,379	5.774	2,107.36
Average Africa	**1,943**	**1,245**	**5.214**	**1,903.20**

	m²/P/a	kcal/P/d	MJ/P/d	MJ/P/a
Southern Asia	1,509	2,360	9.881	3,606.51
South-Eastern Asia	1,278	2,571	10.764	3,928.96
Eastern Asia	1,436	2,941	12.313	4,494.38
Western Asia	2,102	3,055	12.791	4,668.60
Asia	**1,581**	**2,732**	**11.437**	**4,174.61**

	m²/P/a	kcal/P/d	MJ/P/d	MJ/P/a
South America	1,793	2,882	12.066	4,404.22
Central America	1,834	3,023	12.657	4,619.69
Latin America	**1,814**	**2,953**	**24.723**	**9,023.91**

	m²/P/a	kcal/P/d	MJ/P/d	MJ/P/a
North America	**2,364**	**3,690**	**15.449**	**5,638.99**

	m²/P/a	kcal/P/d	MJ/P/d	MJ/P/a
Eastern Europe	2,622	3,301	13.821	5,044.53
Northern Europe	2,127	3,362	14.076	5,137.75
Souther Europe	3,084	3,360	14.068	5,134.69
Western Europe	2,189	3,492	14.620	5,336.41
Europe	**2,506**	**3,379**	**14.146**	**5,163.34**

	m²/P/a	kcal/P/d	MJ/P/d	MJ/P/a
Oceania	**3,555**	**2,388**	**9.998**	**3,649.30**

Tab.4.   FBS (kCal and area) food supply per person comparison (Kastner et al.)

2.2.2. World land masses and the ratio of land used for agricultural production

On a global scale the average cropland needed for a single person (food supply), as we have seen in the previous subchapter is 1,732 m2/a. Land for agriculture use, is divided in arable land[1] (=cropland), land for permanent crops[2] and pastures[3].
The world's surface area is some 510,072,000 km2. 29,10% of this is distributed over the continents, the remaining 70,90 % is the surface of the oceans.[4] The landmass is distributed as follows: forests cover more than a quarter of the continent's surface, pastures around 23 %. The inland water (rivers and lakes) some 3 %. Antarctica and agricultural land have equal dimensions, both being around 15,000,000 km2 each.

On a first glance on the agricultural land in use (both for permanent crops and cropland) and dividing it by the world population of 2007 (6,646,374) when the study of Kastner et al. was published, we can suggest that there is enough agricultural land available. 2,336.57 m2 would be available for a single person, a surplus of around 600 m2 compared to the world media of 1,732 m2. This is a potential expansion in world average to meet future demand for the growing world population, or rather reducing hunger and the percentage of undernourished people.

Now as we know the availability of land per person and the caloric values needed demant a look at the total produce from agriculture. The following comparative data retrieved by the database of the Food and Agriculture Organization are all from the year 2011. This is information on the above statistics (and following on from this) on available agricultural land, FBS, world population numbers and agriculture primary production for food, food manufacture and animal feeds.
The following table shows the FBS of the world. It differentiates between the total production of food, food manufacture, feed and seed-production and also gives information on estimated quantities of food waste and also changes in food stock, and imported and exported food. The methodology and definitions can be obtained from the Nomenclatura.

---

1   Arable land is the land under temporary agricultural crops (multiple-cropped areas are counted only once), temporary meadows for mowing or pasture, land under market and kitchen gardens and land temporarily fallow (less than five years). The abandoned land resulting from shifting cultivation is not included in this category. Data for "Arable land" are not meant to indicate the amount of land that is potentially cultivable. faostat.fao.org/site/375/default.aspx; retrieved 26.08.2015
2   Crops are divided into temporary and permanent crops. Permanent crops are sown or planted once, and then occupy the land for some years and need not be replanted after each annual harvest, such as cocoa, coffee and rubber. This category includes flowering shrubs, fruit trees, nut trees and vines, but excludes trees grown for wood or timber. ibid.
3   Mainly meant as grasslands. ibid.
4   https://www.cia.gov/library/publications/the-world-factbook/geos/xx.html, retrieved 12.03.2012

LANDUSE, BIOCAPACITY AND ENERGY CONSUMPTION

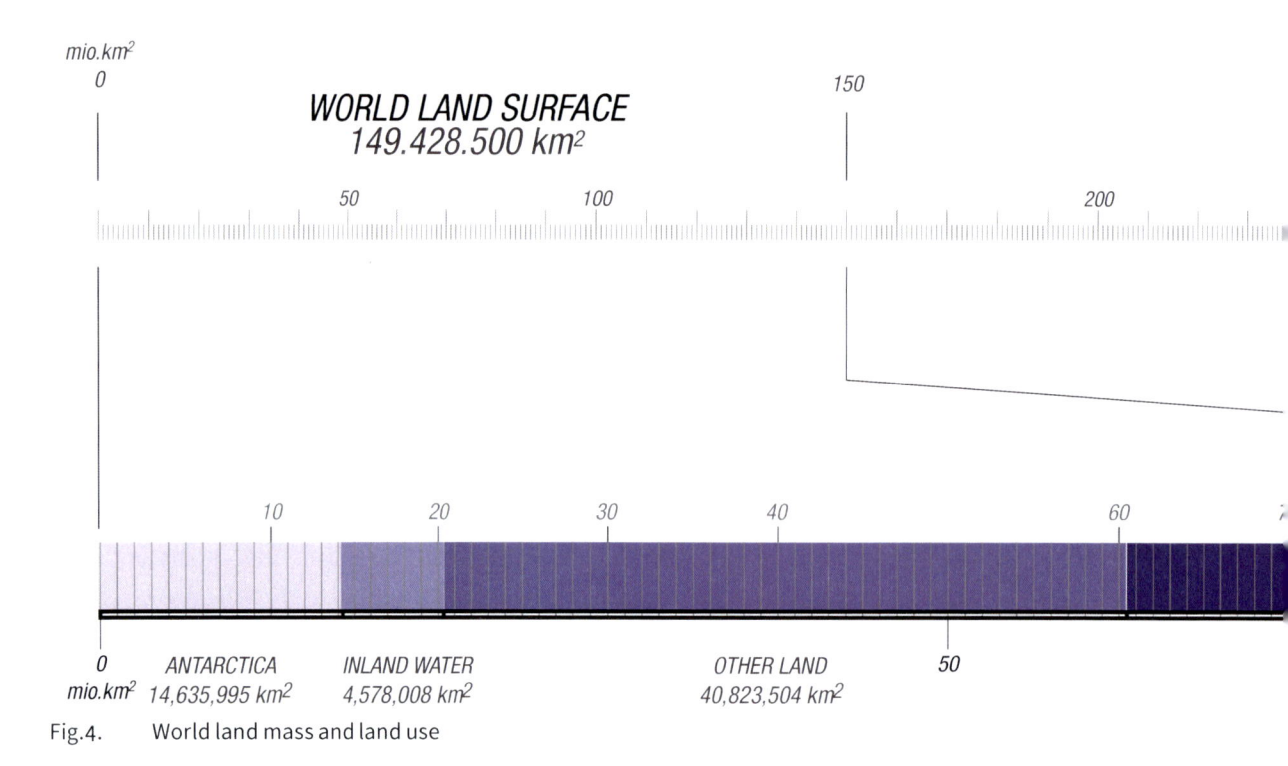

```
mio.km²
0                          WORLD LAND SURFACE                150
                             149.428.500 km²
                    50              100                                      200
```

```
         10          20          30           40                   60
 0    ANTARCTICA   INLAND WATER              OTHER LAND      50
mio.km²  14,635,995 km²  4,578,008 km²       40,823,504 km²
```

Fig.4.   World land mass and land use

Agricultural land of 15,529,767 km2 produced more than 9.5 bn metric tons of primary products in 2011. By assuming a per capita food supply of 900 kg/a (2.4 kg/d) which corresponds to a European high GDP country[1] and dividing it by the total primary production worldwide (9,585,647,000 metric tonnes)[2] every person could be supplied with 1,399.80 kg/a. A number which could lead to the conclusion that there is enough food on earth but unevenly distributed if we consider how some 795,000,000 people are undernourished or suffer from hunger[3][4].

Now subtracting feed, seeds, wastes, other (nonfood/feed) uses and food manufacture, and dividing the result of 4,712,094 by the world population in 2011 the statistic to emerge appears quite different: 688,03 kg/y (1.88 kg/d) are directly related to food supply. Adding food manufacture (which is a consumable part of food production, more specifically explained in the Chapter Nomenclatura) the per capita food supply rises to 997 kg/year. We can assume as a rule of thumbthat world's productivity per km2 (cropland and permanent crops) is 617 kg of which only 50 % is used for direct food production or 70 % when also considering food manufacture.

1   Austria: 1,027.50; Germany: 979.90; Italy: 1,051.90; FBS faostat.org, retrieved 28.08.2015
2   http://faostat3.fao.org/download/Q/*/E
3   http://www.fao.org/docrep/018/i3434e/i3434e.pdf, p.4 retrieved 13.08.2015
4   Undernourishment means that a person is not able to acquire enough food to meet the daily minimum dietary energy requirements, over a period of one year. FAO defines hunger as being synonymous with chronic undernourishment..

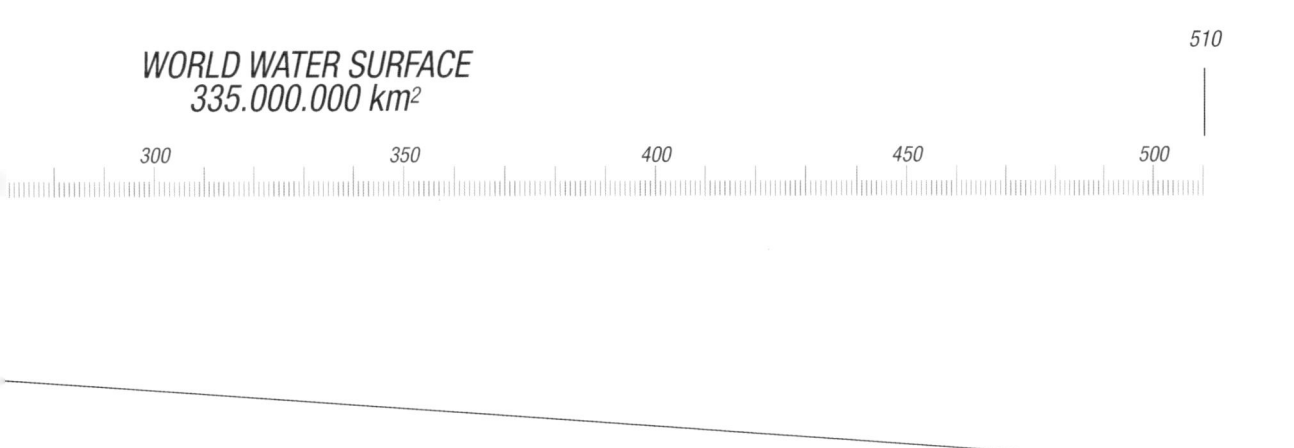

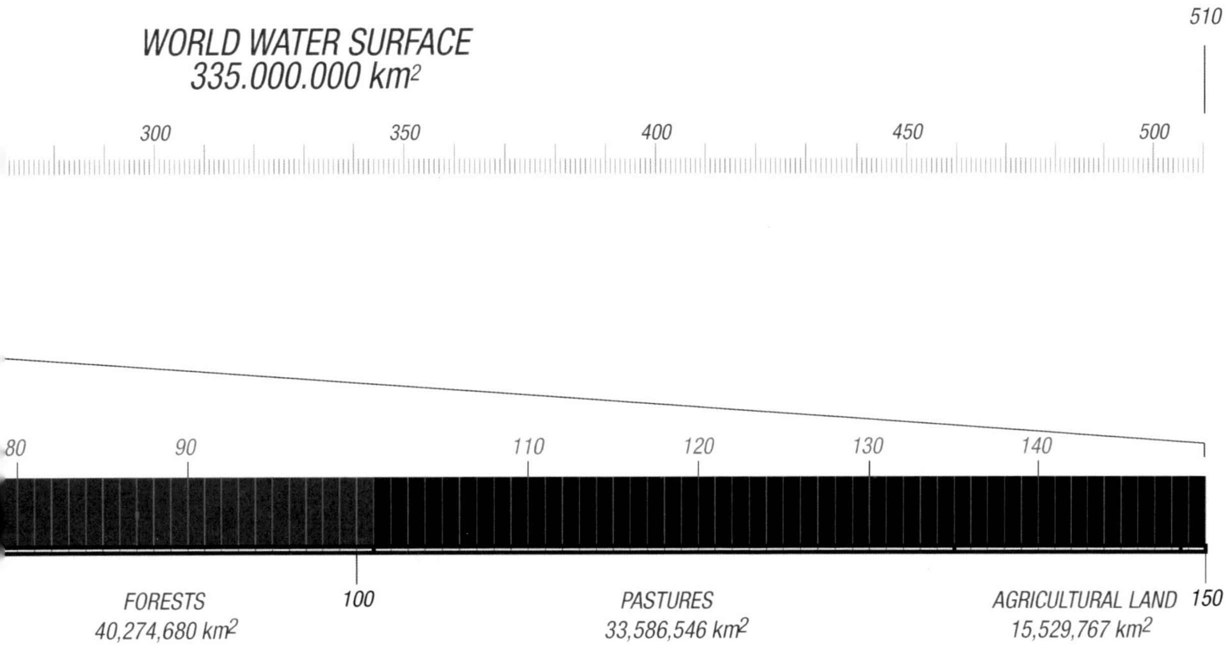

WORLD WATER SURFACE
335.000.000 km²

FORESTS
40,274,680 km²

PASTURES
33,586,546 km²

AGRICULTURAL LAND
15,529,767 km²

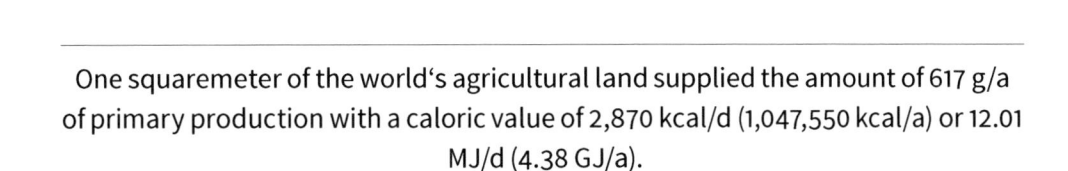

One squaremeter of the world's agricultural land supplied the amount of 617 g/a of primary production with a caloric value of 2,870 kcal/d (1,047,550 kcal/a) or 12.01 MJ/d (4.38 GJ/a).

A total primary production of 9,585,647,000 metric tons must feed 6,847,859,000 people and some hundreds of million pets, about which precise statistics are difficult to obtain.[1]

Faced with these numbers at this stage we can clearly maintain that an increase in agriculture land will almost certainly be a necessity in future. In this context it must first be pointed out that 795 million people are still undernourished or suffer from hunger in 2015.[2] Secondly, a caloric value of 2,870 kcal/day might be enough for most sedentary populations with a high consumption of animal products, but for populations with a higher percentage of labor in agriculture (less technologized) a much higher total energy requirement is necessary. More than 2.6 billion people work in the primary sector[3], more than 70 % of them still till the soil by hand or with animal power. Taking into consideration those countries in a process or rapid growth such as China and India and other emerging economies, it must be borne in mind that their

---

1   http://pets.thenest.com/number-dogs-cats-households-worldwide-8973.html, retrieved 03.09.2105
2   http://www.fao.org/news/story/en/item/288229/icode/, retrieved 28.08.2015
3   http://www.worldwatch.org/asia-and-africa-home-95-percent-global-agricultural-population-0, retrieved 26.08.2015

socioeconomic development is directly related to a change in diet with a move to a consumption of more animal products. Returning briefly to the contemporary situation in 2015: in the time since FAO published the data given above, the world population increased from 6,847,859,000 to more than 7,320,000,000 - in other words half a billion additional people are now on the plantet.

According to the UN-World Population Report we will have to expect a world population of some 8,920,000,000 people in 2050. Furthermore growth will continue and peak in 2075 by some 32 billion before a slight decline is expected to begin.

WORLD	km²	AFRICA	AMERICA	ASIA	EUROPE	OCEANIA
Land area	130,094,732	29,647,666	38,892,312	30,935,392	22,132,815	8,486,547
Inland water	4,578,008	688,038	1,858,200	1,029,554	927,326	74,890
Combined	134,672,740	30,335,704	40,750,513	31,964,946	23,060,141	8,561,437
Agricultural area	49,116,314	11,696,963	12,151,195	16,335,214	4,698,748	4,234,194
Arable land	13,962,790	2,264,531	3,710,517	4,735,722	2,764,979	487,041
Permanent crops	1,539,377	290,901	276,554	800,434	155,591	15,897
Permanent meadows and pastures	33,586,546	9,113,931	8,164,124	10,799,058	1,778,177	3,731,256
Total area equipped for irrigation	3,182,972	139,232	499,335	2,258,164	254,408	31,833
Forest area	40,274,680	6,709,986	15,661,816	5,942,050	10,057,708	1,903,120
Other land	40,823,504	11,360,215	11,079,301	8,658,132	7,376,359	2,349,497

WORLD	%	AFRICA	AMERICA	ASIA	EUROPE	OCEANIA
Land area	100	23	30	24	17	7
Inland water	100	15	41	22	20	2
Combined	100	23	30	24	17	6
Agricultural area	100	24	25	33	10	9
Arable land	100	16	27	34	20	3
Permanent crops	100	19	18	52	10	1
Permanent meadows and pastures	100	27	24	32	5	11
Total area equipped for irrigation	100	4	16	71	8	1
Forest area	100	17	39	15	25	5
Other land	100	28	27	21	18	6

Tab.5. Agricultural land availability and forest land and its distributions over the continents

An additional need for land area for food and feed production now seems to be an unavoidable necessity. The interesting question at this stage now would firstly be to what extent agriculture land needs to increase, in other words, how much land surface - nature - must be converted in to cropland?

And, more importantly what is the total biocapacity of the earth that is - suitable for agricultural production? Does the earth's landmass have the potential to feed more than 9 billion people?

2.2.3. Biocapacity of the Earth

Some 120,000,000 km2 we are defined as biologically productive land and water surfaces in 2011.[1] This corresponds to 1.75 ha (or global hectare [gha]) per person. It is the land and water (sea and inland water) „that supports significant photosynthetic activity and the accumulation of biomass used by humans. Non-productive areas as well as marginal areas with patchy vegetation are not included. Biomass that is not of use to humans is also not included."[2] Also included is the area with the capacity to capture CO_2.

„Land is an indispensable resource for the most essential human activities: it provides the basis for agriculture and forest production, water catchment, recreation, and settlement."[3] For assessing agricultural resources and potentials for the growing world population over the last thirty years FAO together with the International Institute for Applied System Analysis (IIASA) developed the Agro-Ecological Zones (AEZ) methodology which „provides a standardized framework for the characterization of climate, soil and terrain conditions relevant to agricultural production."[4] Five major thematic areas are covered within GAEZ:

- „Land and water resources, including soil resources, terrain resources, land cover, protected areas and selected socio economic and demographic data;
- Agro-climatic resources, including a variety of climatic indicators;
- Suitability and potential yields for up to 280 crops/land utilization types under alternative input and management levels for historical, current and future climate conditions;
- Downscaled actual yields and production of main crop commodities, and
- Yield and production gaps, in terms of ratios and differences between actual yield and production and potentials for main crops."[5]

Of major interest for us are findings regarding the actual potential of suitable land for agricultural production. Major findings of this study are as follows:

---

1 http://www.footprintnetwork.org/en/index.php/GFN/page/glossary/#biologicallyproductivel andandwater, retrieved 28.08.2015
2 ibid. Glossary.
3 FISCHER, G., VELTHUIZEN, H. V. & NACHTERGAELE, F. O. 2000. Global Agro-Ecological Zones Assessment: Methodology and Results. International Institute for Applied Systems Analysis, FAO., Executive Summary, p.x
4 ibid. p.x -
5 http://www.fao.org/nr/gaez/en/#, retrieved 28.08.2015

LANDUSE, BIOCAPACITY AND ENERGY CONSUMPTION

WORLD LAND SURFACE
149.428.500 km²

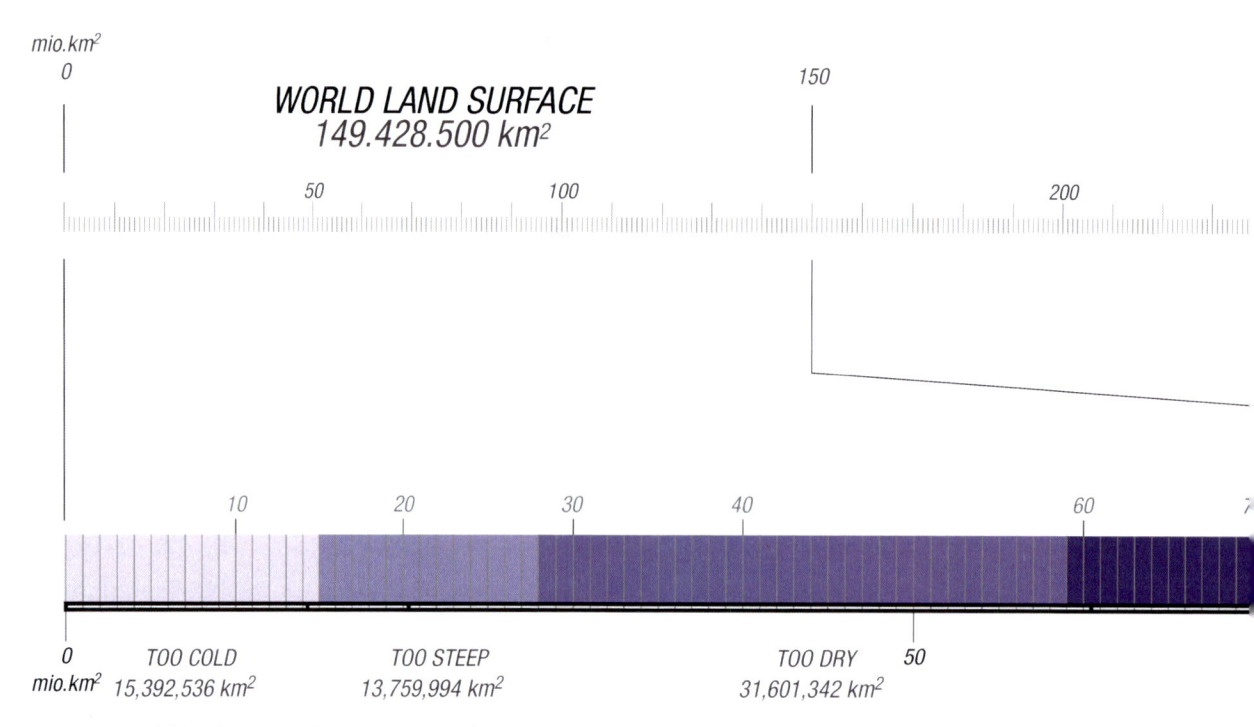

| 0 | TOO COLD | TOO STEEP | TOO DRY | 50 |
| mio.km² | 15,392,536 km² | 13,759,994 km² | 31,601,342 km² | |

Fig.5. World land mass and soil conditions for agricultural production

- Resources (both land and biological) are „sufficient to meet the needs of food and fiber of future generations, and more in particular for a world population of 8.9 thousand million, as projected for the year 2050 by the UN medium variant."[1]
- A closer look at the dataset, however, also leads to „(...)profound concerns. Several regions exist, where the rain-fed cultivation potential has already been exhausted (...)"[2]
- Global warming „may alter the condition and distribution of land suitable for cropping. (...)
- Socioeconomic development may infringe on the current agricultural resource base for want of a concomitant rapidly expanding industrial and service sector. (...)
- Land degradation, if continuing unchecked, may exacerbate regional land scarcities. Concerns for the environment may prevent some resources from being developed for agriculture.[3]
- Roughly two thirds of the total land mass (Antarctica included) „suffer rather severe constraints for rain-fed crop cultivation"[4], an area of 105,000,000 km2.

---

1 FISCHER, G., VELTHUIZEN, H. V. & NACHTERGAELE, F. O. 2000. Global Agro-Ecological Zones Assessment: Methodology and Results. International Institute for Applied Systems Analysis, FAO. Executive Summary, p.xi
2 ibid. p.xi
3 ibid. p.xi
4 ibid. p xi

WORLD WATER SURFACE
335.000.000 km²

BAD SOIL CONDITIONS
75,446,748 km²

SUITABLE FOR AGRICULTURE
32,724,841 km²

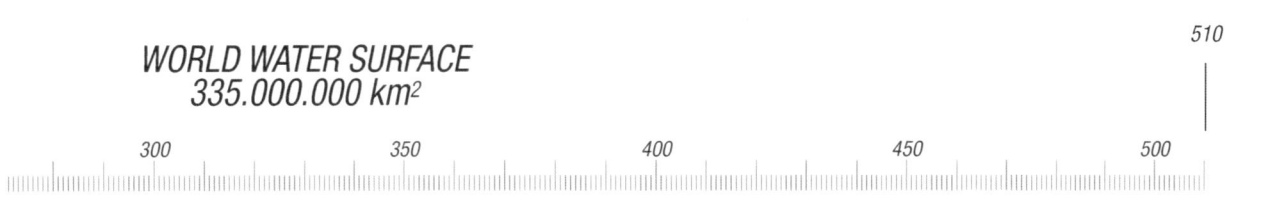

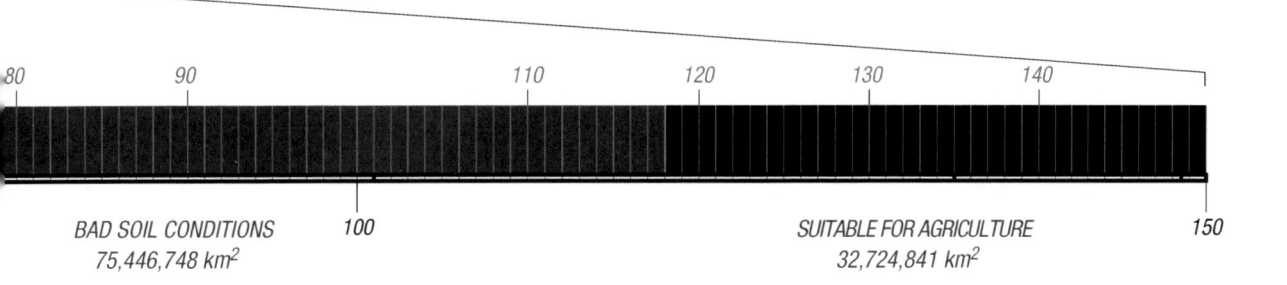

At this point after subtracting the area suffering from severe constraints for rainfed agriculture practices from the total land mass of the earth, we can assume 44,428,500 km2 will be left. This suggests that the potential exists for more than doubling the actual total agricultural area. Estimating the possible extent of land with the potential to grow rain-fed crops is depending on „a variety of assumptions: the range of crop types considered, the definition of what level of output qualifies as acceptable, the social acceptance of land-cover conversion (in particular forests), and the assumption on what land constraints may be alleviated with modern inputs and investment."[1] This explains GAEZ's estimation ranges from 13,000,000 km2 of land „very suitable and suitable for major cereal crops, under high inputs and mechanization, outside current forest areas"[2] to some 33.000.000 km2 which is defined as land „very suitable, suitable or moderately suitable for at least one of the AEZ crop types, within or outside current forest areas."[3]

In absolute numbers we can state in summary that from the 149,428,500 km2 of land mass area:

---

1   FISCHER, G., VELTHUIZEN, H. V. & NACHTERGAELE, F. O. 2000. Global Agro-Ecological Zones Assessment: Methodology and Results. International Institute for Applied Systems Analysis, FAO. Executive Summary, p. xii
2   ibid. p.xii
3   FISCHER, G., VELTHUIZEN, H. V. & NACHTERGAELE, F. O. 2000. Global Agro-Ecological Zones Assessment: Methodology and Results. International Institute for Applied Systems Analysis, FAO. Executive Summary, p.xii

- 32,698,612 km2 is very suitable or suitable
- 116,610,121,566 km is land area with constraints for agricultural production of which
- 75,446,748.65 km2 has bad soil conditions
- 31,601,342.94 km2 is too dry
- 15,392,536.04 km2 is too cold and
- 13,759,994.34 is too steep.

		Potential	VS	S	MS	mS	vmS	NS
Total Land Mass	149,428,500							
Land, Excluding Antarctica	132,950,000	44,950,000	13,150,000	21,870,000	9,930,000	11,110,000	16,270,000	60,610,000
of which in agricultural use	15,590,000	12,600,000	4,420,000	6,160,000	2,010,000	1,200,000	1,040,000	750,000
Gross balance of land with rain-fed potential		32,360,000	8,730,000	15,710,000	7,920,000	9,910,000	15,230,000	
Under forest	37,360,000	16,010,000	4,530,000	8,540,000	2,930,000	3,420,000	5,300,000	12,630,000
Strictly protected land***	6,380,000	1,070,000	300,000	500,000	270,000	390,000	590,000	4,320,000
Built-up land	1,520,000	1,160,000	410,000	610,000	140,000	120,000	100,000	150,000
Net balance of land with rain-fed potential		14,120,000	3,490,000	6,060,000	4,580,000	5,980,000	9,230,000	

Tab.6. GAEZ - Potential for Agricultural production

Tab. 6 gives an overview of the actual results of GEAZ-study. The potential land is subdivided into six classes: VS is prime land (very suitable) „with attainable yields of over 80 % of maximum constraint-free yields. Good land [S] represents suitable and moderately suitable [MS] land with attainable yield levels of 40 to 80 percent of maximum constraint-free yields (...)"[1] Marginally suitable land (mS) has an attainable yield of 20-40 %, very marginally suitable (vmS) from 5-20 % and lastly NS, not suitable land less than 5 %. In this study major crops which were considered were cereals, roots and tubers, sugar crops, pulses and oil-bearing crops.

As a conclusion to this brief perspective of the biocapacity of the earth we could claim on the one hand that there would be enough land usable for agricultural production, considering our above defined world average land use for food production of a single person of 1,732 m2: Counting the potential of land very suitable, suitable and marginally suitable the world could feed more than 25 billion people, 28 billion with a diet corresponding to Asia, 19 billion with a North American diet, 17 billion with a European diet and 12 billion with a diet typical to Oceania.

On the other hand one must be conscious about the fact that some 30,000,000 km2 of natural land has to be converted into agricultural land. Our built up land is currently around one percent of total landmass, but we need 10 % of the total landmass to supply ourselves with food. 25 % of the total land is covered with forests. Nearly a third of it in South- and Central America, or in other words: 75 % of the existing forests are located in developing countries.[2]

---

1 ALEXANDRATOS, N. & BRUINSMA, J. 2012. World Agriculture Towards 2030/2050. The 2012 Revision. [Rome]: Agricultural Development Economics Division, FAO. p. 102
2 FAO Global Forest Resources Assessment 2010, retrieved 28.08.2015

LANDUSE, BIOCAPACITY AND ENERGY CONSUMPTION

World Agriculture is already responsible for 17 - 32 % of total greenhouse gas emissions.[1] Around 47 % of this is related to land conversion to cropland. Considering a media of 300 t/ha of CO2 a rainforest can store which would be released by slash-and-burn methods one can readily imagine that other solutions to increase the total amount of available food for the growing and prospering world population should be considered.

When the area for agricultural production is not to be increased then two options are available to meet future demands: One of these would be to radically change the agricultural system by exclusively delivering every single calorie of crops for human consumption and the second option would be to increase yield per hectare.

GLOBAL CROP PRODUCTION - CALORIES FOR DIRECT HUMAN CONSUMPTION
9,460,000,000,000,000,000 cal

55% DIRECT CONSUMPTION
21.78 EJ

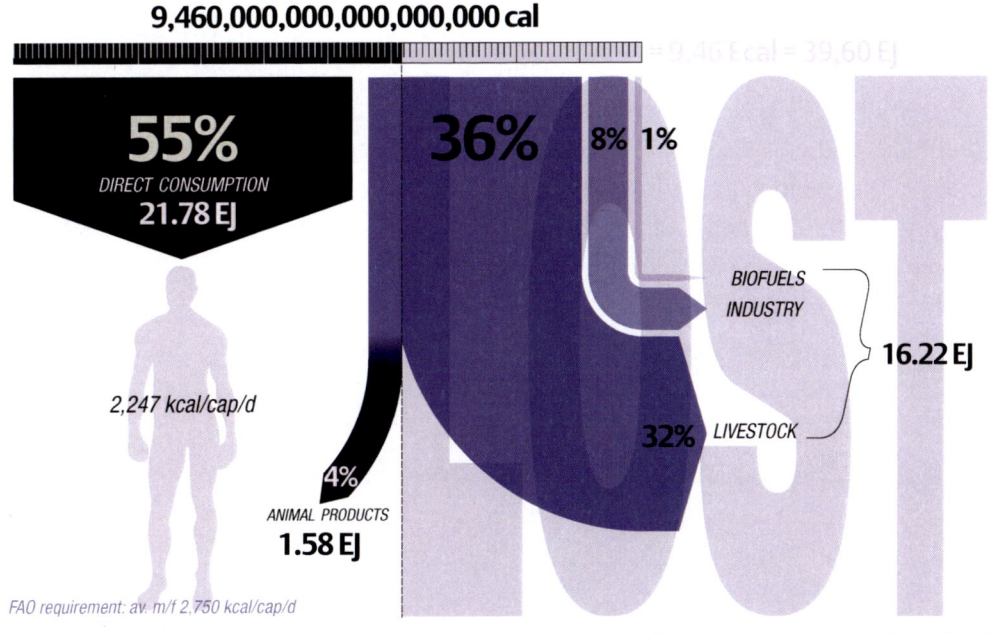

36%   8%   1%

2,247 kcal/cap/d

4% ANIMAL PRODUCTS
1.58 EJ

BIOFUELS INDUSTRY
16.22 EJ

32% LIVESTOCK

FAO requirement: av. m/f 2,750 kcal/cap/d

Fig.6.   World crop production and calories directly consumed by people and losses from food chain

---

1   MILLSTONE, E. & LANG, T. 2008. The Atlas of Food, London, Earthscan. p.62

CROPLANDS EXCLUSIVELY DELIVER CALORIES FOR HUMAN CONSUMPTION?

Higher incomes (Tab.8)[1] are directly connected in the societies where these occur to changes in diets[2], with a move away from more vegetal to more animal products.[3] This consequently increases the footprint related to agricultural production per person. The following table gives an insight about changes of FBS globally from 1961 to 2007 compared to the Gross Domestic Product. China is presented as an exemplary case as a country which greatly changed its FBS - a traditionally vegetal kitchen increased its consumption of animal products fourfold - a change which is most likely to continue and typical for developing countries.

	mass (%)	kcal (%)	proteins (%)
Crops for food	67	55	40
Crops for feed	24	36	53
Industry	6	8	7
Biofuel	3	1	-

Tab.7.   Mass, kCal and proteins lost from FSC

Kastner's et al. findings show that the main drivers „call to dramatically boost global crop production."[4] What would be the potential in radically changing crop production to exclusively direct every calorie to human consumption? What would be the potential?

„Currently, 36% of the calories produced by the world's crops are being used for animal feed, and only 12 % of those feed calories ultimately contribute to the human diet (as meat and other animal products). Additionally, calories edible for humans are used for biofuel production"[5] additionally reducing the available calories for human consumption.

In terms of mass, two thirds are thus produced for direct human consumption. Feed crops „represent 24% of global crop production"[6] on 75 % of all agricultural land, including pastures, which correspond to 36,837,235 km2 of the total of 49,116,313 km2. Crops for industrial uses „including biofuels, make up 9 % of crops by mass, 9% by calorie content (...)"[7]

---

1 GDP-statistics from http://data.worldbank.org/indicator/NY.GDP.MKTP.CD, retrieved 18.08.2015
2 KASTNER, T., IBARROLA RIVAS, M. J., KOCH, W. & NONHEBEL, S. 2012. Global changes in diets and the consequences for land requirements for food. Available: http://www.pnas.org/content/early/2012/04/10/1117054109.full.pdf+html.
3 What also can be observed is that countries with higher GDPs consume less starchy roots or pulses and increase its consumption of sugar and sugar crops and cereals.
4 CASSIDY S. EMILY, WEST C. PAUL, GERBER S JAMES and JOLEY A JONATHAN, 2013. Redefining agricultural yields: from tonnes to people nourished per hectare. Environmental Research Letters, IOP PUblishing, p.1
5 ibid. p.1
6 ibid. p.3
7 ibid. p.3

		1961	2007
World	Animal Products	369	481
	Stimulants	3	8
	Alcoholic beverages	67	70
	Vegetable Oils and Oilcrops	193	325
	Sugar and Sugarcrops	224	233
	Vegetables	50	90
	Fruits	50	76
	Spices	4	8
	Pulses	64	62
	Starchy Roots	178	139
	Cereals	1,174	1,288
	∑	2,376	2,780
	GDP (USD)	8,935,730,000,000	387,627,000,000,000
China	Animal Products	153	622
	Stimulants	-	3
	Alcoholic beverages	26	90
	Vegetable Oils and Oilcrops	116	316
	Sugar and Sugarcrops	58	106
	Vegetables	50	178
	Fruits	10	60
	Spices	2	3
	Pulses	53	13
	Starchy Roots	255	133
	Cereals	1,197	1,417
	∑	1,920	2,941
	GDP (USD)	49,557,050,149	3,523,094,314,821

Tab.8. Changes in food diets related to GDP worldwide and, exemplary China

From the calorie perspective only 55 % of the global crops produced are consumed directly by humans. 36 % of these go to animal feed, „of which 89 % is lost, with the result that only 4 % of crop-produced calories are available to humans in the form of animal products. Another 9% of crop-produced calories are used for industrial uses and biofuels and so completely lost from the food system."[1]

When counting both human-edible crop calories and feed-produced animal calories 59 % of the total production is delivered to the world's food system and 41 % of the total calories „available from global crop production are lost to the food system."[2]

By radically reshaping the production of the food system we could estimate that by not enlarging the agricultural land and by using the actual rate of technologization the calories available for direct human consumption would increase by some 70 %, or in terms of population, it „could potentially feed an additional ~4 billion people."[3]

Addressing challenges for a future food security and considering the actual number of undernourished people „making human consumption a top priority over animal

---

1   ibid. p.4
2   ibid. p.4
3   ibid. p.4

feed and biofuels (...)"[1] emerges as priority. The pressure on the agricultural production is high and will increase. Theoretically there is enough land to feed the population of our world, enough to release 745 million people from undernourishment and hunger and to additionally feed the next 3 billion people. If we again call to mind the numbers of UN-World Population Report statistics and the number of 9.22 billion people at which world population is expected to peak in 2075 before it begins to decline, taking the 2003 statistics (latest reference year of CASSIDY et al.) with a world population of 6,310,000,000 people - we can say that the actual extent of agricultural land can stop world hunger and feed the whole human population until the number will peaks and then it moves to decline.

However, by studying different reports from FAO, e.g. the „World Livestock 2001"-report shows the regrettable inference is a trend going in the opposite direction. The demand for meat and dairy products will increase by more than two thirds, meat consumption will increase by roughly 60 %, biofuel production „has increased sharply in recent years, which has directed more calories away form feed and human food. (...)"[2] Thus land conversion from nature to agricultural land is most likely to continue.

Of particular concern „is the environmental impact of developing new agricultural land. In 1980s and 1990, tropical forests were the source of over 80% of new agricultural land."[3]

The discussion that emerges with the conclusion of the study „Redefining agricultural yields: from tonnes to people nourished per hectare" (exclusively producing food for direct human consumption) is to what extent this might be achievable, whether based on political decisions, which must find social acceptance. It is hard to imagine that the interests of industry and biofuel producers could be stemmed or ignored. Nevertheless the findings of this study explicitly show the dramatic inefficiency of our food system from the perspective of land use but, primarily from that of solar energy conversion to cover the total energy requirement of humans.

Although the potential exists to meet global food demand, the trends go are taking a different way for various reasons.[4] So we could put our question on food security in different terms, for example as: could world food demand be met by increasing yield per hectare?

---

[1] CASSIDY S. EMILY, WEST C. PAUL, GERBER S JAMES and JOLEY A JONATHAN, 2013. Redefining agricultural yields: from tonnes to people nourished per hectare. Environmental Research Letters, IOP PUblishing,. p.6
[2] CASSIDY S. EMILY, WEST C. PAUL, GERBER S JAMES and JOLEY A JONATHAN, 2013. Redefining agricultural yields: from tonnes to people nourished per hectare. Environmental Research Letters, IOP PUblishing,.. p.6
[3] ibid. p.6
[4] ibid. p.6

INCREASING CROP PRODUCTION ON THE CURRENTLY USED AGRICULTURAL AREA TO MEET WORLD FOOD DEMAND

Considering FBS of rich industrialized countries it might be contradictory that the per capita land area needed for food supply is often much lower than that in developing or emerging countries (with the exception of countries with traditional vegetal diets like India or China). But also, for instance, comparing FBS between the USA and Italy one might say that the USA with high meat and other animal product consumption and a high consumption of alcoholic beverages (processed primary product) must have a much higher land use per person. But, in this case it is not so. This is a good example of a comparison between two different countries with a similar GDP but a completely different level of technologization. Whereas the USA is the paragon of the implementation of the Green Revolution principle, Italy's agriculture mostly is industrialized in the north and only marginally in the south. This, together with other reasons, leads to the huge difference in land use per individual food supply of 2,364 m2 (USA) to 3,084 m2 (Southern Europe).[1]

What the developed countries have achieved following World War Two, is to drastically improve yield. Combined with the inventions of the previous century (internal combustion engine, railway and the expansion of the road system) the Green Revolution reshaped the whole agricultural production within a few decades. „(...) [I]nternational centers of agricultural research, financed by large American private foundations (Ford, Rockefeller), selected high-yield varieties of rice, wheat, maize and soya requiring high inputs in fertilizers and treatment products and developed appropriate cultivation methods on experimental stations."[2] The research results, the new cultivated varieties (or optimized varieties) increased yield in many countries. „This large-scale expansion of some elements of the second agricultural revolution (plant and animal selection, mineral and synthetic fertilizers, treatment products, pure culture of genetically homogeneous populations, partial mechanization, strict control of water) to three main grains widely grown in the developing countries was called the „Green Revolution". The benefits, however, hat their focus in the most fertile regions, with higher returns to compensate „the necessary costly inputs."[3]

This brief digression explaining the beginnings of the Green Revolution was necessary to sensitize the reader to an inherent consequence of this: that most likely every intensification of active farming, every intention of increasing productivity, increasing yield per hectare is connected to an increase in energy inputs for (whatever is

---
1 KASTNER, T., IBARROLA RIVAS, M. J., KOCH, W. & NONHEBEL, S. 2012. Global changes in diets and the consequences for land requirements for food. Available: http://www.pnas.org/content/early/2012/04/10/1117054109.full.pdf+html. Table
2 MAZOYER, M. & ROUDART, L. 2006. A History of World Agriculture, London, Earthscan. p. 450
3 ibid.

missing or necessary) building up infrastructure, watering systems (canals or sprinkling systems), greenhouse constructions, mechanization (tractors, trolleys, etc.) or the production of hydrocarbon-based macronutrients, pesticides, herbicides, etc.

> „In general, the sustainability of the food production system is being questioned. Doubts are cast on the possibility to continue doing more of the same, that is, using high levels of external inputs in production, increasing the share of livestock in total output, expanding cultivated land and irrigation, and transporting products over long distances."[1]

The following subchapter aims to examine the actual energy dependency of world agriculture. In addition a short presentation is made concerning which intersection of the food supply chain consumes most of the energy.

---

[1] ALEXANDRATOS, N. & BRUINSMA, J. 2012. World Agriculture Towards 2030/2050. The 2012 Revision. [Rome]: Agricultural Development Economics Division, FAO. p.8

2.3. Energy Consumption

2.3.1. Looking back a century

Approximately 119 EJ a year of all energy used in the world is consumed by the food system, a quarter of this within the farm gate. The green revolution radically changed the production method of the primary sector. Energy inputs per ha soil based agriculture exploded in the last decades.

The world cultivation area nearly doubled from 1900 to 2000, world population increased from 1.500.000.000 to over 6 billion, while the agricultural land only had to be expanded between 80 to 100%. Energy subsidies and technologization increased the yield output up to 600% to feed the exploding human population. In other words, by relating to the question of increasing productivity to the existing land used for crop production, the Green Revolution clearly brought about a great improvement and led to a productivity increase which is sixfold compared to that at the beginning of

CHANGES IN ENERGY CONVERSION IN AGRICULTURE - THE 20th CENTURY

+ 300 %   + 80-100 %   + 600 %   + 8,500 %

WORLD POPULATION | WORLD CULTIVATION AREA | ENERGY HARVESTED | ENERGY SUBSIDIES

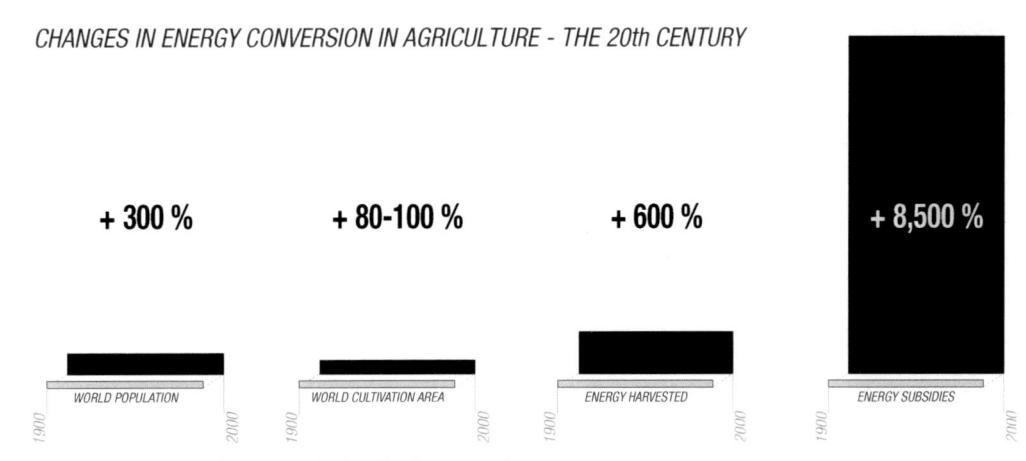

Fig.7.   Increase in Energy Subsidies in the 20th century

the 20th century. The price for achieving this was the enormous intensification of energy subsidies by 8,500 %.

Breaking that global scenario down to wheat cultivation in the U.S., „[i]n 1945 average subsidies of nearly 6 GJ/ha helped to produce about 2.2 t/ha of grain (...). By 2003 subsidies of about 18 GJ/ha aided in harvesting 9 t/ha (..). The energy subsidy rate had tripled, but the efficiency of converting solar radiation into harvested grain had more than quadrupled."[1]

---

1   SMIL, V. 2008. Energy in Nature and Society, Cambridge, Mass. MIT Press. p.304

This increase in crop yield led to the situation that in 1945 one single hectare of agricultural land was able to provide enough grain for 1.5 people, covering 10 MJ/cap/d. Today it is possible to produce grain for up to six people on the same area, by means of an even higher energy supply - 15,7 MJ/cap/day - before losses.[1]

WHEAT ENERGY CONVERSION - THE 20th CENTURY

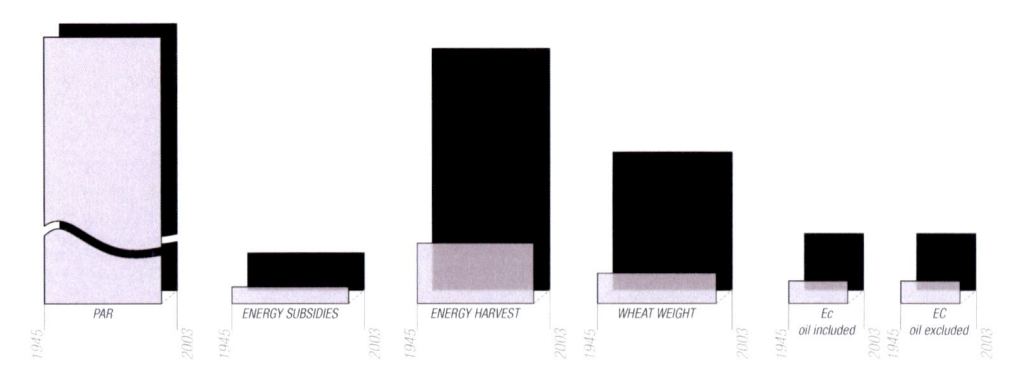

Fig.8.   Energy conversion coefficient change since 1945 using the example of wheat

If higher energy subsidies result in higher productivity, then the conversion efficiency of a cropping system is clearly increasing.[2]

These examples show the obvious success of modern agriculture in the context of increasing productivity, yield and also, the in energy output per hectare.

PEOPLE SUPPLIED PER HECTARE - THE 20th CENTURY

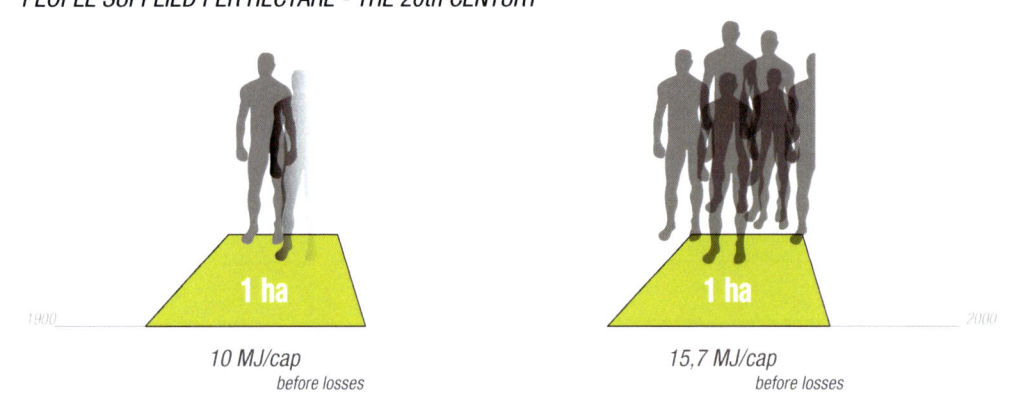

Fig.9.   Increase in productivity per area from 1900 to 2000

On one hand in the context of energy efficiency, it must be said that „[t]he overall magnitude of agricultural energy subsidies is insignificant compared to the input of solar energy."[3] On the other hand, on a global scale, we have now created an agricultural system which is directly dependent upon hydrocarbon energy, not only to

---

1   SMIL, V. 2008. Energy in Nature and Society, Cambridge, Mass. MIT Press. p.301
2   ibid. p.303 and 304
3   ibid. p.300

maintain yields at the levels of today, but there might also be a considerable need to increase the use of fossil fuels in the field of traditional soil based agriculture.

Where exactly, looking throughout the food supply chain, is the most energy input needed? How much is used directly on the farm and how much is needed to process food, to transport and cook it? And, lastly, by considering Vertical Farming as an alternative, what parts of the food sector might have the potential to minimize energy subsidies? The following subchapters aim to establish the status quo for traditional soil based agriculture.

2.3.2. Energy and the Food Supply Chain

The findings of the Food and Agriculture Organization (FAO) study " 'Energy-Smart' Food for People and Climate"[1] confirm the large share of global energy supply required and the strong reliance on fossil fuels „to meet production targets and contribute to greenhouse gas emissions. The study concluded that agrifood systems will have to become „energy smart" to meet future food and energy challenges, and recommended establishing a major long-term multipartner program on energy-smart food systems (...)."[2] One of the main aims and objectives of this study is to „evaluate how the fossil fuel dependency of the transport and processing components of the food sector can be reduced together with energy costs and GHG emissions."[3]

The key findings of this study are the following:

- The agrifood chain consumes 32 percent of the world's available energy - with more than 70 percent consumed beyond the farmgate.
- The agrifood chain produces about 20 percent of the world's greenhouse gas emissions.
- More than one-third of the food we produce is lost or wasted, and with it about 38 percent of the energy consumed in the agrifood chain.

The total world primary energy consumption in 2011 was 549,02 EJ [4] 32 % of world end-energy-consumption is related to the food sector, 24% of it until the farmgate.

The food system is heavily dependent on fossil resources. A study from 2014 claim to a drastic reduce its dependencies, especially by industrialized countries, such as the UK. „Both direct energy use for crop management and indirect energy use for fertilizers, pesticides and machinery production have contributed to the major increases in food production (...)."[5]

The achieved results relate as a principle to „increasingly volatile fossil fuel prices. (...)

---

1   In all quotations used in this dissertation „food sector", „food systems" and „food chain" are used interchangeably. These terms refer to stages from the production on-farm, through manufacturing to consumption.
2   FOOD AND AGRICULTURE ORGANIZATION OF THE UNITED NATIONS 2011.Energy-Smart Food for People and Climate, Issue Paper, Rome:FAO
3   ibid, p.7 and 8
4   http://www.iea.org/publications/freepublications/, retrieved 05.04.2014
5   FOOD AND AGRICULTURE ORGANIZATION OF THE UNITED NATIONS 2011.Energy-Smart Food for People and Climate, Issue Paper, Rome:FAO, p.1

- Fossil fuel prices, particularly those of oil-derived products, will increase significantly over the coming decades and will become more volatile.
- Prices, on a unit energy basis, between oil, gas and coal, are likely to diverge with the possibility of a break in the traditional linkage between gas and oil prices emerging Unless substantive agreements emerge from the UNFCCC's[1] intergovernmental negotiations that limit access to coal, its large and widely distributed reserves will mean that it is the least vulnerable of the fossil fuels to price increases; a switch to coal away from oil and natural gas is probably where that is possible e.g. for processing and nitrogen fertilizer production.
- The world's major crops are dependent on different shares of their energy inputs from oil, gas and coal. Thus, relative changes in fossil fuel prices will affect each crop type differentially."[2]

Major areas of concern are identified in
- Fuel use for tillage, transport from farmgate to storage to proecessing and end use will be directly affected by increasing oil prices.
- Nitrogen fertilizer prices are immediately affected by increasing natural gas prices.
- „Coal is still used for nitrogen fertilizer production, particularly in China, and is likely to be least affected by worries about reserve depletion. From a GHG perspective, a switch away from oil and gas to coal, rather than to renewable, would be detrimental.
- Increased costs for direct and indirect energy inputs for agriculture may lead to lower yields for the world's major agriculture commodity crops. In turn, this is likely to bring an expansion of land areas unter these crops, leading to increased GHG emissions, as a result of LUC, and increased prices owing to less efficient production.
- Significant land expansion will also have detrimental effects on biodiversity and possibly on water resources."[3]

---

1   The United Nations Framework Convention on Climate Change is an environmental treaty on an international level, a climate policy venue with a broad legitimacy due its universal membership, negotiated at the Earth Summit, Rio de Janeiro (3.-14.6.1992), the UNCED (United Nations Conference on Environment and Development) intentionally to stabilize greenhouse gass concentrations in the atmosphere at a level to prevent „anthropogenic interference with the climate system".

2   FOOD AND AGRICULTURE ORGANIZATION OF THE UNITED NATIONS 2011.Energy-Smart Food for People and Climate, Issue Paper, Rome:FAO, p.1

3   WOODS, J., WILLIAMS, A., HUGHES, JOHN K., BLACK, M., MURPHY, R. 2010, Energy and the Food System, Philosophical Transactions of the Royal Society. 2991-3005. Available: http://rstb.royalsocietypublishing.org/content/royptb/365/1554/2991.full.pdf, retrieved 14.10.2015

An issue that should be questioned is how the situation may develop by 2050. The use of fossil fuel in agriculture will in all probability not decrease over the coming decades. Future prospects on world population growth, changes in per capita-income, especially in developing countries and changes in diets suggest the assumptions that dependencies between the primary sector and fossil fuel will increase, not least because an intensification in farming practices to increase yield per hectare might only be achievable by increasing the use of macronutrients, supplemental watering systems and additional production and use of machinery.

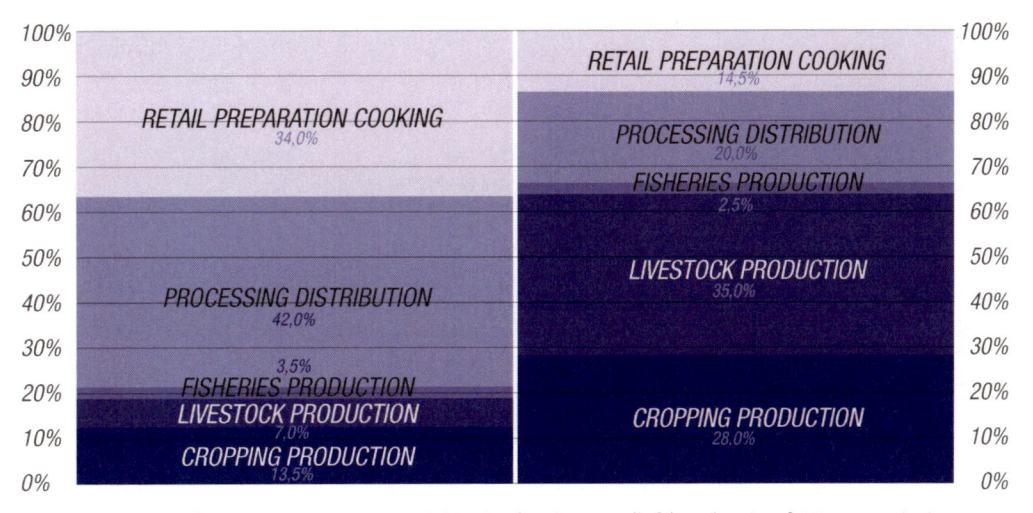

Fig.10.   Ratio of energy consumption within the food sector (left) and ratio of CO2eq. emissions

ENERGY CONSUMPTION FROM FARMGATE TO FARM GATE

„Once conventional oil and gas flows reach a peak as is predicted, the food sector's continued reliance on these non-renewable resources for production, processing and transportation activities will lead to greater business risks, especially from unpredictable price spikes."[1]

Fossil fuel is needed through the entire food supply chain. But where is most of the overall energy consumed within the food sector? Directly on the farm, or up to the farm gate (without the inclusion of energy consumption for products or media used and needed for cultivation), and human and animal power excluded, the world energy demand on farm is estimated with 6 EJ per year. Around 50% of that energy is consumed by OECD-countries. Fig. 10 shows the ratios of energy consumption within the food sector and the related CO_2-emissions.[2]

Indirect energy demand for food production is 50% higher, namely 9 EJ per year. This value includes energy demands for boats, tractors and other farm machinery - operations and fertilizer manufacturing.[3]

ENERGY DEMAND FOR PRIMARY PRODUCTION

The biggest variations in energy demand for primary food production are at the level of the farm and the crop itself. „The energy demand for the production of similar food products under different production systems can be used to compare fossil fuel dependency. For example, the direct energy inputs of an extensive, unsubsidized, grazing enterprise in Australia (2-3 GJ/ha) can be compared with intensive, subsidized, dairy farming systems in the Netherlands (70-80 GJ/ha).[4]

In terms of energy consumption in 2005 tractors and other agriculture machinery consumed around 5 EJ of diesel for land development, transport and field operations. „A further 1.5 EJ per year was used for the manufacture and maintenance of tractors and farm implements."[5] Exact numbers of two-wheel design agricultural machinery, primarily used by small-scale farms is difficult to gather and therefore not covered by the 5 EJ. Additional machinery such as balers, combined harvester-threshers, manure spreaders, fertilizer distributors, milking machines, ploughs, root and tuber harvesting machines, threshing machines, seeders, planters and trans-

---

1　FOOD AND AGRICULTURE ORGANIZATION OF THE UNITED NATIONS 2011.Energy-Smart Food for People and Climate, Issue Paper, Rome:FAO, p.9
2　ibid. p.III
3　ibid. p.13
4　FOOD AND AGRICULTURE ORGANIZATION OF THE UNITED NATIONS 2011.Energy-Smart Food for People and Climate, Issue Paper, Rome:FAO, p.13
5　FOOD AND AGRICULTURE ORGANIZATION OF THE UNITED NATIONS 2011.Energy-Smart Food for People and Climate, Issue Paper, Rome:FAO

planters are not included in this calculation. FAO is currently updating and extending the lists of machinery used for primary production. Current data are only available for parts of the above mentioned items, but only in import and export data and not for those units in working use.[1]

FISHERIES AND AQUACULTURE

World fish production per year is approximately 130,000,000 t. In 2012 136,000,000 t were produced for direct human consumption (86 %) and roughly 22,000,000 t were destined for non-food uses such as fishmeal or fish oil, ornamental purposes, for culture, bait pharmaceutical uses, etc.[2] Some 2 EJ are consumed directly by the global primary production. This figure is mainly associated with fish aeration, water pumping and diesel propulsion for vessels and boats. 400 PJ or 0,4 EJ per year „of indirect energy is embedded in aquaculture feedstuff."[3]

IRRIGATION

As we have seen in the former subchapter, of the potential 44,950,000 km2 only a third (14,120,000 km2) is a net balance of land with rain-fed potential. If pressure on agricultural land in existence at present increases further and land conversion for agriculture is needed, irrigation demand will most likely also increase. The current consequences of world climate change already affect vast landfills with water shortages, as can be observed in various parts of the world. Around 17% of all agricultural land in use is irrigated. These 2,760,000 km2 and the additional water consumption for agricultural production together consume 70% of all freshwater withdrawals. By comparison Industry and domestic use consume 22% and 8% respectively.[4]

As a result of increasing the agricultural area during the past century demand for irrigation also increased, made possible by developing more efficient technologies for watering and by using hydrocarbon energy to operate the watering plants. „Traditional agriculture provided the needed water by simple open-ditch irrigation fed by gravity flows or by a variety of human- or animal-powered devices (...). Modernizing agricultures retain the inefficient ridge-and-furrow arrangements and supply them with simple mechanical pumps. (...) The global dependence on irrigation has trebled since the end of World War II, when about 75 million ha of cropland were watered. A

---

1 http://faostat3.fao.org/download/I/RM/E, retrieved 12.10.2014
2 FAO. The state of World Fisheries and Aquaculture 2014, Rome. p. 42
3 FOOD AND AGRICULTURE ORGANIZATION OF THE UNITED NATIONS 2011.Energy-Smart Food for People and Climate, Issue Paper, Rome:FAO, p.9
4 http://www.ifad.org/english/water/key.htm, retrieved 06.09.2015

generation later the total was 140 million ha, and by 2000 the figure topped 275 million ha, with three-fifths in Asia and nearly one-fifth in China alone."[1]

Some 0,225 EJ per year are needed to power the pumps for irrigation. The mechanical pumping of water on approximately 10 percent of the world arable land area (around 300 Mha) consumes around 0.225 EJ/a to power the pumps. In addition, another 0.05 EJ/a of indirect energy is required to manufacture, deliver equipment for irrigation.[2] Irrigated lands produce higher yields than rainfed systems and allow for yield increases of up to threefold. These lands areas for example, provide 40 % of the global cereal supply.[3]

MACRONUTIRENTS

The synthesis of nitrogenous fertilizers, also known as macronutrients such as nitrogen, phosphate and potassium account for approximately 7 EJ/year (accounting approximately 5 % of the world gas consumption per year).[4] „Chemical fertilizers represent the largest indirect energy subsidy in nonirrigated farming. No other innovation has contributed so much to increased yields as the three macronutrients (...)[5]- phosphate, nitrogen and potassium.
The production of phosphates requires more than 50% of the total embodied energy consumption for macronutrient production (or 4-5 GJ/t), 2,8 EJ, production of nitrogen 1,85 EJ (or 55 GJ/t) and Potassium 0,25 EJ (or 5 to 20 GJ/t). Every year 198,5 million tonnes of macronutrients are produced. The biggest share is phosphates with 140 Mt, followed by nitrogen with 33,5 Mt and potassium with 25 Mt.
By dividing the area of arable land referred to above (15,529,767 km2) with the annual total use of nitrogenous fertilizers (198,5 million tonnes) on one single hectare on average roughly 130 kg of macronutrients is used. This, of course, is unevenly distributed worldwide - with no use at all in Sub-Saharan Africa and up to 500 kg/ha/a in China.[6]

PEST CONTROL

The explosion in macronutrient production was „accompanied by increasing use of herbicides to control weeds, and insecticides and fungicides to raise the yields of new high-yielding varieties. (...) Like most of the changes in modern agriculture we find its application origins after World War I and World War II."[7] Starting around 1945 the first

---

1   SMIL, V. 2008. Energy in Nature and Society, Cambridge, Mass., MIT Press. p. 294
2   ibid. p. 294
3   http://www.fao.org/docrep/004/y3557e/y3557e08.htm, retrieved 28.08.2015
4   ibid. p.9
5   SMIL, V. 2008. Energy in Nature and Society, Cambridge, Mass., MIT Press. p.294
6   ibid, p. 295
7   ibid. p.295

weed and insect controls were used on agricultural fields, today more than 50,000 different types of products used as pesticides have been registered „to fit thousands of specific applications, but the bulk goes to only handful of corps. Pesticide applications are highly effective and economically rewarding. Although the compounds are derived by energy-intensive processes from petrochemical feedstocks, their low application rates translate to only minor subsidies in absolute terms."[1]

With around 500 PJ, or 0.5 EJ per year production of herbicides and pesticides contributes to the energy consumption of the world agriculture system. These data relate to the year 2000. „Synthesis of common active ingredients atypically requires 100 - 200 MJ/kg, and total energy costs, including formulating, packaging, and marketing, are mostly 200 - 300 MJ/kg (...)."[2]

Making allowance for a conservative average of 150 MJ/kg as the sum of energy subsidies for synthesis including the embodied energy for formulating, packaging and marketing today's energy consumption we can assume that the global energy consumption for pest control has increased by around 10 % and accounted for 0.55 EJ for the year 2011.

CROPS AND ENERGY SUBSIDIES

Standardized methods for energy calculations on crops are still lacking in the world today. Compared to mass-produced industrial goods, it is extremely difficult to achieve uniform data standards. „Published energy costs of individual crops are not readily comparable because of the arbitrary and non-uniform choice of analytical boundaries and sometimes substantial differences in input equivalents."[3] Energy analysis for crop productions sometimes calculate machinery and irrigation, others include embodied energy of macronutrients and pest-control products. The system border is not always precisely described. Some analysis stops at the farmgate, others include transportation energy etc.

„Typical annual rates are 8-15 GJ/ha for dryland cereals, 20-25 GJ/ha for rain-fed, and more than 40 GJ/ha for irrigated Corn Belt corn and California rice. Nitrogen-fixing soybeans need no more than 8-15 GJ/ha, but potatoes, vegetables and tree crops, with heavy fertilization and irrigation, need 50-100 GJ/ha, and orange groves require about 120 GJ/ha and even these rates are dwarfed by hydroponic cultivation."[4]

1　SMIL, V. 2008. Energy in Nature and Society, Cambridge, Mass., MIT Press. p. 297
2　ibid. p.296
3　ibid. p.296
4　ibid. p.296

GREENHOUSES

Fruit and vegetable production in greenhouses is characterized, when compared to soil based agriculture, above all by the need for seasonal heating. In some cases, especially in rich industrialized countries, seasonal light requirements are partly provided by artificial lighting in order to enlarge the productive season and secure the crop rotation.

„In warm climates operation costs are dominated by irrigation, fertilization, and cultivation needs. Turkish rates (with some supplemental heating range from 2.6 GJ/t of tomatoes to 4.3 GJ/t of peppers[1], but in heated Dutch greenhouses the same crops may consume as much as 40 GJ/t and the heating rate may be several TJ/ha. These subsidies translate to about 3-4 GJ/t of Manitoba spring wheat, 5 GJ/t of Iowa corn, and up to 7 GJ/t of rice. Although vegetables and fruits need much higher energy subsidies per hectare, their high yields translate into GJ/t rates that are similar to those for cereals."[2]

Volatile fuel prices and increasing demand for greenhouse crops through increasing pressure on food security and food supply on traditional soil based agriculture are caousing a greenhouse explosion in terms of both numbers and the area cultivated. Due to the fact that greenhouses are inherently energy inefficient production entities (U/V-ratio), especially in temperate climate zones, different strategies for energy reduction get applied and developed. Although greenhouses in specific cases need more energy for cultivation than traditional soil based agriculture, their use is inevitable if fresh food is to be provided for certain areas with constraints on the available arable land, disadvantageous climate or water scarcity. Increasing demand for organic and, locally grown food in the past few decades has also prompted a remarkable growth in urban greenhouses.

Energy efficient lighting, the insertion of energy screens, computer controlled greenhouse climates, air leaks reduction and, efficient ventilation systems are required to reduce the energy demand of greenhouses.

---

1   SMIL, V. 2008. Energy in Nature and Society, Cambridge, Mass., MIT Press. p.296
2   ibid. 297

Greenhouse crop production is now a swiftly expanding area and a fact of life throughout the world with an estimated 405,000 ha of greenhouses spread over all the continents. The degree of sophistication and technology depends on the specific local climatic conditions and socio-economic environment.[1]

A yield of 300t/ha fruit or vegetables in the Mediterranean countries is not unusual. A fact relevant for the following simulation model, is that by comparing the soil based agricultural yield of tomatos in Austria of 27.2 kg/m2/crop rotation to the yield achieved in greenhouses in the same country with some 43 kg/m2/crop rotation it emerges clearly that a remarkable production increase can be expected from this method.[2]

As a rule of thumb calculation to estimate the total energy requirement for world greenhouses (without embodied energy) the following might apply:

estimation: approx. 40 MJ/kg[3], 405,000 ha = 4,050 km2 of greenhouses[4]:

405.000 ha x 150 t/ha (50% = OECD)[5] x 20 MJ/kg (=50% - high GDP/low GDP) = 60.750.000 t (greenhouse crops world / a) x 20 MJ/kg = 1.215.000.000.000 MJ = **1.215 EJ**[6]

An estimated sum of 1.2 EJ is thus defined for operating world greenhouses.

---

1 FAO Good agriclutural practices for greenhouse vegetable crops.pdf, p.9
2 www.wien.gv.at/statistik/wirtschaft/tabellen/gemüseernte-anbauflaeche.thml, retrieved 26.10.2104 and www.statistikaustria.at, Ernteerhebung, retrieved 26.10.2104
3 FOOD AND AGRICULTURE ORGANIZATION OF THE UNITED NATIONS 2011.Energy-Smart Food for People and Climate, Issue Paper, Rome:FAO p.15
4 http://www.fao.org/docrep/018/i3284e/i3284e.pdf, p.vii, retrieved 21.6.2014
5 OECD, 2008. Environmental Performance of Agriculture at a Glance, p.24, http://www.oecd-ilibrary.org/docserver/download/5108091e.pdf?expires=1446819166&id=id&accname=guest&checksum=5CC696AEB345D9B622DE8C6987704E98, retrieved 12.09.2015
6 own calculation estimation

FOOD AND TRANSPORT

Food and transport is a much discussed and controversial issue, the opportunities or problems of which are completely dependent on the point of view taken. While more and more people (especially the urban population) ask for locally or regionally grown products[1], which are associated with a low CO2-footprint, the findings of several studies and their resulting recommendations for policies ask by contrast, for a stronger food allocation related to the best fitting climatic condition. In fact, in very specific situations, a locally grown food that is consumed could have a larger CO2-footprint than an imported one. Locally produced apples, for example, can do better in terms of ecology. „If an apple, grown in the southern hemisphere, is offered for sale during the spring time in a German supermarket, it will generally appear more appetizing than a local apple from last year's crop that has been stored in the interim period."[2] It is springtime, the local apple was harvested about six month ago. Since then huge amounts of energy have been expended for washing, storing and cooling the apple. This is why regional food is „seriously challenged"[3] from the ecological perspective when compared to the fresh Chilean apple. Harald von Witzke also refers to a detailed environmental assessment of 150 of the most popular products in UK-supermarkets. „A team from the Manchester Business School (...) examined the entire production process end-to-end, from the harvest all the way though to retail packaging."[4] Key findings are that it is often true that imported goods have a worse CO2-balance, compared to locally grown food.

The other side of the picture, once again in a brief reference to a UK context, is that more than 80% of all food consumed in London is imported from outside the UK. The data in this paragraph summarize major findings of the study on „The Validity of Food Miles as an Indicator of Sustainable Development."[5] Referring to Harald von Witzke's point, we can confirm that transport-emissions related to seas (and overseas) only contribute some 12 % to the total CO2-emissions associated with UK food transport.[6]

Likewise we can assume that the higher the dependency on food imports, the bigger the percentage of food miles on land will be. A quarter of all driven miles in the UK

---

1   The terms „local" or „regional" are not strictly defined. „Local" is often described from 100 - 250 miles. „Locavore: A consumer who primarily eats minimally processed, seasonally available food grown or produced within a specifi ed radius from his or her home, commonly 100 or 250 miles", Local Food SystemsConcepts, Impacts, and Issues vSteve Martinez et al., USDA, May 2010
2   WITZKE, H. V. 2011. Bananas from Bavaria?, Augsburg, Ölbaum-Verlag. p.17
3   ibid. p.18
4   ibid. p.18
5   WATKISS, P., SMITH, A., TWEEDLE, G., MCKINNON, A., BROWNE, M., HUNT, A., TRELEVEN, C., NASH, C. & CROSS, S. 2005. The Validity of Food Miles as an Indicator of Sustainable Development: Final report produced for DEFRA. AEA Technology.
6   ibid. p.iii

are food related. A further point to be underlined here is the strong impact this has on social costs, too. The total costs generated by the UK food transport system, including costs related to CO2-emissions, air quality, congestion, accidents, the infrastructure and its maintenance and also animal welfare are estimated at over 9 billion pounds/a.[1]

There is a strong movement for re-implementing local social and economic interdependencies, related to food production, retail sales and consumption. „The growth of market share of the supermarkets with the associated decline in local shops and markets, and the increase in international food trade, have led to a move away from locally produced food in the UK."[2] This issue is addressed by several national campaigns (Eat the View), locally run farmers' markets, farm shops and, community-led initiatives such as community growing projects, to name only a few. A new culture is rapidly growing - not only in Great Britain - arising from the necessity of urban population of consuming locally grown food.[3]

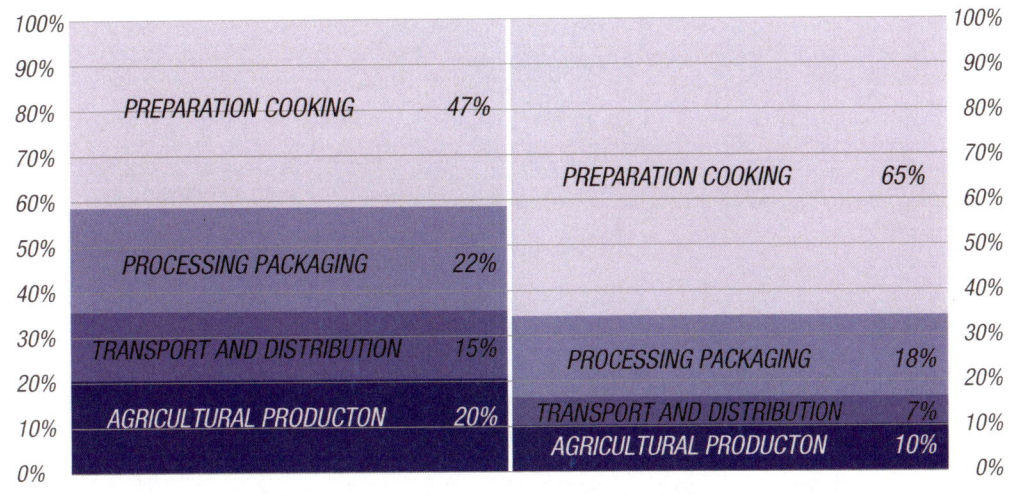

USA		Africa	
PREPARATION COOKING	47%	PREPARATION COOKING	65%
PROCESSING PACKAGING	22%	PROCESSING PACKAGING	18%
TRANSPORT AND DISTRIBUTION	15%	TRANSPORT AND DISTRIBUTION	7%
AGRICULTURAL PRODUCTON	20%	AGRICULTURAL PRODUCTON	10%

Fig.11.   Energy consumption within the food sector comparing USA (left) and Africa

In the context of FAO's Issue Paper transport is included under the heading „processing and distribution". In order to make an assumption on what the percentages for processing and distribution (transport) are, we make a comparison between a high-GDP-country such as USA and the average statistics for the continent of Africa. (Fig.12) In this we can see that agricultural production in USA has twice the weighting

---

1   WATKISS, P., SMITH, A., TWEEDLE, G., MCKINNON, A., BROWNE, M., HUNT, A., TRELEVEN, C., NASH, C. & CROSS, S. 2005. The Validity of Food Miles as an Indicator of Sustainable Development: Final report produced for DEFRA. AEA Technology. p.41 ff.
2   ibid. p.15
3   http://www.theguardian.com/sustainable-business/2014/jul/02/next-gen-urban-farms-10-innovative-projects-from-around-the-world, retrieved 28.08.2015

compared to Africa in terms of energy use of the total energy inputs, but it comes in up to a third lower for preparation and cooking.

Processing and packaging is similar, namely 22% (USA) vs. 18% (Africa).[1] Transport and distribution although again has twice the volume in the United States compared to the situation in Africa. Drawing an inference for the world on this basis to the itemization of the overall world energy consumption in the food sector, we thus roughly assume that a third of the 42% related to processing and distribution mentioned falls to transport, and two thirds to processing.

This number might be not far from the real situation, considering the FAO estimate on global food miles.[2] Some 800 million metric tonnes of food were shipped worldwide in the year 2000. In 2011 world food imports- and exports totaled up to 1.2 billion metric tonnes. By using the shares of global t/km as presented here related to different means of transportation and comparing the extrapolated values of total energy consumption based on the sum of the transported food items, and comparing the estimated 33.3% energy related to the 42% share of processing and distribution mentioned by the FAO-Issue Paper, we carrive at a figure of 24.64 EJ (of 550 TPES). This would lead to the conclusion that all transported food, consignments will all have travelled more than 600 km on a global average.

	share of global transport (% of total t/km)	[t]	Global shares of local distribution transport (% of total t km)	Energy intensity of travel mode (MJ/t km)	Energy intensity of travel mode av. (MJ/t km)	av.10 km (MJ)	av.100 km (MJ)	av.1000 km (MJ)
RAIL	29	348,000,000	16	8-10	9	31,320,000,000	313,200,000,000	3,132,000,000,000
MARINE SHIPPING	29	348,000,000	not applicable	10-20	15	52,200,000,000	522,000,000,000	5,220,000,000,000
INLAND WATERWAY	13	156,000,000	19	20-30	25	39,000,000,000	390,000,000,000	3,900,000,000,000
ROAD TRUCKS	28	336,000,000	62	70-80	75	252,000,000,000	2,520,000,000,000	25,200,000,000,000
TROLLEY, CYCLE, TRACTOR	data not available	data not available	3	variable				
AVIATION	1	12,000,000	0	100-200	150	18,000,000,000	180,000,000,000	1,800,000,000,000
	100	1,200,000,000	World FBS 2011			392,520,000,000	3,925,200,000,000	39,252,000,000,000
		800,000,000 Vaclav Smil 2008			EJ	0.39	3.93	39.25

world population in 2000:	6,055,049,000
world population in 2011:	6,847,859,000

Tab.9.   World food transport

Before we combine the overall data on world energy consumption within the food system, we first wish to highlight and summarize the previously mentioned findings on energy consumption within the farm gate according to the FAO-Issue Paper, which refers largely to the study by Vaclav Smil on „Energy in Nature and Society"[3].

Fig. 12 on the next page shows the overall energy consumption up to the farm gate. The embodied energy involved is also included in this calculation. 7 EJ is used to produce nitrogen, phosphate and potassium, while 6 EJ is required in the food sector consumption directly on on the farm for water pumping, housing livestock, cultivation practices, harvesting, heating, drying and storing crops.

---

1   FOOD AND AGRICULTURE ORGANIZATION OF THE UNITED NATIONS 2011.Energy-Smart Food for People and Climate, Issue Paper, Rome:FAO, p.13
2   Estimates based on the statistics published by FOOD AND AGRICULTURE ORGANIZATION OF THE UNITED NATIONS 2011.Energy-Smart Food for People and Climate, Issue Paper, Rome:FAO, p.16
3   SMIL, V. 2008. Energy in Nature and Society, Cambridge, Mass., MIT Press.

The indirectly consumed 9 EJ refers to tractors, use of additional farm machinery, operating boats and also for fertilizer manufacturing. The energy involved for the total fuel consumption and maintenance is included in the 7 EJ of machinery. 300 PJ (Petajoule) are needed for irrigation and additional 500 PJ for pesticides. Fishery and aquaculture worldwide consume 2.35 EJ, while roughly 2 EJ are needed for breeding, raising and keeping livestock and 1.25 EJ are required for the energy consumption of greenhouses.

In summary the energy consumption of the food sector exceeds 33 EJ before leaving the farm gate. Comparing this with the findings of the FAO Issue paper where 24% of the energy consumed by the food sector is used, we reach a number of 42.24 EJ, by using the number of 550 EJ TPES, representing a difference of approximately 9 EJ.

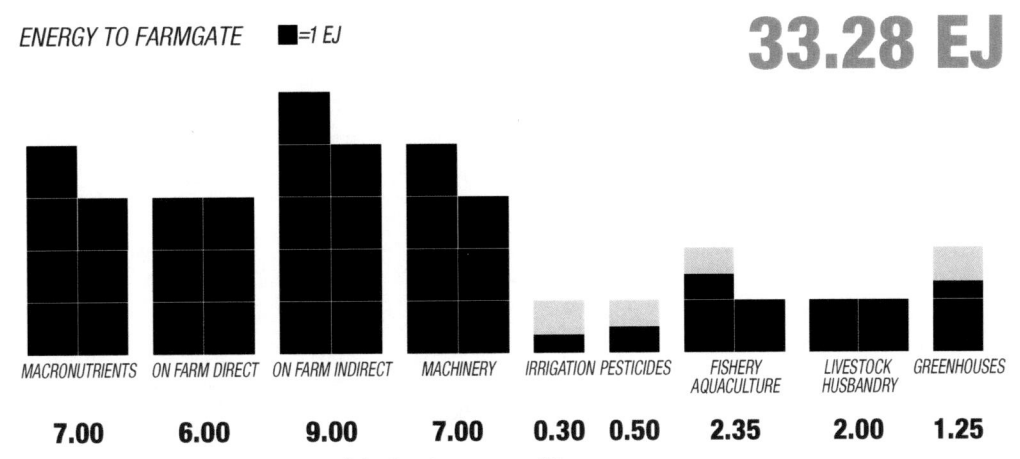

ENERGY TO FARMGATE ■=1 EJ **33.28 EJ**

MACRONUTRIENTS	ON FARM DIRECT	ON FARM INDIRECT	MACHINERY	IRRIGATION	PESTICIDES	FISHERY AQUACULTURE	LIVESTOCK HUSBANDRY	GREENHOUSES
7.00	6.00	9.00	7.00	0.30	0.50	2.35	2.00	1.25

Fig.12. Energy consumption of the food sector until farmgate

LANDUSE, BIOCAPACITY AND ENERGY CONSUMPTION

FOOD LOSSES AND WASTES AT DIFFERENT FSC STAGES, EXAMPLE FOR FRUIT AND VEGETABLE

CONSUMPTION
DISTRIBUTION
PROCESSING
POST HARVEST
AGRICULTURE

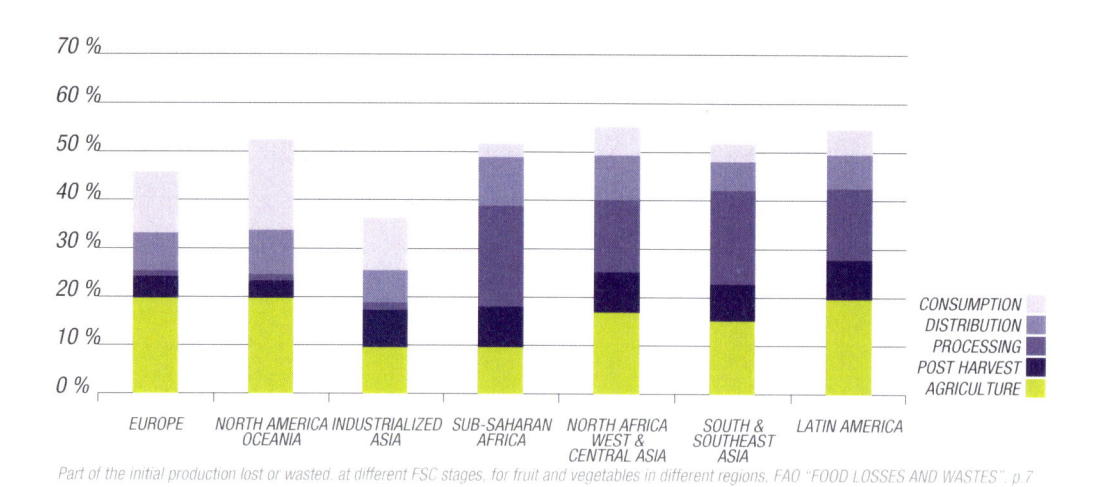

Part of the initial production lost or wasted, at different FSC stages, for fruit and vegetables in different regions. FAO "FOOD LOSSES AND WASTES", p.7

FOOD LOSSES AND WASTES PER CAPITA

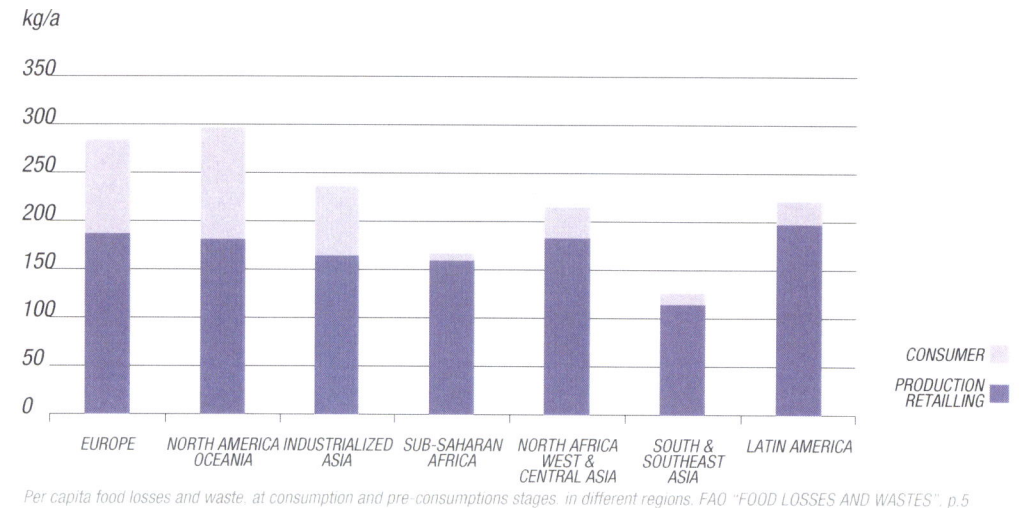

kg/a

CONSUMER
PRODUCTION RETAILLING

Per capita food losses and waste, at consumption and pre-consumptions stages, in different regions. FAO "FOOD LOSSES AND WASTES", p.5

PRODUCTION VOL. PER REGION (million tonnes)

EUROPE
N-AMERICA, OCEANIA
INDUSTRIALIZED ASIA
SUB-SAHARAN AFRICA
N-AFRICA, W& CENT. ASIA
S& SE ASIA
LATIN AMERICA

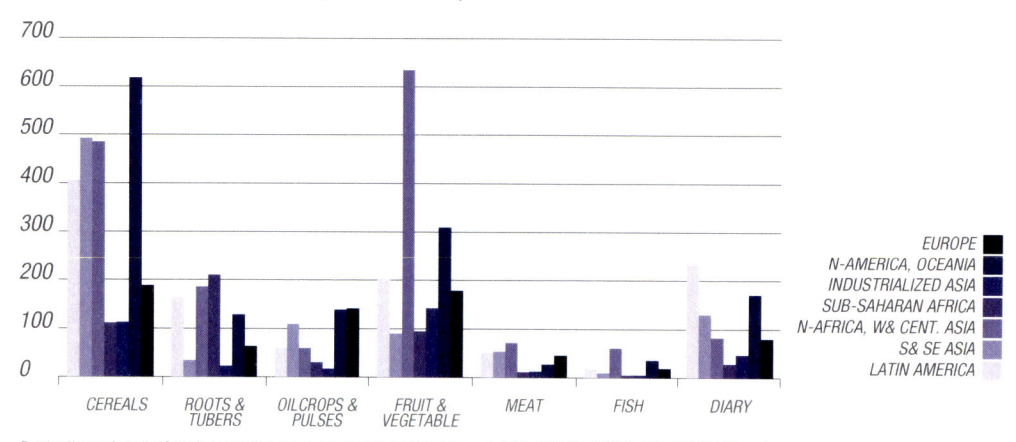

Production volumes of each comodity group, per region (million tonnes). FAO "FOOD LOSSES AND WASTES", p.4

Fig.13. Food losses and wastes related to the food supply chain, per capita and Food production quantity per world regions

FOOD LOSSES AND FOOD WASTES

A serious problem of our food system is the fact that we do not produce our food where we consume it. The fragmented food supply chain has the result that calories "leading to edible products destined for human consumption"[1] gare lost or wasted. "Food losses", as defined by the FAO encompass all losses from the farm to food processing, whereas "food wastes" refer to losses of the consumer – from the retail and catering trades through to private households.[2]

Findings of a representative study for Germany[3] sketch a typical picture for an industrial country, where the percentage – specifically here 25-31% of all the calories lost for human consumption are allocated as "food losses" and the rest (two thirds to three quarters) are assigned to "food wastes". This study also points out that there is a lack of comparable methodological standards to provide a solid basis for comparison between the different studies relating to this issue.

Some 6.6 million tonnes of food are wasted in Germany, every year, which correspond to approx. 80 kg/capita.[4] In this context the attempt has been made to answer another interesting question: What portion of the total calories lost are inevitable losses, and what portion is system immanent? On a consumer level this study refers to Cofresco[5] a study which estimates that around 59% of all the food waste (which in the case of Germany would amount some 3.6 metric tonnes of food) could be saved from being lost to the food supply chain.

The last study published by the FAO can be referred to for an estimation of "food losses" and "food wastes" statistics on a global scale. This is an ongoing research project in which standardized methodologies are now being developed. Until now questions on how much food is lost or wasted prompted responses such as "impossible to give precise answer, and there is not much ongoing research in the area."[6]

Nevertheless some assumptions relevant to our work should be presented:

- Nearly one third of food produced for human consumption is lost or wasted every year.
- The food losses and wastes in the industrialized countries are ten times higher than food losses and food wastes in developing countries.
- In low income countries the main drivers of food losses and food wastes are connected to "financial, managerial and technical limitations in harvesting techniques,

---

1 GUSTAVSON et al. 2011. „Global food losses and food wastes", Study conducted for the International Congress ‚SAVE FOOD!', at Interpack 2011, Düsseldorf, Germany, FAO, p.2
2 NOLEPPA, S. & WITZKE, H. V. 2012. Tonnen für die Tonne. Berlin: WWF. p. 20
3 ibid.
4 ibid. p.23
5 COFRESCO. 2011. Vermeidbare Lebensmittelverschwendung in europäischen Haushalten: Erkenntnisse und Lösungsansätze. Minden: Cofresco
6 NOLEPPA, S. & WITZKE, H. V. 2012. Tonnen für die Tonne. Berlin: WWF. p. 20

storage and cooling facilities in difficult climatic conditions, infrastructure, packaging and marketing systems."[1]

- In high GDP countries, however, there are different main reasons cited for the problem. There is also a lack of "coordination between different actors in the supply chain."[2] Food is wasted because of defined quality standards. Food which is to all appearances healthy and fresh can be thrown away, because it is lacking some pre-defined aesthetic properties.
- On a consumer level people tend to buy larger quantities in rich countries, which if not consumed in time tends to be wasted.
- The "best-before-date" practice also leads to vast amounts of food waste.
- In highly industrialized countries food is also lost "when production exceeds demand.[3]

The diagrams in Fig. 13 on page 82 show exemplary food losses and wastes at different FSC stages for fruit and vegetables in different regions of the world. A striking fact here from comparing industrialized regions with developing areas of the world is how different types of calories get lost from the food supply chain. The same is also the case on a comparisons of the ratio between food losses and wastes based on a consumer level and production/retailing level. The last diagram represents the production volumes of each commodity group per region in million tonnes.

If food is lost or wasted throughout the production and distribution processes in these huge amounts, then as a matter of course energy is also wasted. And considering the quantity of energy consumed in the food sector, it must be pointed out that the issue of food losses and food wastes is also a serious energy problem. In addition to this it is also an ethical problem. Food losses and waste amounting to 30% of the food produced in 2011 could feed some 1.5 billion people, calculating an average consumption of 900 kg/year, and assuming that all the food items required for healthy nutrition are lost or wasted in equal quantities.

---

1   NOLEPPA, S. & WITZKE, H. V. 2012. Tonnen für die Tonne. Berlin: WWF. p. 20
2   ibid.
3   Global food losses and food wsastes, executive summary, p.2

LANDUSE, BIOCAPACITY AND ENERGY CONSUMPTION

Agricultural area for food production per person and year vs. ecological food footprint per person and year (world average)

Agricultural area for food production per person and year (world average)

1,700 m²
8,000 m²
1,700 m²
2,100 m²

area needed for kg consumption

WORLD AGRICULTURAL AREA IN USE 15,000,000 km2
WORLD BIOCAPACITY 44,000,000 km2
WORLD LAND MASS 150,000,000 km2

Energy consumption within farmgate: about 24 % of the overall energy consumption of the food system

The food sector's energy consumption is about 30 % of the world's overall end-energy consumption

110 EJ
95 EJ
373 EJ
550 EJ
177 EJ

WORLD END-ENERGY CONSUMPTION 2011

ENERGY LOSS

WORLD TOTAL PRIMARY ENERGY SUPPLY (TPES)

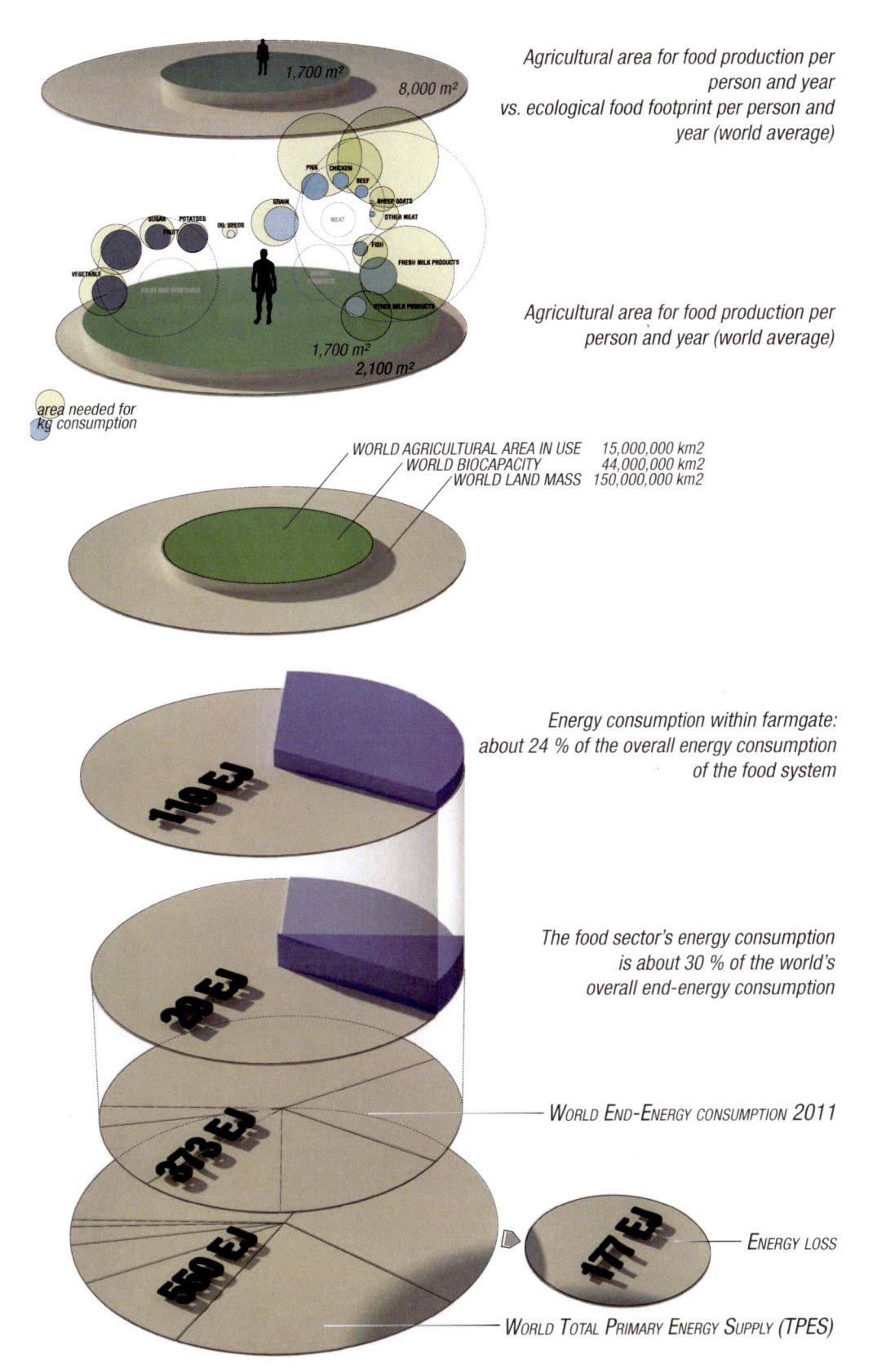

Fig.14.   World energy consumption, energy consumption of the food sector, food footprint per person and world biocapacity

2.3.3. Sketching the Big Picture of World Energy Consumption by the Food System

By bringing the different studies together to sketch the big picture of world energy consumption in the food system, and overlaying it with the world energy consumption for the year 2011 we can come up with the statistics for current end energy consumption. By the end of this chapter the TPES will be generated, but due to the lack of data the percentage of energy consumption within the food sector will be the same as the total primary energy consumption.[1]

29 EJ are used for agricultural production, 24% of the overall consumption of the food sector. The biggest portion is used for crop production, followed by livestock production and fisheries. 14% or 16.6 EJ has been assigned to transport, as mentioned in the subchapter above and one third of the energy or 33.3 EJ. Is needed for world food processing 40 EJ can be assigned for retail, preparation and cooking.[2]

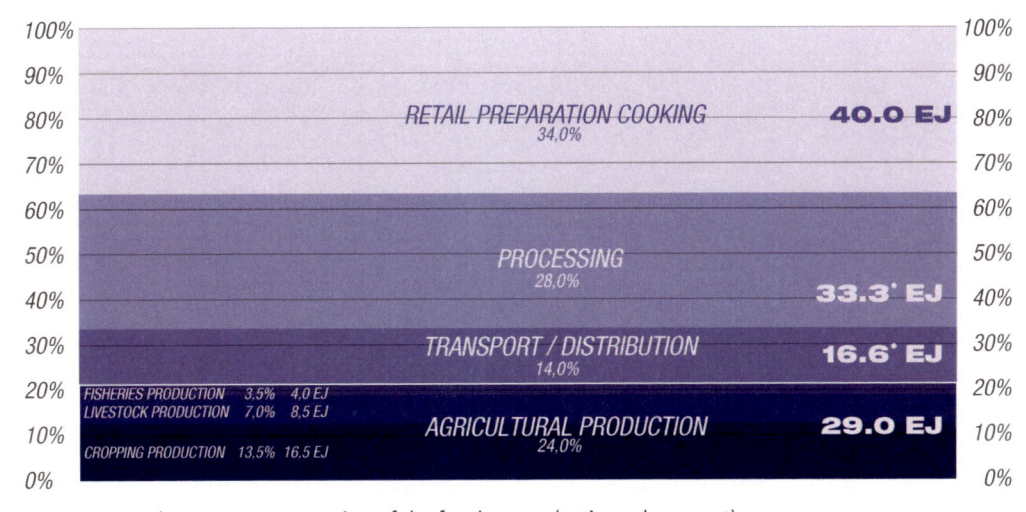

Fig.15. End energy consumption of the food sector (ratio and amount)

Fig.15 represent the percentages and energy consumption for the different items of the food supply chain and highlight that portion of the energy consumption which will then be relevant for a direct comparison of the ratio of the total primary energy supply for traditional soil based agriculture (the food supply system currently in operation) and Vertical Farming food production, assuming that the percentage for pre-

---

[1] It is self-explaining, that the numbers of TPES of the food sector are only rough-sketches. E.g. for agricultural production and processing a much higher electricity consumption can be assumed than for transport and distribution, and therefor a deeper breakdown of these items would be necessary to apply the right conversion factor from end-energy to total primary energy supply.

[2] FOOD AND AGRICULTURE ORGANIZATION OF THE UNITED NATIONS 2011.Energy-Smart Food for People and Climate, Issue Paper, Rome:FAO, p.11; Numbers in Fig. 15 and ff. are applied in the same percentage to a world energy consumption of 550 EJ

paration and cooking is not face with change, although energy supply for retail also could vary as a result of the shorter distances food will be travelling and therefore fewer structural elements for refrigerating, storage or building entities for ethylene-ripening processes would be necessary.

2.3.4. Landuse, Energy Consumption and Biocapacity - Resumée

The Chapter "Landuse, Biocapacity and Energy Consumption" presents some good news in terms of potential solutions for how to feed the world population in future and specifically in the next few generations – the biocapacity of the earth has the strength and the capacity to provide an adequate agricultural yield to cover the total daily energy requirement of every single human being on the planet.[1]

Urbanization will continue to make strides in coming decades. Some 1% of the total land mass today is defined as built-up land. An area more than ten times the size of this is cultivated to provide the people with food.[2] The past few decades have seen the detachment of the spatial union between food production areas and urban areas and also the implementation of a global food production network, with the consequence that the food supply chain has grown substantially larger, more complex and more energy intensive.

Enough potential remains on different levels for traditional soil based agriculture to provide human kind with food. Furthermore the area for agricultural land can be tripled in size, but with the consequence that natural landscape will diminish further as it is changed into agricultural land. Land conversion releases CO_2, especially through slash-and-burn clearing methods applied to wildland and, above all to the rainforests. The latter has provided by far the biggest share of new agricultural land since the 1980s.[3]

There is enough potential in terms of increasing the productivity of existing agriculture land. Looking back to the last century, energy conversion from sun to food increased many times.[4] But we also see the consequences in terms of energy consumption by investing in infrastructure, machinery, macronutrients and other upgrades for the pace of technologization - and a glance at the short history of agricultural practice since the Green Revolution makes the scenario likely that the dependency of agriculture on hydrocarbon energy will also increase many times - a problem of the current situation the FAO wishes to avoid and to reduce.[5]

---

1 FISCHER, G., VELTHUIZEN, H. V. & NACHTERGAELE, F. O. 2000. Global Agro-Ecological Zones Assessment: Methodology and Results. International Institute for Applied Systems Analysis, FAO., Executive Summary, p.xi
2 FAO Global Forest Resources Assessment 2010, retrieved 28.08.2015
3 FISCHER, G., VELTHUIZEN, H. V. & NACHTERGAELE, F. O. 2000. Global Agro-Ecological Zones Assessment: Methodology and Results. International Institute for Applied Systems Analysis, FAO., Executive Summary, p. xii
4 SMIL, V. 2008. Energy in Nature and Society, Cambridge, Mass., MIT Press. p.304
5 FOOD AND AGRICULTURE ORGANIZATION OF THE UNITED NATIONS 2011.Energy-Smart Food for People and Climate, Issue Paper, Rome:FAO. p.3

FOOD SECTOR END ENERGY CONSUMPTION
119 EJ

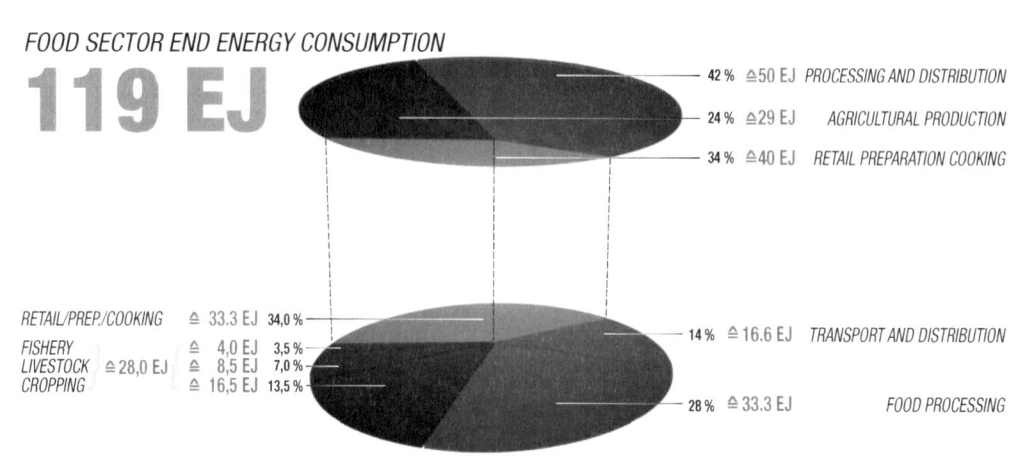

- 42 % ≙ 50 EJ PROCESSING AND DISTRIBUTION
- 24 % ≙ 29 EJ AGRICULTURAL PRODUCTION
- 34 % ≙ 40 EJ RETAIL PREPARATION COOKING

RETAIL/PREP./COOKING ≙ 33.3 EJ 34,0 %
FISHERY ≙ 4,0 EJ 3,5 %
LIVESTOCK ≙ 28,0 EJ ≙ 8,5 EJ 7,0 %
CROPPING ≙ 16,5 EJ 13,5 %

- 14 % ≙ 16.6 EJ TRANSPORT AND DISTRIBUTION
- 28 % ≙ 33.3 EJ FOOD PROCESSING

Fig.16.   End energy consumption within the food sector

Only 55% of all the calories in the agricultural process are produced for direct human consumption (Fig. 6 on page 59).[1] The rest is provided for animal feed, biofuel and industrial production. An enormous potential to establish food security is to be found in this ratio. Unfortunately the trends clearly indicated that this ratio is not going to change in coming decades. Socio-economic changes lead to changes in FBS with a move away from a vegetal to a diary based and a meat diet which will increases the per capita footprint for food.

In the light of these facts producing food there where it is consumed, appears to be a meaningful postulate. A legitimate question can well be asked at the conclusion of this section: can Vertical Farming truly provide a viable remedy for the problems ahead of us in the current situation of world agriculture? Does its potential include bringing an unavoidable increase in energy dependency, in particular on hydrocarbon sources? Is there a potential to drastically reduce the pressure on land and the need to convert natural landscapes, above all the rain forests, for agricultural uses?

---

1   CASSIDY S. EMILY, WEST C. PAUL, GERBER S JAMES and JOLEY A JONATHAN, 2013. Redefining agricultural yields: from tonnes to people nourished per hectare. Environmental Research Letters, IOP PUblishing, p.1

LANDUSE, BIOCAPACITY AND ENERGY CONSUMPTION

FOOD SECTOR TOTAL PRIMARY ENERGY SUPPLY

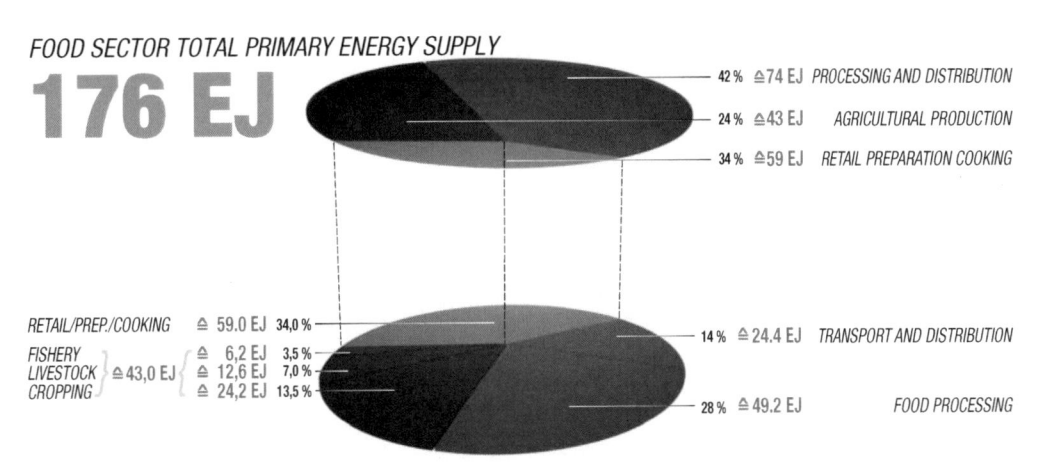

176 EJ

- 42 % ≙ 74 EJ PROCESSING AND DISTRIBUTION
- 24 % ≙ 43 EJ AGRICULTURAL PRODUCTION
- 34 % ≙ 59 EJ RETAIL PREPARATION COOKING

RETAIL/PREP./COOKING   ≙ 59.0 EJ 34,0 %
FISHERY   ≙ 6,2 EJ   3,5 %
LIVESTOCK ≙ 43,0 EJ ≙ 12,6 EJ   7,0 %
CROPPING   ≙ 24,2 EJ 13,5 %

- 14 % ≙ 24.4 EJ TRANSPORT AND DISTRIBUTION
- 28 % ≙ 49.2 EJ FOOD PROCESSING

Fig.17.   Total primary energy consumption (est.) within the food sector

One question can already be answered intuitively: producing food at the location where it is consumed will bring about a shrinking of the food supply chain and it will make the infrastructure including buildings for storage, wholesale, refrigeration and packaging tasks (on a large scale) if not redundant, then certainly less essential. A remarkable number of food travel miles will simply vanish. This intuitively answered question comprises highly interesting research fields in many areas which can readily be conceived, but which are beyond the scope of this dissertation.

Quantification of the land area reduction and of the energy demand for Vertical Farming is the focus of the next chapters:

Can Vertical Farming reduce land use and actually increase the overall energy efficiency of cities?

3. The Vertical Farm Reference Models

3.1. Goals and Process

This chapter examines the status quo of component availability for food production adaptable for Vertical Farming. In this context Vertical Farming components are defined as elements of an available technology adaptable for food production in vertical, enclosed building systems, which are directing food production away from natural agro-ecological systems and envisaging drastic scaling of biomass output compared to classical greenhouses.[1]

Design proposals for Vertical Farms have been published exponentially over the past few years. The foremost typology applied by architecture studios and students around the world is based on the typology of the skyscraper. The author defines the skyscraper accepted by most of the design proposals, as the most challenging typology in terms of daylight supply for the well-defined reasons explained in Chapter 4.

Irrespective of the typology, however, this chapter will introduce the available production method alternatives, which go beyond using soil as a "physical support system"[2], soil as a "solid base of operations into which they can spread their roots."[3] Contrary to popular belief, plants do not necessarily require soils. What they need is space for root development and water with "dissolved minerals, and a source of organic nitrogen."[4]

---

[1] MITCHELL, C. 1994. Biogenerative life-support systems. The American Journal of clinical Nutrition 60 (5), p. 820 - 824. Available on: http://www.ajcn.org/content/60/5/820S.abstract 2.2; To produce foot for an individual in space, estimated 28m2 are needed for a daily output of 2.000 kcal. Retrieved 21.03.2014

[2] DESPOMMIER, D. 2010. The Vertical Farm, New York, St. Martin's Press, p. 163

[3] ibid, p. 162

[4] ibid, p. 163

VERTICAL FARM - REFERENCE MODELS

Of principal interest are systems of different soil-less cultivation methods such as

hydroponic,
aquaponic and
aeroponic systems

and production methods in terms of

horizontal layers (bedding and stacking practices),
horizontal rotating elements (horizontal dynamic systems) or
vertical rotating elements (vertical dynamic systems) and
3D-conveyor belts.

An existing or planned Vertical Farm has been chosen for each of the production method to exemplify and demonstrate the implementation of the cultivation methods. These farms are used for research purposes, e.g. Suwon Vertical Farm in Suwon, South Korea, for animal food production in Devon, UK by the Vertical Farm in Paignton Zoo, United Kingdom and for marketable food production in Singapore by SkyGreens Vertical Farm, Plantagon Vertical Farm in Linköping, Sweden and Vertical Harvest in Jackson, Wyoming, USA.

3.2. Evaluation of components - Modelling a Vertical Farm - Prototype

Research results and experimental approaches in the practice of several studies and of institutions with a focus on high-tech greenhouses and experimental food production for and in space[1], focusing on optimization of energy efficiency and maximization of edible biomass output compared to traditional soil based agricultural practices have encouraged follow-up work along this route of implementing food production entities, which contribute to relieving the current situation of increasing pressure on land, water and energy consumption. The following subchapters give an overview of the already established practices for industrially optimized food production.

3.2.1. Cultivation methods - Overview and Evaluation of the advantages and challenges

Hydroponics refers to the cultivation method for plant growing in nutrient solutions. The roots of the plants can grow with or without the use of substrate (gravel, rockwool, peatmoss, cocopeat etc.) Two distinct systems can be differentiated within hydroponic plant growing practice:

• Liquid hydroponic systems without a supporting medium for the plant roots and "The roots are hanging into the nutrient solution which can be either in the form of a liquid or a mist."[2]
• Aggregate hydroponic systems where plants get supported by a solid growing medium and „irrigated with a complete nutrient solution."[3]

„Historically, water culture has been undertaken in research since the 17th century, with considerable publicity engendered by Gericke in the 1930s."[4] „Hydroponics, developed (...) by Dr. William Frederick Gericke at the University of California, Davis, is the method of Choice used routinely by nurseries to get seeds to germinate and sprout roots before they are transplanted into some form of potting soil."[5]

---

1 http://www.nasa.gov/mission_pages/station/research/experiments/863.html, retrieved 11.02.2015
2 RORABAUGH, P.A., 2014. Introduction to Hydroponics and Controlled Environment Agriculture. Tucson, University of Arizona, Controlled Environment Agriculture Center, p. 5-4
3 ibid.
4 HANAN J.J. 1998. Greenhouses: advanced technology for protected horticulture, Boca Raton, CRC Press, p. 315
5 DESPOMMIER, D. 2010. The Vertical Farm, New York, St. Martin's Press, p. 163

„The setup for a hydroponic facility is constrained by the crop itself, especially determined by the root system of the plant. „The liquid portion of the operation is pumped slowly through a specially constructed pipe (…)"[1] avoiding or reducing evaporative water loss. Of principal interest are hydroponic production methods in enclosed environments, where evaporated water can be recaptured, recycled and again enriched with nutrients. Furthermore enclosed environments increase the potential for better control of diseases and pests to the extent of eliminating the necessity for pesticide and herbicide-use.

> *„A major advantage of hydroponics as compared to growth of plants in soil is the isolation of crops from the soil, which often has problems associated with diseases, salinity or poor structure and drainage. Costly and time consuming soil preparation is unnecessary in hydroponic systems and a rapid turnover of crops is readily achieved as replanting can be done within a day or two after harvesting. The principal disadvantages of hydroponics are the cost of capital and energy inputs relative to conventional open-field production. A high degree of competence in plant science and engineering skills is also required for successful operation of the system. Because of its significantly higher costs, successful application of hydroponic technology is limited to crops of high economic value."* [2]

The following is a summarized overview of basic principles and advantages of using hydroponics, published by Patricia A. Rorabaugh, Ph.D., Arizona University[3]. Some points of this work are not quoted, because they are not relevant to the goal of this analysis, some have been extended and augmented by results of similar papers for the topic of this work.

Hydroponic food production has the capacity to grow food without being dependent on soil fertility. Crops can be grown on land which is unsuitable for conventional soil based agriculture. "Land with poor soils, and contamination (i.e., high heavy metal and salinity levels)"[4], bad climate conditions etc. land, as we defined it in Chapter 2, is an endless resource, whereas land with fertile soil conditions is scarce. Cities with reduced open (and fertile) areas, can benefit from this cultivation method to "make themselves more independent of food imports through a widespread use of hydroponic systems, whether in a climatically fully controlled environment or outdoors.

---

1   DESPOMMIER, D. 2010. The Vertical Farm, New York, St. Martin's Press, p. 165
2   Arizona University, Gene Giacomelli on hydroponic tomato production; http://ag.arizona.edu/hydroponictomatoes/overview.htm, retrieved 11.02.2015
3   RORABAUGH, P.A., 2014. Introduction to Hydroponics and Controlled Environment Agriculture. Tucson, University of Arizona, Controlled Environment Agriculture Center.
4   ibid. p.5-1

Isolation from diseases or insect pests usually found in the soil

The plant roots are contained in continuous pipes, substrates, bags, etc. and do not grow through soil that might contain diseases or other pests such as insects and nematodes.[1]

Direct and immediate control over the rhizosphere

- Since the roots are either growing in water or growing through an inert medium, whatever is in the nutrient solution is bathing the roots. Therefore, nutrient concentrations and pH can be adjusted quickly.[2]

High planting densities are possible which minimize use of land area

- A typical planting density for field tomatoes is 4,000 to 5,000 plants per acre [12.500/ha, Ed.). Greenhouse hydroponic tomatoes can be 10,000 to 11,000 plants per acre [27,500/ha, Ed.]. 23,672 tomatoes can be planted by using an optimized bed size of 60 x 70 cm.[3]
- Plants can be grown closer together because of the use of indeterminant ("vining") varieties that take up less area than the bush varieties usually used for field cropping. They also need less root room – the plants are "spoon fed" the nutrient and water they need and do not have to grow a large root system to find these, as field tomatoes do in the soil.[4]

For the Vertical Farm Simulation Model the planting bed size will be defined by an accurate study which examined different distances between tomatos (*Lycopersicon Esculentum* (Mill.) aiming the ideal distances between the plants.[5]

Higher yields are possible

- As a result of higher planting densities, higher yields are also possible. The indeterminant varieties bred for the greenhouse, can also produce over 6-12 months.
- Since most commercial hydroponic production takes place within greenhouses (or other CEA[6])production, through the use of interplanting, can continue year around.

---

[1] RORABAUGH, P.A., 2014. Introduction to Hydroponics and Controlled Environment Agriculture. Tucson, University of Arizona, Controlled Environment Agriculture Center. p.5

[2] ibid.

[3] LUITEL, B. P., ADHIKARI, P. B., YOON, C.-S. & KANG, W.-H. 2011. Yield and Fruit Quality of Tomato (*Lycopersicon esculentum* Mill.) Cultivars Established at Different Planting Bed Size and Growing Substrates. Horticulture, Environment, and Biotechnology; 53(2), 102-107.

[4] ibid.

[5] ibid.

[6] Controlled environment agriculture, see also „list of abbrevations". For further informations visit http://ceac.arizona.edu/, retrieved 15.08.2015

- Yields are also greater due to better control over water, nutrition, EC, pH and diseases (see above).
- The yields for field grown tomatoes are 10-40 tons per acre compared with 300 tons per acre or more for tomatoes grown using greenhouse hydroponics (equates to 75 kg/m2 (750t/ha); this figure has recently risen as high as 90-104kg/m2 [900-1,040t/ha]).[1]

Efficient use of water and nutrients

- In soil culture water may be lost in wetting the soil beyond the reach of the plant roots or from the surface through evaporation.
- In hydroponic culture, since the nutrient solution is transported within an en-closed pipe system (or similar) water loss and water stress on plants can be drastically reduced if not eliminated.
- Nutrients (...) are also not lost to the soil but retained in the root zone and, in closed systems, are replenished and recycled.[2]

- *No weeding or cultivation is needed*

This reduces if not eliminating completely the necessity for using all sorts of herbicides.

- Transplanting of seedlings is easy – No transplant shock
- In soil culture the root mass can be easily disturbed during transplanting causing root breakage, plant stress and stunted growth for up to a week.
- In hydroponic culture seeds are started in Rockwool cubes or plugs, and then transplanted into larger cubes with holes made for that purpose. There is no disturbance of the root mass, little or no root breakage and therefore minimal plant stress and transplant shock.

- *Fruit of hydroponically grown plants can have more flavor*

Field tomatoes are mostly harvested before they ripen. Long transportation routes make this necessary to dovetail the crop with the buying time. Tomatoes then gas treated with ethylene for lycopene[3] formation.

---

1 LUITEL, B. P., ADHIKARI, P. B., YOON, C.-S. & KANG, W.-H. 2011. Yield and Fruit Quality of Tomato (*Lycopersicon esculentum* Mill.) Cultivars Established at Different Planting Bed Size and Growing Substrates. Horticulture, Environment, and Biotechnology ; 53(2), 102-107.
2 RORABAUGH, P.A., 2014. Introduction to Hydroponics and Controlled Environment Agriculture. Tucson, University of Arizona, Controlled Environment Agriculture Center p. 5-2
3 Lycopene is a chemical compound which gives tomatoes the red color. More detailed information see Nomenklatura

- "Hydroponically grown tomatoes (...) are picked after they have begun to ripen, which includes the typical red color formation of the fruit (lycopene), the formation of gel within the locules and the characteristic taste."[1]

It's not the greenhouse that distroys the taste of tomatoes.

Tasteless tomatoes which nevertheless have an attractive appearance to the buying public are commonly associated with the greenhouse tomato. It is a matter of fact that in the early years of mass-produced tomatoes in Europe, especially in the Netherlands, and also in America fifteen years ago, the greenhouse industry concentrated on aesthetics, especially on the skin-quality of tomatoes. Today after studying specific outdoor cultivation conditions researchers know how and why a tasty vegetable develops, "they concluded that some stress was necessary in order to elicit flavonoids (complex organic molecules specific to plants). These molecules are the essence of why most vegetables have distinctive flavors and aromas. In addition, restricting the water a plant receives increases its sugar content, heightening the flavor even more"[2] Electrical conductivity can also be raised within hydroponic production. "This tends to stress the plant and enhance fruit flavor"[3], too.

ADVANTAGES OF GREENHOUSE CULTURE OVER FIELD CULTURE

Virtual indifference to the seasons

- Crops can be grown year around in any climate zone, from tropical regions in deserts and on to the Polar regions on earth. This cultivation method enables fresh food production in space, too.[4] The VEGGIE-research program, launched by NASA, will start fresh vegetable production on the International Space Station within 2015.[5]

---

1   RORABAUGH, P.A., 2014. Introduction to Hydroponics and Controlled Environment Agriculture. Tucson, University of Arizona, Controlled Environment Agriculture Center p. 5-2
2   DESPOMMIER, D. 2010. The Vertical Farm, New York, St. Martin's Press, p. 166
3   RORABAUGH, P.A., 2014. Introduction to Hydroponics and Controlled Environment Agriculture. Tucson, University of Arizona, Controlled Environment Agriculture Center, p. 5-2
4   During the lecturing phase of this Doctoral Thesis nasa pronounced first vegetable consumption of astronauts on ISS, completely grown in space. https://www.nasa.gov/mission_pages/station/research/news/meals_ready_to_eat, retrieved 10.08.2015
5   http://www.nasa.gov/mission_pages/station/research/news/veggie.html, retrieved 27.02.2015

More efficient use of space

Hydroponic production methods reduce the depth for root development compared to soil based agriculture. If light energy supply is guaranteed, supplementary stackings of one plant above another is possible.

Control over the aerial (upper) portions of the plant to achieve higher yields

- Air temperature and relative humidity can be regulated within enclosed environments. Both can be adapted to the specific crop and to the actual growing- and maturation stage.
- CO_2 concentrations can be regulated. If light, nutrient, and water supplies are optimal, with higher CO_2 concentrations photosynthesis can be boosted. Plant and fruit growth will accelerate and higher sugar concentrations will result. Normal CO_2 concentrations of between 330 and 380 ppm the can be doubled or tripled.
- "Light levels and quality (wavelengths) can be controlled by choosing an appropriate shade cloth or glazing. In certain high light regions, shade cloth can be used during the summer to protect the crop. Furthermore certain forms of glazing can block UV radiation, which can harm plants and plant productivity."[1]

---

1 RORABAUGH, P.A., 2014. Introduction to Hydroponics and Controlled Environment Agriculture. Tucson, University of Arizona, Controlled Environment Agriculture Center, p. 5-3

CHALLENGES OF CLOSED ENVIRONMENT AGRICULTURE

Regardless of whether hydroponic, aeroponic or aquaponic methods are imple-mented in an enclosed environment, each of these alternatives to soil based agriculture shares similar, if not the same disadvantages.
The following outline can also be read as common challenges of classical greenhouse food production, and, of course for future implementation of Vertical Farms. For the sake of completeness all the different lines of thought are listed here, even though most of the following points influence neither the configuration of the simulation model as defined in Chapters 4 and 5, nor the architecture and design of Vertical Farms. Referring to Rorabaugh, Arizona University[1], the following challenges are listed such as:

Large capital, energy and labour input

- Any size of commercial operation (including injector irrigation systems, computer controls, etc.) will cost about $600,000 per acre with the land itself costing $1000 - $2000 per acre or more (depending on location).
- Energy costs can be high and include those for heating (usually burning natural gas), cooling (usually through use of evaporative cooling) and electricity to run equipment (injectors, computers, motors, sorting/packing/storage equipment, etc.).
- Labor is essential on a daily & intensive basis with significant wage costs for year-around workers plus associated benefits.

The grower needs a high level of competence in plant science, engineering, computer control systems and marketing

If not available, experts in these fields needs must be hired. This is an intensive form of agriculture where a small problem can very quickly escalate into a major disaster.

The technology is limited to crops of high economic value

- Since the initial cost of a large commercial facility is so high it would not be profitable to grow anything but crops of high economic value including tomatoes, colored bell peppers, cucumbers and even lettuce which, in a hydroponic greenhouse, can yield multiple crops per year.

---

[1] RORABAUGH, P.A., 2014. Introduction to Hydroponics and Controlled Environment Agriculture. Tucson, University of Arizona, Controlled Environment Agriculture Center, Chapter 5, p.3

Plant diseases and insect pests may prove more difficult to control

- Root pathogens that produce water-borne spores (e.g., zoospores of Pythium) can be devastating to plants growing in a recirculating system since infected solution could circulate to all plants. For treatments see Chapter 4.
- The greenhouse, with its controlled environment, is a perfect habitat for many types of insects, good (beneficials) and bad (white flies, aphids, thrips, spider mites, shore flies and fungus gnats). Although IPM [integrated pest management, Ed.] and biological control are available, the plants will require constant vigilance and swift action.

In summary the following plant needs can be listed, as the requirements which are critical for implementing hydroponic systems within a controlled environ-ment:

- "Water – Critical for metabolic processes, for transport of substances throughout the plant body (phloem and xylem) and for transpirational cooling.
- Light – Critical for photosynthesis. (Where you put your system is important.)
- Inorganic mineral nutrients – at the correct concentrations (EC[1]) and pH levels.
- Carbon dioxide – Critical for photosynthesis (needed at the leaf surface).
- Oxygen – Critical for respiration (needed by all parts of the plant including the roots, as a result aeration of the nutrient solution may be required).
- The appropriate temperatures and relative humidity (specific to type of plant).
- The proper temperature and relative humidity (specific to type of plant)
- Support systems for the roots and shoots. For plants where the roots hang directly into the nutrient solution and do not provide any support for the plant, mechanical support may be needed. For an indeterminant tomato plant, support for the stem will be needed in the form of hooks, twine and vine clips."[2]

---

1   Electrical Conductivity of Water estimates the solids dissolved in water. See List of Abbreviations.
2   RORABAUGH, P.A., 2014. Introduction to Hydroponics and Controlled Environment Agriculture. Tucson, University of Arizona, Controlled Environment Agriculture Center, Chapter 5, p.4

AEROPONICS

"The aeroponics systems allow the growth of plants in an air/mist environment without the use of soil or an aggregate media. This high performance food production technology will rapidly grow crops using 99% less water and 50% less nutrients in 45% less time."[1]

Richard Stoner[2] invented this technology in 1982. In principle it is an advanced hydroponics technology, soilless with water and nutrient supply for the plant. By contrast with standard to hydroponics, the roots of the plants no longer "swim" (or are planted in a substrate, Ed.) in a nutrient solution, but are sprayed with a "nutrient-laden mist onto the roots, supplying them with everything they need (...)."[3] Small nozzles cover the plant roots with a continuous mist.

Enclosed root systems which are supplied with water and nutrients with a mist system have now been under observation for decades. The technical requirements for aeroponics are far more advanced than those for classical hydroponics. This might be one of the principal reasons why aeroponics have so far failed to gain widespread use. "Water quality requirements are very high, and the system must be very reliable. The method (...) is not applicable to products sold that require a substrate. For (...) vegetables, the supporting system must be rearranged. As a rule, the method has been used mostly for research. (...) Apparently the improvement in net return makes it difficult to warrant the initial investment and operating costs."[4]

The advantage of drastic water (and therefor weight) reduction facilitates food production in space. Research with aeroponic systems is thus strongly supported by NASA. This cultivation methods provides "clean, efficient and rapid food production. Crops can be planted and harvested in the system all year round without interruption, and without contamination from soil, pesticides and residue. Since the growing environment is clean and sterile, it greatly reduces the chances of spreading plant disease and infection commonly found in soil and other growing media."[5]

---

1 http://sbir.gsfc.nasa.gov/SBIR/successes/ss/10-026text.html, retrieved 12.10.2014
2 Also visit http://www.nasa.gov/offices/ipp/centers/kennedy/success_stories/Inflatable_Aeroponic_System_BBlinds_prt.htm, retrieved 15.08.2015
3 DESPOMMIER, D. 2010. The Vertical Farm, New York, St. Martin's Press, p. 165 - 166
4 HANAN J.J. 1998. Greenhouses: advanced technology for protected horticulture, Boca Raton, CRC Press, p.339-340
5 http://www.nasa.gov/vision/earth/technologies/aeroponic_plants.html, retrieved 12.01.2015

AQUAPONICS

The intent of aquaponics is to create a symbiotic relationship between classical aquaculture and hydroponics. Basically it is a hybrid of hydroponics (plants in growing substrates, without soil) and aquaculture[1]. Aquaponics enables the possibility to grow food by creating an "ecosystem in which the wastes of one process become resources for another. Fish waste feeds the plants, and plants clean the water which is returned to the fish tanks. Beneficial bacteria in our filters break down fish waste so that it is easily absorbed by the plant roots, and beneficial insects and organic foliar feeding help control pest populations naturally and safely."[2]

A promising implemented concept within the context of this doctoral thesis is "The Plant", an aquaponic Vertical Farm in Chicago with a production area of roughly 2.500 m2 where the aquaponics farm also houses its own breeding system. The main fish cultivated is the tilapia. Additional aquaponics systems will be installed in the next future with prawns. For more information in the Appendix "The Plant" is described and classified in more detail.

The biomimicry of nature can be defined as a helpful cultivation method to change the principle of cradle to grave by traditional industrial farming to the principle of cradle to cradle. Nutrient delivery by fish for plants has the potential to additionally reduce necessary macronutrients for plant cultivation within fully controlled environments. In addition to other operations the conversion of ammonia released into water in the excreta of fish, into nitrates, is the central one. The byproduct of fish metabolism becomes the nutrient supply for crops. Plants are able to absorb ammonia from water to some degree, but nitrates are assimilated more easily. This drastically reduces the toxicity of the water for the fish.[3]

---

1   Also known as aquafarming, is the production of aquatic organisms in both sweet- and saltwater under controlled conditions. The Food and Agriculture Organization (FAO) defines aquaculture the „farming of aquatic organisms including fish, molluscs, crustaceans and aquatic plants. Farming implies some form of intervention in the rearing process to enhance production, such as regular stocking, feeding, prodtection from predators, etc. Farming also implies individual or corporate ownership of the stock being cultivated.", http://www.fao.org/fishery/statistics/global-aquaculture-production/en, retrieved 12.01.2015
2   http://www.plantchicago.com/non-profit/farms/plantaquaponics/, retrieved 12.01.2015
3   For more information on the nitrogen cycle visit: http://www.britannica.com/EBchecked/topic/416271/nitrogen-cycle, retrieved 01.04.2015

3.2.2. Production methods - Building footprint, cultivation area and soil-based-area-equivalent

Food production within a fully controlled environment aims to maximize yields in order to compete with the traditional food production system. Beyond the economic pressure on food prices it is of central interest due to the potential it offers for relieving the current and future stress situation on available additional farmland. The Vertical Farm "design must make optimum use of its internal space by accommodating the largest possible growing area. (...) It is also of vital importance that the building design allows for the most efficient method for workers to tend the plants (...)." But considering the different production methods explained in more detail below and based on existing or planned Vertical Farms, the ratio between the production footprint and the horizontal circulation on the ground floor level is of basic interest and as a consequence in order to make different production methods in Vertical Farms comparable one must consider the ratio between the production volume to the circulation volume (horizontal and vertical).[1]

Different design strategies using specific production methods aiming a maximization of yield, lead to a maximization of operation costs. The high light requirements of most plants s an issue of central interest. The higher the plant density per level or building volume, the lower the light accessibility of plants will be. This means, "[t]he dense growing configurations distributed throughout a Vertical Farm's multiple floors create far too many physical barriers for sunlight to penetrate. (...) [E]ven if sunlight could somehow bend around these obstacles, the sum of solar energy cast on a particularly dense Vertical Farm may be less than the farm's total energy needs."[2]

Greenhouses using technology adaptable for future Vertical Farms, but also Vertical Farms in construction mostly implement high technology to maximize yields. Great efforts are being made in this to ensure sunlight reaches the crops. These two competing targets are crucial in design and have to lead to new Vertical Farm typologies[3]

---

1 For an advanced building analysis additional information for construction and HVAC must be considered.
2 ibid.
3 Gordon Graff stated the problem in a similar way: „The battle between maximizing yields and maximizing solar penetration thus becomes the most important design consideration for architects, and ultimately will be the primary criteria from which the efficacy of a design will be determined." GRAFF, G. 2011. Skyfarming. Master of Architecture, Waterloo, Ontario, Canada, p.76

- We can differentiate typologically between four different food production methods within conventional or stacked greenhouses:
- horizontal layers
  - - single layer
  - - multiple layer
- horizontally rotating elements
- vertically rotating elements and
- 3D-conveyor belt.

SINGLE HORIZONTAL LAYERS

Single horizontal cultivation layers are used by all greenhouses, irrespective of whether the unit is for soil-based or soil-less cultivation. Greenhouses for food production, are a "support in controlling temperature, sunlight and [the, Ed.] availability of water during the growing season. To increase the fertility of an ecosystem, one can also act on the temperature (possibly with heated greenhouses), on sunlight (providing shade), on the water supply and its organization (irrigation, drainage, windbreaks, soil covering that minimizes evaporation), and even on the carbon dioxide content of the air (...).[1] Today, artificial lighting enables an additional increase in yield output per square-meter: Stacking the crop.

Fig.18. Single horizontal layer

By stacking the growing substrate automatically the available sunlight for plants decreases.
Multiple, stacked non-moving layers are possible and dependent on specific plant- and greenhouse heights, as on light requirements for plant growth, photomorphosis and photosynthesis. The smaller the light requirement of plants is, the easier it is to additionally stack the cultivation footprint within a greenhouse. Specially designed troughs for plant cultivation already enable an optimization of the light supply, and can be used both for greenhouses and outdoor cultivation. This could be achieved by perforating the horizontal substrate supporting element for enabling light to pass through the structure or simply by stacking only the hydroponic substrate and watering pipes. By contrast with the other three production methods, this one can be considered as the only static, non-moving production method. The plants continue in the same position throughout crop rotation. The main advantage of this is in

---

1   MAZOYER, M. & ROUDART, L. 2006. A History of World Agriculture, London, Earthscan, p. 63-64

the greatly reduced material and investment cost for additional technical, mechanical and computerized elements.

Single horizontal layer-cultivation is used today in classical greenhouses for crops ranging from tomatoes with their high light dependence, to mushrooms with their low-light requirement. The very low light requirement for mushrooms enables use of a cultivation principle on multiple layers. The low height of the mushrooms also makes multiple stacking possible. The cultivation method for most mushroom production is hydroponic with the use of a growing substrate by adding water and nutrients, or is soil-based using fertile soil.

MULTIPLE HORIZONTAL LAYERS

Extending beyond classical greenhouses and approaching closer to the typology of a Vertical Farm, the Research Center in Suwon, South Korea located just outside the capital Seoul (ca. 32 km from center to center), should to be mentioned for the purpose of this work.

Here computers and robots are cultivating leafy crops within a fully controlled environment using LED light and reducing the time to harvest by half. The aim is to identify the optimum wavelength for specific crops in order to reduce necessary artificial lighting. "With a plant factory the environment can be artificially controlled.

Fig.19.   Suwon Vertical Farm, Multiple stacked horizontal layers

The operator can change conditions such as temperatures, humidity and carbon dioxide to provide an optimal environment for plants to grow in."[1]

The Vertical Farm, built in 2010 can now look back to four years of empirical data. Although specific statistical data has still not been published, the main challenge that is being faced is, as expected, high energy costs. "This factory consumes a lot of energy and electricity. (...) In particular you need to cool the temperature down for some plants through air conditioning. Technological development to reduce energy is [the key for] success of plant factories."[2]

Compared to the following examples, this kind of stacking is fixed and non-rotational. This cultivation and production method as currently implemented in Suwon, South Korea, is thus classified as a production method that is suboptimal for Vertical Farms having crop productions with high light re-quirements.

The Suwon Vertical Farm is fully dependent on artificial lighting, although it can conceivably be used for cultivation of crops with low or no light requirements within a potentially fully controlled environment and with the potential for heating, ventilation and cooling. The methods for use here are hydroponic, aeroponic, aquaponic or soil-based cultivation.

---

1 Lee Sang-Woo, Gyeonggi Province Agricultural Research, In a June 5 report entitled "South Korea moving towards Vertical Farming," Al Jazeera's Wayne Hay described the scene at a Vertical Farm in Gyeonggi-do (Gyeonggi Province). http://www.korea.net/NewsFocus/Sci-Tech/view?articleId=103942, retrieved 03.11.2015
2 ibid.

HORIZONTALLY ROTATING ELEMENTS

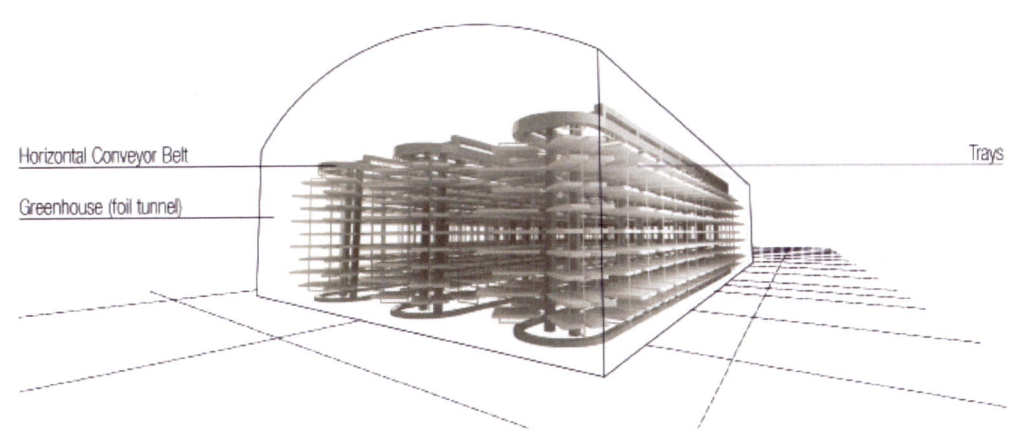

Fig.20. Paignton Zoo: Rendering of the production entity

In the Southwest of England, Devon[1], Paignton Zoo Environmental Park[2], a combined zoo and botanical garden is producing vegetables to feed the animals within a specially constructed greenhouse on 120 m2. It has now been running since 2009.

While horizontal layers are the cultivation method used in most greenhouses producing food hydroponically, horizontally rotating elements are used at Paignton Zoo for production by the horticultural company Valcant. The columnar design enables the stacking of plants within the entire floor height of 3m. The plant columns are con-

Fig.21. Paignton Zoo, Devon: Photo of the production volume

1   The following daylength and sunlight hours got taken from the weatherfiles from Kiev, Ucraina, for unavailability of Devon's weather files. Devon's latitude is the closest to a city available from the official energyplus weather file-site (http://apps1.eere.energy.gov/buildings/energyplus/).Latitude comparison: 50°43'18"N Devon and Kiev 50°27'13"N.
2   Paignton Zoo Environment Park Totnes Road Paignton Devon TQ4 7EU

nected to a conveyor track and are turned around the z-axis along the conveyor belt on top of the trays to equalize the available daylight coming through the transparent building skin. A pair of eight layers (or trays) is moved along the conveyor reaching the starting point in 40 minute cycle.

This production unit for Paignton Zoo began pilot operation with VertiCropT™ [1] at Paignton Zoo in Devon. „The pilot project grows 11,200 plants in a greenhouse of 1[2]0 square meters, using a conveyor driven stacked growing system various micro greens, lettuce and salad mixes have been planted sequentially to provide a regular supply of fresh green leaves (...) to improve animal feeding regimes at the zoo."[2]

The production method consists of horizontally rotating elements, which make use of the full floor height. The rotation enables to equal distribution of the available natural light to all the plants - principally leafy vegetables – on the production line. The trays are "suspended from an overhead track and rotate on a closed loop conveyor."[3] The principle advantage of this production method consists in allowing "centralized locations for irrigation and the loading and unloading of crops"[4] The environment is fully (computer) controlled providing plants with ideal growing conditions. The plants get provided with water and nutrients at regular interval by the trays. "(...) [I]ntegrated advanced hydroponic technology supplies water and nutrients at the correct pH and EC levels automatically via a central feeding station."[5] A monitoring operation controls the environmental temperature, humidity, water and nutrient solution composition. "UV filters ensure the re-circulated water is clean and free from potential plant pathogens."[6]

The production method used by Paignton Zoo, implemented by VertiCropTM is adaptable to all existing and new stacked building typologies. It must be accepted that the high density of plants in the z-axis behind the facade drastically reduces light penetration in the deeper zones of the levels and makes artificial lighting necessary. The construction of the horizontally rotating elements also reduces the Choice of crop types. Only those vegetables and fruits can be considered that have a low growth in the z-axis.

1	http://www.verticrop.com/, retrieved 12.03.2015
2	IHC 2010 Lisbon S10 220 verticrop 1Fin 2011 (SHS Acta Horticulturae (3)-2.pdf, retrieved 12.01.2015, p.1
3	IHC 2010 Lisbon S10 220 verticrop 1Fin 2011 (SHS Acta Horticulturae (3)-2.pdf, retrieved 12.01.2015, p.2
4	http://www.zoolex.org/publication/frediani/feeding_time_frediani_horticulturist2010.pdf, p.14, retrieved 12.01.2015
5	ibid.
6	ibid.

A much higher output per m2 compared to traditional soil-based agriculture or greenhouse practice is certainly conceivable. "Scaling up, a 6 m high VertiCropTM requires 87% lower land and building footprint than conventional hydroponic systems to grow the same quantity of plants."[1]

This cultivation and production method in Devon is thus classified as a Vertical Farm within a potentially fully controlled environment with the potential for heating-, ventilation and cooling. The cultivation method used is hydroponic. The greenhouse construction and the production method is optimized for natural sunlight, but artificial light, if necessary, could be adapted without negatively influencing the production procedure. The production volume is a monovolume with potential for additional level stacking.

---

1 http://www.zoolex.org/publication/frediani/feeding_time_frediani_horticulturist2010.pdf, p.14, retrieved 12.01.2015

VERTICAL FARM - REFERENCE MODELS

VERTICALLY ROTATING ELEMENTS

Two additional relevant Vertical Farms should be mentioned at this point:

• SkyGreens Vertical Farms in Singapore.
• Vertical Harvest, a Vertical Farm in Jacksonville, Wyoming.

One of the most efficient and economically relevant Vertical Farms is managed by SkyGreens in Singapore, specialized on lettuce production. The chosen production method is thus a vertically rotating system.[1]

SKY GREENS, SINGAPORE

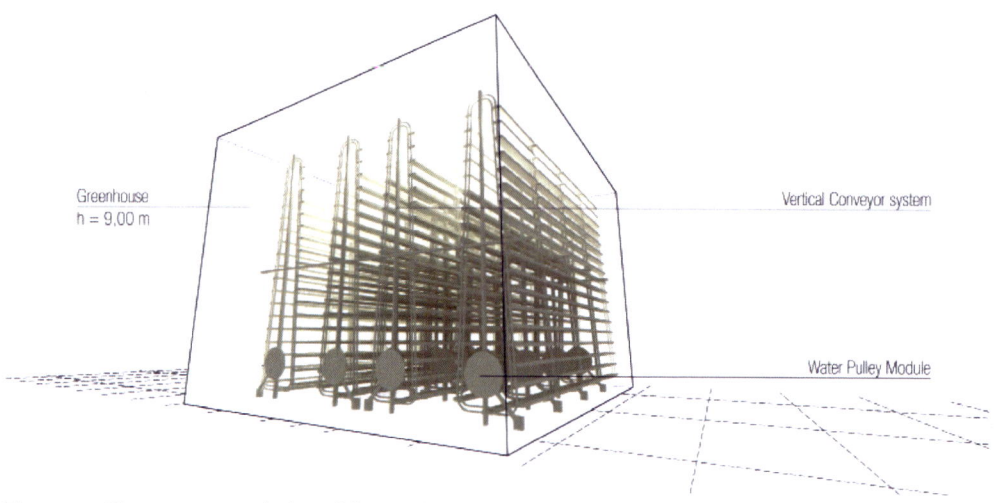

Fig.22.   Skygreens: Rendering of the production entity

This unit is typologically a tall greenhouse (9 meters high) with growing troughs mounted on an aluminum form. This form consists of a modular structure which is customizable and scalable, it allows different heights (in absolute terms) and between the troughs. Different crops can thus be cultivated. Troughs are currently optimized and produced for the following medium- to high PPFD requirement plants xiao bai cai, naibai, cai xin, Chinese cabbage, mao bai, bayam, kai lan kang kong, spinach.[2]

---

1   http://www.skygreens.com/technology/, retrieved 01.02.2015
2   ibid.

The plants rotate twice a day following an A-shaped path (cross section), which guarantees the same light exposure for each plant. The rotation itself is powered "by a unique patented hydraulic water-driven system which utilizes the momentum of flowing water and gravity to rotate the troughs. Only 60W electricity (...) is needed to power one 9m tall tower."[1] Hydroponics is the cultivation method used, although this system would also work with a soil-based cultivation method.

While passing through different points in the structure, the troughs are irrigated by a nutrient solution. According to information from SkyGreens, the yield output is ten times higher per unit land area equivalent. Higher yields through the control of the enclosed greenhouse environment are not declared.

Fig.23.   SkyGreens, Singapore: Production Volume

The SkyGreen production method is to work with natural light. The principle of the rotating trough-construction is based on using natural light, although this system makes it possible to apply artificial light fixtures.

Water usage, according to SkyGreens, depends on an underground reservoir system and all the water is fully recycled and reused. The drainage system is based on the flooding method (which drains and fertilizes the plants). This eliminates electricity wastage from sprinkler systems as well as water run-offs.[2]
The specific water amounts data per m2 of cultivation area or per crop rotation for this work were not available and it is thus difficult to make comparisons with soil-based equivalents.
The patented and modular A-shape-production method allows an expansion in the x- and y-axis. The distance between the production structures allows an optimizati-

---

1   http://www.skygreens.com/technology/, retrieved 01.02.2015
2   ibid.

on of the horizontal circulation area, which are necessary in every greenhouse. The ratio between the crop growing area and the horizontal circulation area is reduced approx. tenfold compared to other greenhouses with the same crops.

This cultivation and production method in Singapore is thus classified as a Vertical Farm in a potentially fully controlled environment with the potential for heating, ventilation and cooling. The cultivation methods that can be used are hydroponics and soil-troughs. The greenhouse construction and the production method are optimized for natural sunlight, but artificial light could be adapted without negatively influencing the production procedure if necessary. The production volume is a monovolume.

VERTICAL HARVEST, JACKSON, WYOMING, USA

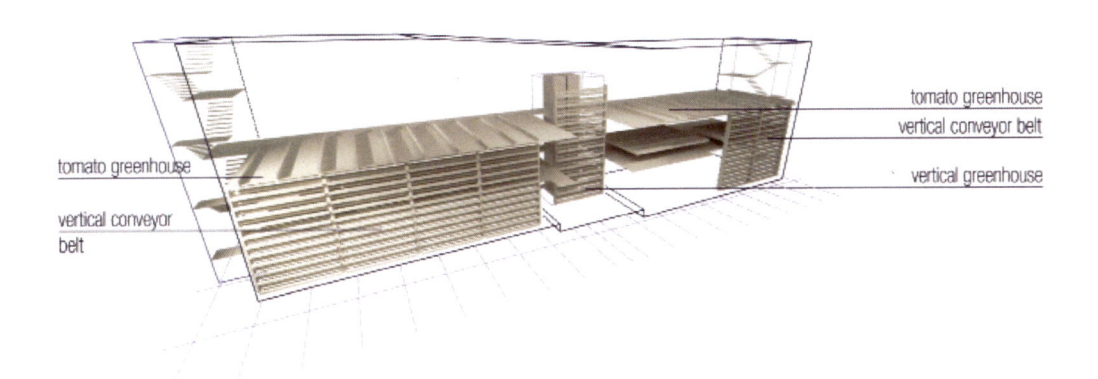

Fig.24.   Vertical Harvest: Rendering of the production entity

The Vertical Farm in Jackson, Wyoming, USA, was initiated by the architect Nona Yehia and environmental consultant Penny McBride.

Vertical Harvest is designed for natural daylight use. The facade was developed to include the highest daylight transmission possible. It is oriented to the south.
"This natural lighting not only keeps our energy use down, but is a free source for photosynthesis. As a result of our preliminary feasibility report, we determined that in our location there are periods of the year where the natural daylight is not suffici-

Fig.25.   Vertical Harvest, Jackson, WY: Vertically rotating cultivation conveyor

ent for growing certain produce. For example, it is impossible to grow tomatoes in the heart of winter in Jackson Hole without the use of supplemental artificial lighting."[1]

Supplemental growing lights were fixed due to cope with the long winter periods with low light availabilities, based on provided information of operators of Vertical Harvest. The company calculates that it must use approx. 3,000 artificial lighting hours per year to meet the production targets. HPS (high pressure sodium) - lightbulbs are used for tomato production and LED- fixtures for lettuce varieties, microgreens and for the propagation areas of the vertical greenhouse.

---

1   www.verticalharvest.org, retrieved 12.02.2015

3D-CONVEYOR BELT - PLANTAGON, LINKÖPING, SWEDEN

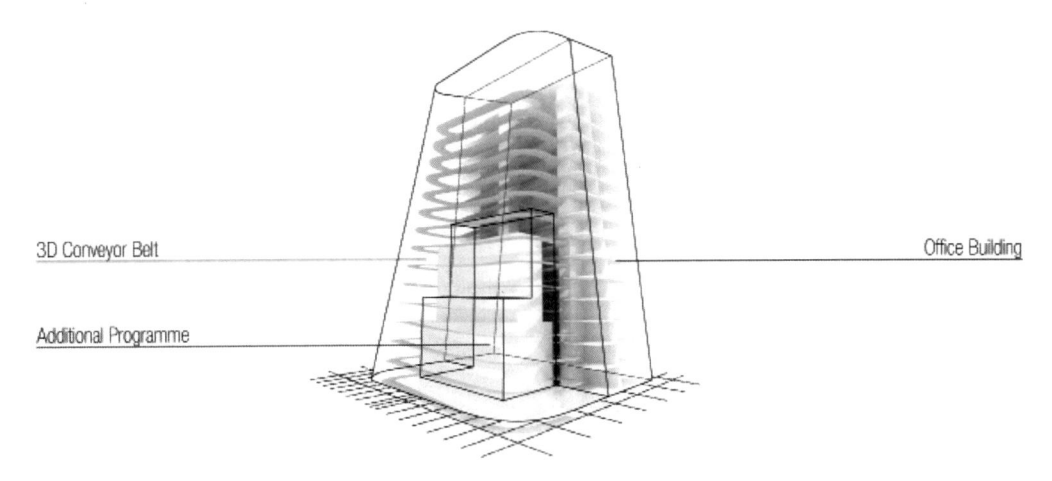

Fig.26.   Plantagon Linköping: Rendering of the production entity

The most advanced and promising production method is the 3D-conveyor belt. The most developed system is patented by Plantagon . Information on this has been comprehensively published and the system is thus investigated in greater depth for this doctoral thesis. Furthermore the author participated at the GUA-Summit[1] in Linköping, Sweden, organized by Plantagon, at the end of January 2013 and thus had the opportunity to discuss in greater detail the state of the art in Vertical Farming using this very specific production method.
Besides different urban and Vertical Farm typologies Plantagon International[2] is also developing, a project which is of particular interest for this doctoral thesis: this is the Vertical Farm project for Linköping, Sweden. The ground-breaking ceremony for this was on February 9th 2012.[3]

A continuous movement from the top of the helix to the bottom (from the top to the ground floor of the building) of the growing crop throughout its crop rotation guarantees a similar amount of sun exposure. Light penetration is certainly less the more distant the plants are positioned from the facade. Additional LED lighting equalizes the differences.

---

1   http://www.urbanagriculturesummit.com/summits/urban-agriculture-summit-2013
2   To their own description Plantagon International „is the global innovation leader in the sector urban agriculture. Plantagon's resilient food systems minimize the need for land, water, energy and pesticides. The environmental impact is very low, and if the products are delivered directly to consumers in the city, the transportation costs are also minimized. The Plantagon concept is simple and appealing: fresh, local vegetables delivered daily directly to consumers. No middle hands, no yesterday's food. We develop innovative solutions to meet the rising demand for locally grown food in cities all around the world. We minimize the use of transportation, land, energy and water – using waste products in the process but leaving no waste behind..", http://plantagon.com/about, retrieved 13.03.2015
3   In addition information exchange and discussions between the author and Prof. Dickson Despommier, one of the keynote-speakers on that Summit, was possible.

The plant begins the journey as a seedling or a young plant at top of the building and throughout the growing- and ripening phase it moves to the building entrance level, ready to be harvested. "Food is harvested in batches using an automatic harvesting machine. After the harvest the trays and pots are disinfected, and the pots are then sorted, separated and replanted with a fresh seed for the next round in the cultivation loop. After germination, the pots are recombined with the trays and elevated to the top of the growing helix to repeat the process."[1]

Fig.27.   Plantagon, Linköping: Production helix

„The trays are equipped with a light sealed nutrient solution reservoir, and the pots are irrigated about three times per day using an ebb-and-flow technique. A capillary mat at the bottom of each tray protects the individual plants from drought. Excess nutrient solution is collected and reused after disinfestation."[2] Regardless the system design of the Vertical Farm, conceived by Plantagon, this production method by using the 3D-conveyor belt with moving trays from the top to the basement level, follow the same production flow.

The depth of the trays used at the farm in Linköping is three meters and can be increased to six meters for use in other projects. The multifunctional structure in Linköping is a combination of the vertical greenhouse with an office building, which is situated to the north of the building. In addition the production unit contains spaces for germination, processing, washing and packaging. A restaurant, a visitor's area and a conference room are also planned.

The greenhouse volume of this Vertical Farm has a truncated form, developed for optimizing sunlight penetration of the cultivation surface for Pak Choi[3] (also known as bok choy) production.

This patented conveying system moves the trays (containers) on an inclined track and a conveying device. This device is "arranged to travel down the track and comprises a container moving unit which after passage below a container moves the

---

1   http://plantagon.com/urban-agriculture/cultivation-systems, retrieved 07.04.2015
2   ibid.
3   Brassica rapa chinensis is a Chinese cabbage, very popular in Southern China and Southeast Asia. This winter-hardy vegetable is increasingly grown in Northern Europe. Due to its tolerance to colder temperatures and relatively low light requirements (220 µmol/m-2/s-1) this product presumably got chosen as crop in Linköping with harsh climate and low light intensities.

container one step up the track. (...) The invention also relates to a tower structure comprising a container conveying system and a method for conveying containers (...)".[1] The railing or the "inclined track", inspired by the structural logic of roller coasters, allows the implementation of this production method in principle to theoretically all the potential geometric forms of three dimensional greenhouses.

"Plantagon integrates the building on site and adapts it to site specific light conditions which vary from location to location in the existing urban fabric of cities. The verticality is important in order to optimize the production of food and the functionality of the building.Plantagon Greenhouse. The patented transportation helix (also referred to as the spiral or the ramp) is the most important concept behind the industrial process of all Plantagon systems. It has been developed based on three main optimisation factors: 1. maximise the footprint usage ratio of the helix; 2. minimize the use of water; 3. minimise the demand for artificial lighting and to gain as homogeneous light levels as possible."[2] The principle of the helix, beyond the implementation into the Vertical Farm in Linköping, with all its geometrical (and climatic) limitations, was conceptually developed strictly for use in two standalone Vertical Farms, as published on the organization website.

The example here mentioned (Fig. 27) shows a closed basement, for industrial process and HVAC and a vertical circulation element, containing additional building services in the North (or the South)[3]. The building geometry with leaning extrusion axis of the greenhouse thus permits full sunlight penetration throughout the day and the year, based on the initial premise that the Vertical Farm in this picture is based on a concept devised for equatorial regions.

---

[1] Excerpt from the patent of Ake Olsson, registered 27.05.2009, published 15.03.2012, "Conveying system, tower structure with conveying system, and method for conveying containers with a conveying system, US 20120060414 A1", http://www.google.je/patents/US20120060414, retrieved 07.04.2015
[2] http://plantagon.com/urban-agriculture/vertical-greenhouse, retrieved 13.03.2015
[3] Dependent on the location wether it's on the Northern or Souther Hemisphere.

3.3. Typological comparison of existing Vertical Farms

Should it prove true that the biocapacity of the earth no longer is able to compete with world population growth; urban agriculture (especially Vertical Farm farming) is an issue of greatly increased relevance today. If cities are in most cases dependent on food imports with all the consequences this entails[1], then food markets and the food supply chain also need to be rearranged. Food supply of city dwellers from traditional agriculture as we now have it, which has long been completely dependent on hydrocarbon energy, should be expanded by food production practices which release the current food provision system fundamentally and in three ways from the pressures to which it is subjected:

- by increasing the food independencies of cities,
- by decreasing the (economic and political) pressures and stresses resulting from the exploitation of natural habitats[2],
- decrease the dependencies of hydrocarbon energy.

A closer look at the five Vertical Farms, presented in Chapter 3.2 shows that food production (which is one item within the food supply chain) is the focal task here and it currently not only obtains its components from outside the system but also demands extensive structures for dealing with the tasks that follow-on after the harvest. At the present time a Vertical Farm, should it happen to be embedded in an existing system (urban, infrastructure, program), is an idiosyncratic building and the following typological comparison clearly indicates the differences the concept brings into the food provision arena, based on the primary design decisions it incorporates, all of which have the background aim of making cities more independent and resilient.

The issues of fundamental interest here are the production output, or the output of edible biomass by the Vertical Farms, as also the ratio between the building footprint and the equivalent of traditional soil based agriculture production, the A/V-ratio, the ratio between transparent and opaque facade area of the greenhouse and the embodiment of additional programs to reduce the spatial impact on the food supply chain.

---

1. WATKISS, P., SMITH, A., TWEEDLE, G., MCKINNON, A., BROWNE, M., HUNT, A., TRELEVEN, C., NASH, C. & CROSS, S. 2005. The Validity of Food Miles as an Indicator of Sustainable Development: Final report produced for DEFRA. AEA Technology.
2. BRUINSMA, J. 2009. The Resource Outlook to 2050. By how much do land, water and crop yields need to increase by 2050? [Paper presented at the FAO Expert Meeting, 24-26 June 2009, Rome on „How to Feed the World in 2050".] [Online]. Available: ftp://ftp.fao.org/agl/aglw/docs/ResourceOutlookto2050.pdf.

VERTICAL FARM - REFERENCE MODELS

Before we go into detail, the production types of the quoted Vertical Farms up to here must be classified more precisely to make the comparison replicable.

The world map on the next page provides an overview of selected relevant world Vertical Farm types (existing buildings, high-tech-greenhouses with stacked production methods or design proposals, relevant for this discussion) . The geographic position of the buildings compared in this chapter can be found by reference to the numbers.

- 5 for Vertical Harvest, Jackson, Wyoming, United States,
- 10 for Plantagon Vertical Farm in Linköping, Sweden,
- 11 for Paignton Zoo, Devon, United Kingdom,
- 15 for Suwon Vertical Farm, Suwon, South Korea and
- 20 for SkyGreens, Singapore.

VERTICAL FARM - REFERENCE MODELS

WORLD VERTICAL FARM PROJECTS[1]

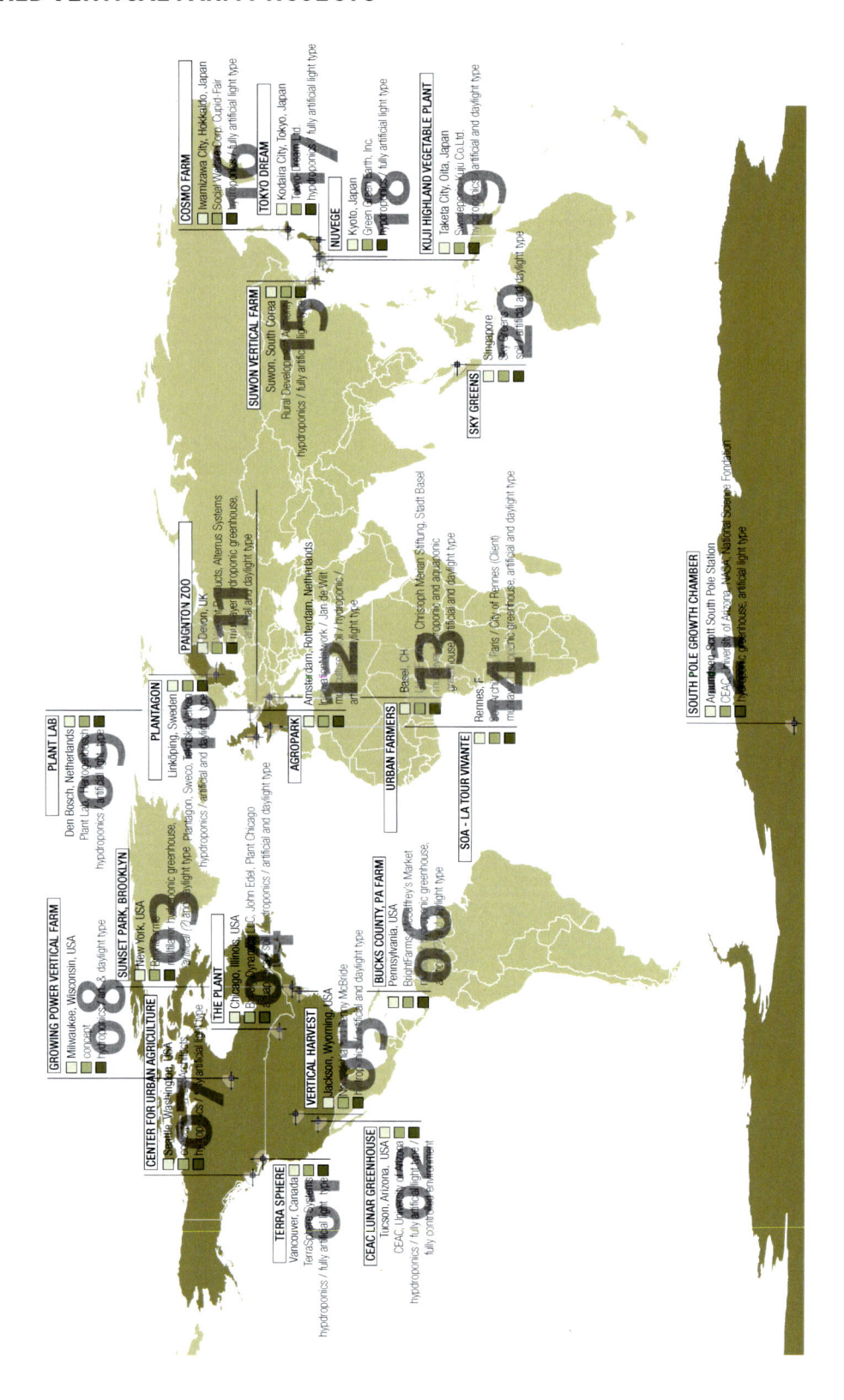

TERRA SPHERE — Vancouver, Canada — TerraSphere Systems — hydroponics / fully artificial light type — **01**

CENTER FOR URBAN AGRICULTURE — Seattle, Washington, USA — Mithun — hydroponics / artificial and daylight type — **07**

GROWING POWER VERTICAL FARM — Milwaukee, Wisconsin, USA — concept — hydroponics / artificial and daylight type — **08**

SUNSET PARK, BROOKLYN — New York, USA — BrightFarm Systems — hydroponic greenhouse, hydroponics / artificial and daylight type — **03**

THE PLANT — Chicago, Illinois, USA — Bubbly Dynamics LLC, John Edel, Plant Chicago — hydroponics / artificial and daylight type — **04**

VERTICAL HARVEST — Jackson, Wyoming, USA — Nona Yehia, Penny McBride — hydroponic greenhouse, hydroponics / artificial and daylight type — **05**

BUCKS COUNTY, PA FARM — Pennsylvania, USA — BrightFarms / Jeffrey's Market — hydroponic greenhouse, hydroponics / artificial and daylight type — **06**

CEAC LUNAR GREENHOUSE — Tucson, Arizona, USA — CEAC, University of Arizona — hydroponics / fully artificial light type, fully controlled environment — **02**

PLANT LAB — Den Bosch, Netherlands — Plant Laboratory — hydroponics / artificial and daylight type — **09**

PLANTAGON — Linköping, Sweden — Plantagon, Sweco, SweGreen — hydroponics / artificial and daylight type — **10**

PAIGNTON ZOO — Devon, UK — Valcent Products, Alterrus Systems — multi-layer hydroponic greenhouse, hydroponics / artificial and daylight type — **11**

AGROPARK — Amsterdam-Rotterdam, Netherlands — Dutch Research Center / Jan de Wit — multi-layered greenhouse / hydroponic / artificial and daylight type — **12**

URBAN FARMERS — Basel, CH — Christoph Meier Stiftung, Stadt Basel — greenhouse, hydroponic and aquaponic, artificial and daylight type — **13**

SOA - LA TOUR VIVANTE — Rennes, France — SOA Architects, Paris / City of Rennes (Client) — multi-layered hydroponic greenhouse, artificial and daylight type — **14**

SUWON VERTICAL FARM — Suwon, South Corea — Rural Development Administration — hydroponics / fully artificial light type — **15**

COSMO FARM — Iwamizawa City, Hokkaido, Japan — Sopaz Wataru Corp. Cupid-Fair — hydroponics / fully artificial light type — **16**

TOKYO DREAM — Kodaira City, Tokyo, Japan — Terra Dream Ltd — hydroponics / fully artificial light type — **17**

NUVEGE — Kyoto, Japan — Green Earth, Inc. — hydroponics / fully artificial light type — **18**

KUJI HIGHLAND VEGETABLE PLANT — Taketa City, Oita, Japan — Suntory Kuju Co. Ltd. — hydroponics / artificial and daylight type — **19**

SKY GREENS — Singapore — Sky Greens — greenhouse, artificial and daylight type — **20**

SOUTH POLE GROWTH CHAMBER — Amundsen-Scott South Pole Station — CEAC, University of Arizona, NASA, National Science Foundation — hydroponics, greenhouse, artificial light type

---

1   Classifications of the projects not mentioned in this chapter can be found in the Appendix.

CLASSIFICATIONS FOR EXISTING VERTICAL FARM PROJECTS

PRODUCTION SPACE

- stacked layers within a mono volume
- stacked layers on multiple building levels
- food production behind the facade
- enlarged double skin depth or atriums
- under existing building programme
- over existing building programme

PRODUCTION TYPE

- **G^{PRO}** high tech greenhouse with crop stackings
- **VF** crop stacking embodying additional food supply programme
- **VF+** VF + additional urban functions

BUILDING SERVICES

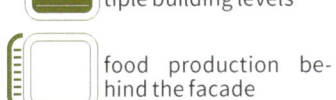

- **H** heating
- **V** ventilating
- **AC** air conditioning

CULTIVATION METHOD

- hydroponics
- aeroponics
- aquaponics
- **CO$_2^+$** elevated CO2-system

LIGHTING SYSTEM

- fully artificial lighting
- full natural lighting
- combined artificial and natural lighting

PRODUCTION METHOD

- horizontal layers
- horizontal rotation
- vertical rotation
- 3D conveyor belt

VF SUWON Vertical Farm, SUWON, SOUTH KOREA

This facility uses high-technology within a fully controlled environment. It is run specifically on an electric light dependence concept using LEDs, the indoor climate is fully computer controlled, all production spaces are either heated or cooled on demand and the facility is artificially ventilated. The research function of this facility means it is not intended to produce marketable food and thus no additional program for the food supply chain is considered. Linear programmatic expansions such as research spaces are likewise not considered in in this case as with of the Plantagon Vertical Farm in Linköping.
This production type is thus is classified as a Vertical Farm.

G^PRO PAIGNTON ZOO, DEVON, UNITED KINGDOM

Although feed production within this type proceeds in a stacked manner, the classification for this building is thus as a Vertical Farm. The building envelope takes the same form as a foil tunnel of the kind that is well known and widely distributed in Central Europe. The production method with its horizontal conveyor system, the computer controlled watering and plant nutrition provision, temperature and humidity control are considered in the light of their potential for possible implementation in Vertical Farms. The production type is thus not classified as a Vertical Farm, but as a greenhousePRO.

SKYGREENS, SINGAPORE

SkyGreens produces food in tall greenhouses. Vertically rotating troughs make full use of the total height of the greenhouse and the exposure to natural sunlight, entering from all sides of the building envelope. No additional programs other than food production take place within the building volume. Computer controlled indoor climate such as temperature and humidity and the rotation of the vertical rotating elements make this a high-tech-greenhouse that achieves remarkable increase in the biomass output per m2. SkyGreens is classified as greenhousePRO.

VERTICAL HARVEST, JACKSON, WYOMING

Vertical Harvest not only produces fresh food, but also enlarges its program linearly to include those spaces, which are normally separated by the current food supply chain such as preparation, retail and offices. In addition public spaces are implemented for guided tours and educational purposes. Lighting fixtures and computer controlled HVAC-systems are implemented. These components make a Vertical Farm + out of this production type, which is functionally comparable with the next Vertical Farm.

PLANTAGON, LINKÖPING, SWEDEN

The Vertical Farm in Linköping combines an office building with a monovolumetric vertical greenhouse. This design decision enables an energetic optimization in terms of the synergy potential resulting from the combination of the two different uses. Additional spaces are integrated within the Vertical Farm for those functions which are normally separated spatially by transport routes within the food supply chain such as washing, preparation and packaging of the harvested. In addition, the computer controlled 3D conveyor belt and building systems make a Vertical Farm+ out of this production type.

Suwon Vertical Farm
Suwon, South Corea

http://www.rda.go.kr/foreign/eng

Suwon Vertical Farm
Rural Development Authority

Fig.28. Suwon Vertical Farm, South Korea

Fig.29. Suwon Vertical Farm, South Korea - Cultivation Room

VERTICAL FARM - REFERENCE MODELS

CLIMATE DATA, Suwon, South Corea: 37°15'4"N 37°15'48"W

hours/year	8,760
dayhours	4,392
sunshine hours	(49,24%) 2,163
kWh/m2/a	1,156.64
diffuse sunlight (kWh/m2/a)	702.69
direct sunlight (kWh/m2/a)	453.95
GJ/m2/a	4.16
kCal/m2/a	994,531
l Oil Eq./m2/a	112.19
av. kWh/m2/a	0.26
av. MJ/m2/a	0.95
av. kCal/m2/a	226.44
av. l Oil Eq./m2/a	0.03

The climate, similar to Seoul, is a humid continental/subtropical transitional climate with strong differences between relatively cold winters (daily mean -2,4°C in January) and hot summer month (daily mean 25,7°C in August). Winters generally are dry and summer month have a high relative humidity. Two thirds of the total radiation are diffuse light, and approximately 50% of the dayhours are sunshine hours.

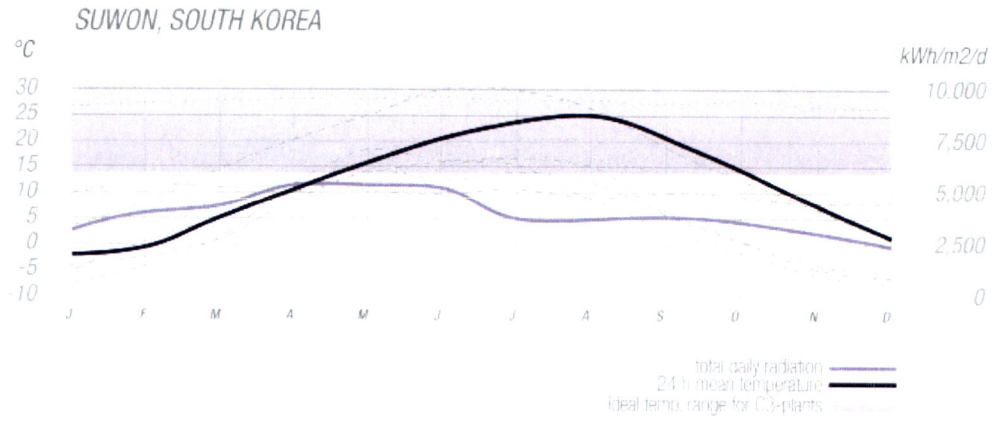

Fig.30.   Total daily radiation and 24 h mean temperature, Suwon, South Korea

SUWON Vertical Farm, SUWON, SOUTH KOREA

This Vertical Farm in Suwon, South Korea, is mentioned here, because it embodies the principle of completeness. Regrettably the author found it was not possible to obtain sufficient information regarding the dimensions of the building and the biomass output yield to contribute significantly to the typological comparisons made within this doctoral thesis.[1]

Nevertheless this Vertical Farm is of central interest. The Korean government, particularly the Rural Development Administration, RDA, have paid close attention to urban and Vertical Farming since 2009 when this research institution was established with the primary focus on optimization of the light supply for crops by investigating the ideal individual wavelength for the specific growing periods.

This Vertical Farm produces primarily leafy vegetables on multiple layers in a horizontal stacking manner. The building itself keeps natural light out. Crops are grown completely through use of artificially light provided by LED fixtures, which produce primarily red and blue light.

The cultivation method is hydroponics with nutrient enriched water.

The building itself requires an enormous amount of energy. Since within the climatic differences between day and night and throughout the seasons it must be heated, ventilated and also air conditioned in summer.[2]

---

1 http://www.korea.net/NewsFocus/Sci-Tech/view?articleId=103942, retrieved 31.10.2015
2 http://www.spiegel.de/international/zeitgeist/vertical-farming-can-urban-agriculture-feed-a-hungry-world-a-775754.html, retrieved 13.03.2014

Paignton Zoo
Devon, UK

http://www.paigntonzoo.org.uk

Paignton Zoo Environmental Park
Totnes Road
Paignton
Devon TQ4 7EU

Fig.31.   Paignton Zoo, Devon: Photo of the production volume

Fig.32.   Paignton Zoo, Devon: Photo of the greenhouse envelope

CLIMATE DATA, Devon, England, Great Britain: 50°43'18"N 3°32'01"W

hours/year	8,760
dayhours	4,387
sunshine hours	(33,37%) 1,464
kWh/m2/a	1,004.90
diffuse sunlight (kWh/m2/a)	586.71
direct sunlight (kWh/m2/a)	418.19
GJ/m2/a	3.62
kCal/m2/a	864,058
l Oil Eq./m2/a	97.48
av. kWh/m2/a	0.26
av. MJ/m2/a	0.82
av. kCal/m2/a	196.96
av. l Oil Eq./m2/a	0.02

Devon's climate is influenced by the North Atlantic Drift. Its relatively mild climate for its latitude is relatively seldom, only compareable e.g. with Vancouver, Canada. The temperature throughout the year change between 8°C and 20°C. Only a third of the dayhours in Devon are sunshine hours. The total solar radiation is around 1.000 kWh/m2/a, the diffuse and direct part are roughly 50% to 50%.

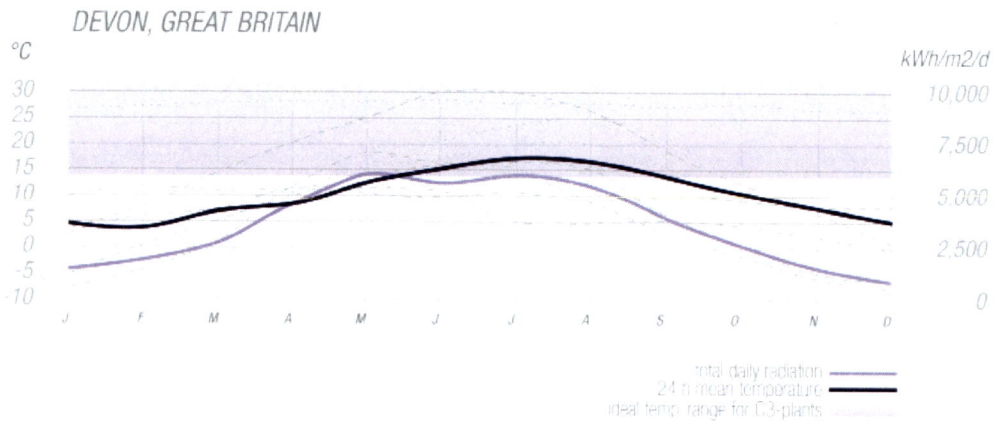

Fig.33. Total daily radiation and 24 h mean temperature, Devon, United Kingdom

VERTICAL FARM - REFERENCE MODELS

PAIGNTON ZOO, DEVON, UNITED KINGDOM

In the same year 2009, the Vertical Farm in Devon started to produce leafy vegetables for zoo animals. In the context of food production within enclosed environments this typology is defined as GPRO, and classified as a high tech greenhouse with crop stackings[1]. The building envelope is the same as that used for foil tunnels (steel construction with transparent and translucent foils), but the production method is conceived for horizontal tray stackings moving along a conveyor belt. For this reason alone it is of basic interest. This production method also can be adapted to existing buildings with stacked levels, which light only reaches from the side. The horizontal transportation offers all the plants an equal exposure to natural light.

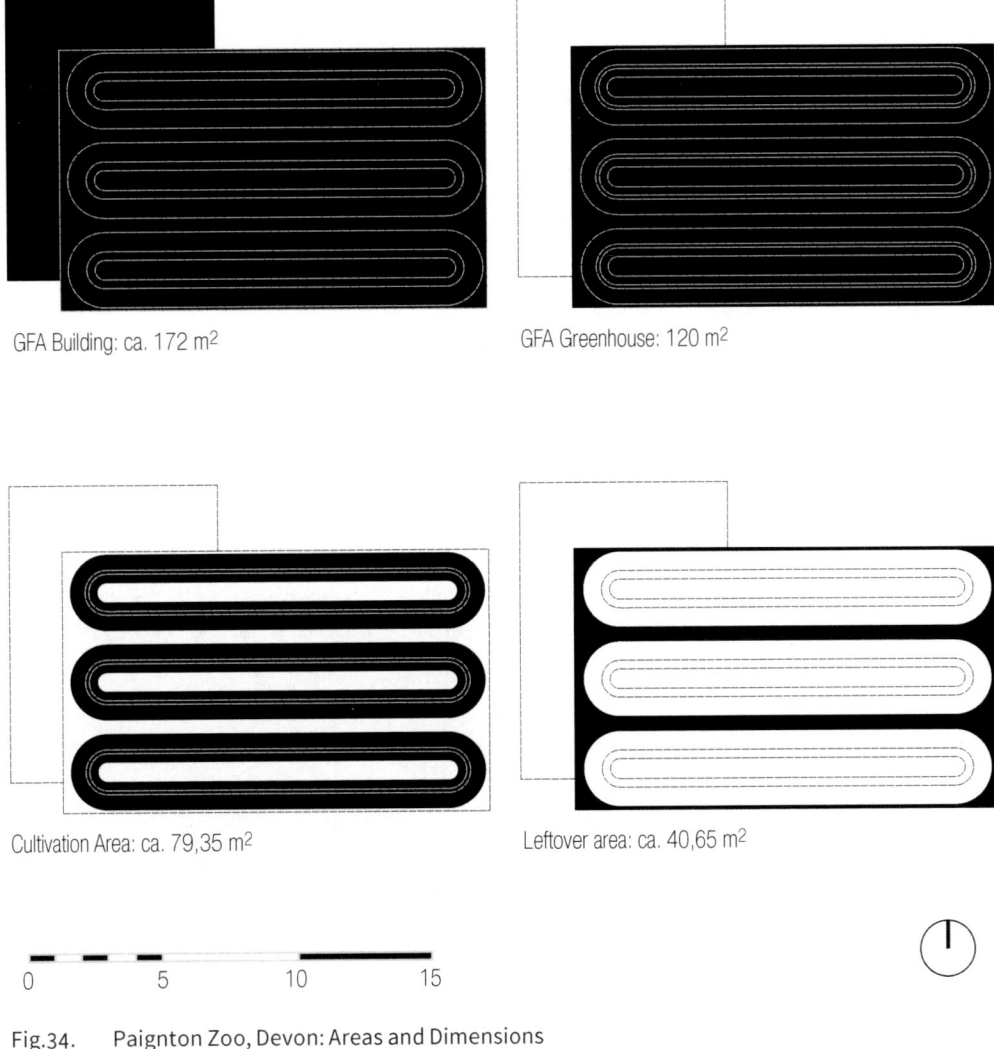

GFA Building: ca. 172 m²

GFA Greenhouse: 120 m²

Cultivation Area: ca. 79,35 m²

Leftover area: ca. 40,65 m²

0   5   10   15

Fig.34. Paignton Zoo, Devon: Areas and Dimensions

---

[1] This system, developed by VertiGrow, is currently implemented at the center in Vancouver, Canada, on the rooftop of a building which is ten times the area of the greenhouse building footprint in Devon.

With this 3m-pilot system Paignton Zoo produces around 112 lettuces/m2/p.a. Soil-based yielded lettuces can reach up to 500g/head. Greenhouse lettuces normally reach between 100 to 250g/head. Based on studies by VertiCrop, however, the forecasts for this model "suggest annual yields of lettuce around 50 times higher than typical for field grown crops (...)."[1]

GREENHOUSE VOLUME: 1.968.92 m3
Greenhouse +
Building Footprint: 144,34 m2
Cultivation Area: 388,32 m2

Soil Based Equivalent: 1.587,00 m2

9,09 %

Fig.35.   Paignton Zoo: land use reduction, production volume and A/V-ratio

Based on average central European data for biomass output, the production per m2/a is roughly ten times higher than for soil based agriculture according to the author's calculation.[2]. Considering that lettuce in a non-protected environment can be cultivated only six month in the year, and Paignton Zoo produces for 365 days, the ratio between soil-based and greenhouse output can be doubled, the ratio is therefore 1/20.

In absolute numbers, the 79.35 m2 cultivation area guarantees an output where 1,587.00 m2 of soil based agricultural land would be needed, or compared to the GFA-building footprint the land requirement is 10.8%.

The VertiCrop[3] conveyor system was developed to support heights up to 6m with 250 lettuces/m2/a. Based on these figures the yield would be 29. 20 times higher compared to soil-based agriculture, or calculated throughout the year, 58.40 times higher. This result comes very close to the estimates of Bayley et al. for the system involved. In this case the GFA of Plantagon's greenhouse would be 2.45% of traditional soil based agriculture land equivalent.

---

1   BAYLEY J.E. 2010. Sustainable Food Production Using High Density Vertical Growing (VertiCropTM), Valcent Products EU Ltd., Cornwall, BAYLEY SUSTAINABLE FOOD PRODUCTION PAIGNTON ZOO.pdf, p. 4
2   Considering that one lettuce weighs 250g at the time when it is harvested and the expectation is 2,14kg/m2/a during a 6month season in middle-Europe of traditional soil based agriculture land equivalent.
3   http://www.verticrop.com/, retrieved 31.10.2015

The building footprint for the greenhouse is 69.76%. A visitor's and educational area is situated together with an area for packaging the crops for immediate distribution is located to the west and the north. HVAC facilities are situated under the 0-level. Due to the fact that the crops are fed to animals immediately after harvesting, the plants do not have to be prepared as with a subsequent conventional food supply chain.

Vertical Harvest
188 South Millward Street, Jackson, WY, USA

www.verticalharvest.org

Vertical Harvest
Penny McBride, Project Administrator
PO Box 7290
Jackson, WY 83002

verticalharvest@verticalharvest.org

Fig.36. Vertical Harvest, Jackson, Wyoming - Rendering of vertical rotating production

Fig.37. Vertical Harvest, Jackson, Wyoming

CLIMATE DATA, Jackson Hole, Wyoming, USA: 43°29'39"N 110°45'54"E

hours/year	8,760
dayhours	4,363
sunshine hours	(68,98%) 3,010
kWh/m2/a	1,568.18
diffuse sunlight (kWh/m2/a)	545.70
direct sunlight (kWh/m2/a)	1,022.48
GJ/m2/a	5.65
kCal/m2/a	1,348,392
l Oil Eq./m2/a	152.11
av. kWh/m2/a	0.36
av. MJ/m2/a	1.29
av. kCal/m2/a	309.05
av. l Oil Eq./m2/a	0.03

There are extreme differences not only between day and night temperatures, but also through the seasons (-46°C to 37°C). The high elevation above sea level (1.901 m) results in high total radiation. More than two thirds of the day hours throughout the year are sunshine hours.

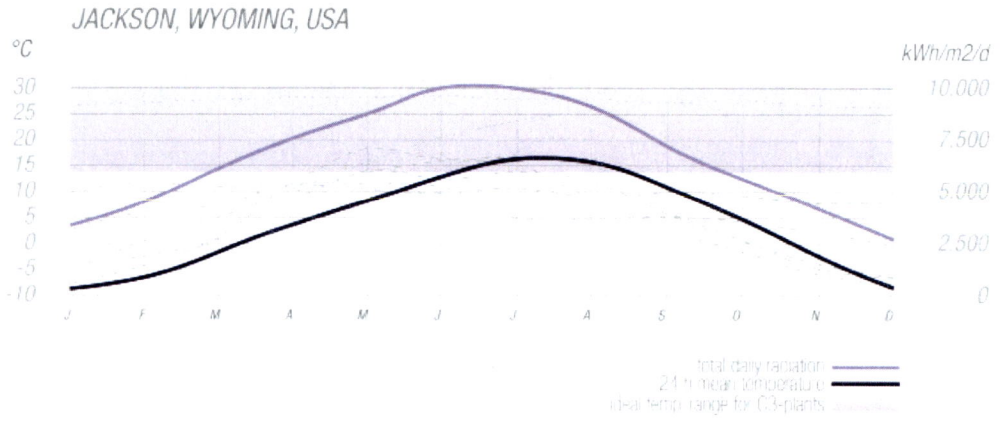

Fig.38.   Total daily radiation and 24 h mean temperature, Jackson, WY, USA

VERTICAL HARVEST, JACKSON, WYOMING, USA

In the classification of this doctoral thesis, this building is a Vertical Farm +. An existing multi-storey car park has been spatially enlarged by adding an enclosed greenhouse structure. The three-storey Vertical Farm has the cultivations of different plants are implemented on each storey. A largely educative public growing-showroom is on the ground floor, while above this the first floor is used mainly for the production of leafy vegetables. The third floor is covered with a glass and tomatoes are produced on this level.
Produce is sold directly from the retail area within the building, after it is got washed, prepared and packaged.

The varied usage involved here means that besides embodying a locally disconnected program of the traditional food supply chain in this production entity, Vertical

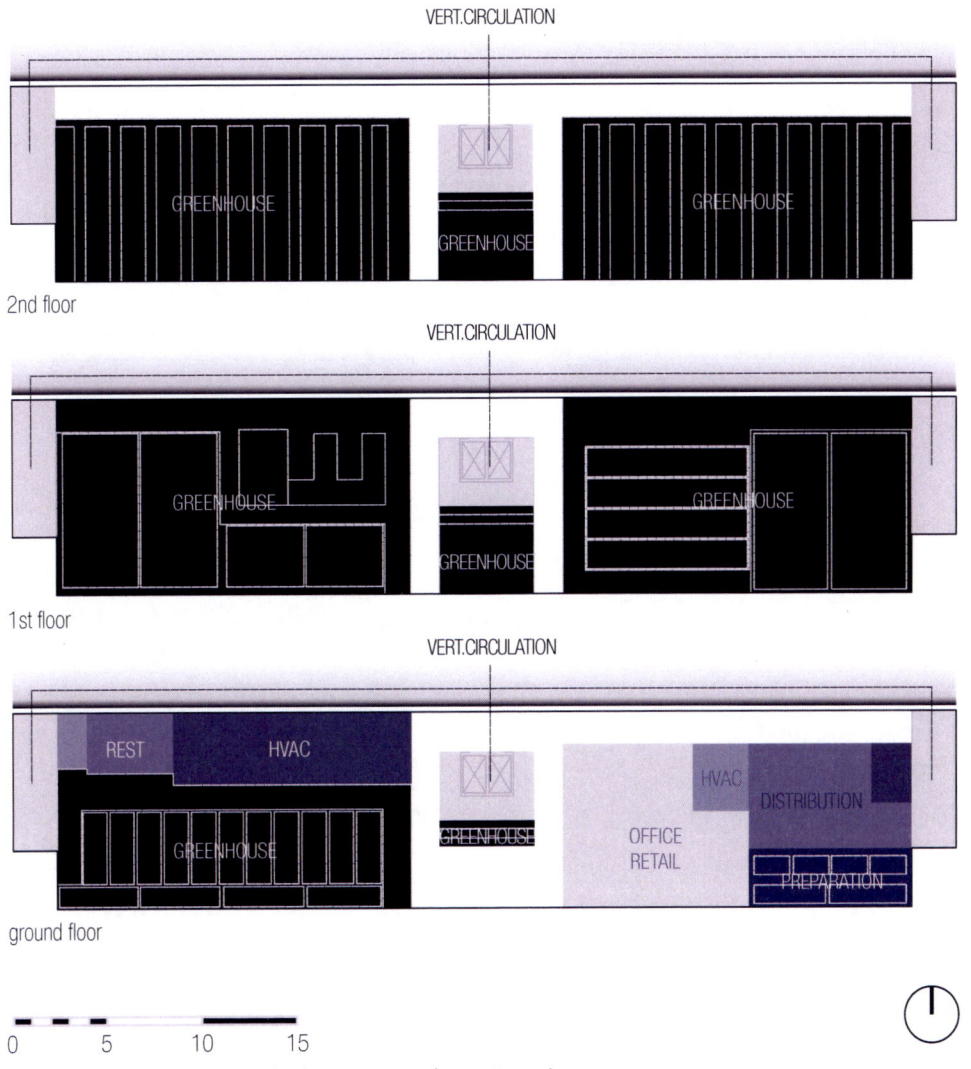

Fig.39.   Vertical Farm Vertical Harvest, Jackson: Functions

VERTICAL FARM - REFERENCE MODELS

GFA Building: ca. 488,44 m²

GFA Greenhouse: 229,48 m²

Cultivation Area: ca. 194,08 m²

Leftover area: ca. 92.68 m²

0   5   10   15

Fig.40.   Vertical Farm Vertical Harvest, Jackson: Areas and Dimensions

Harvest also has the additional objective of involving the community with a dual approach: offering visitor spaces within the building for guided tours and education and also involving people from the community with physical or mental disabilities in the work of growing food. Sensitizing people to the potential local food production and involving the community distinguishes this Vertical Farm from the other examples mentioned in this chapter.

> „Vertical Harvest has a dedicated central public atrium that is physically separated by glass walls from the major growing areas in the building to encourage visitors to visually experience the greenhouse without risk of contaminating the crops. On the ground floor Vertical Harvest will have a small but functional 'living classroom' where we can grow a limited amount of specialty crops while at the same time incorporating educa-

tional initiatives. Here, because there is no soil used and we will growing in mobile containers that are up off the ground, making potential pests and diseases much easier to avoid."[1]

The chosen cultivation method is hydroponics. The climate in Jackson Wyoming makes heating the glasshouse necessary. No air conditioning is needed. The greenhouse is naturally ventilated.
Additional artificial lighting is required, although the exterior facade „is specified to include the highest light transmission possible (...)"[2] Approximately 3,000 hours/a are expected to support the growing- and ripening process of the chosen crop. HPD (high Pressure sodium lamps) are used for tomatoes and LED fixtures for "lettuce varietals,

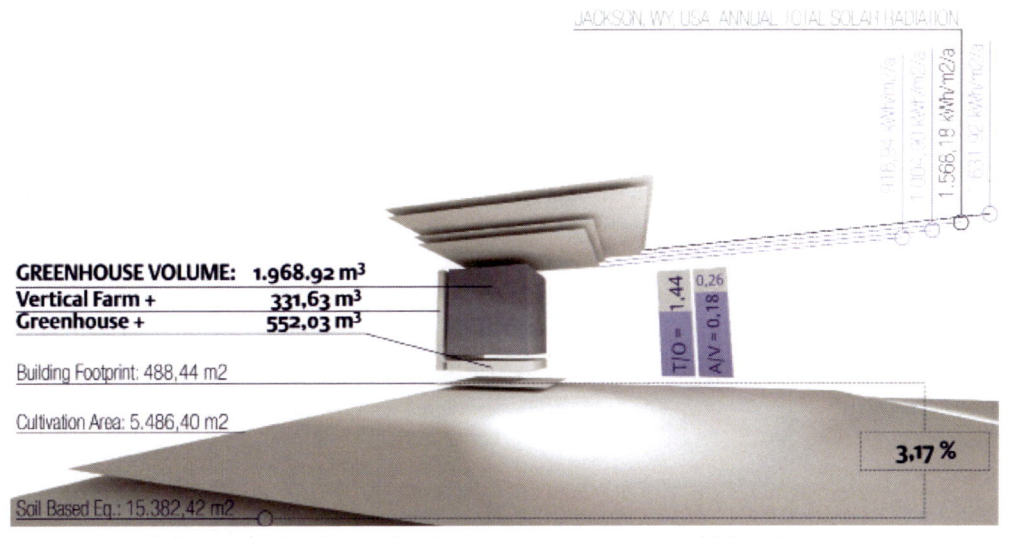

Fig.41.   Vertical Harvest: land use reduction, production volume and A/V-ratio

microgreen and propagation areas."[3]
Tomatoes are cultivated throughout the year on a 216,91m2 area. A comparison to traditional soil based agriculture is most likely irritating, because the months with the right temperature for tomato growth are too few for the necessary crop rotation in Jackson. But we can approach using estimates for an expected yield output. Greenhouse tomatoes can be sold at a profit for cultivation values above 45 kg/m2/a. An ordinary greenhouse season in central Europe is about 9 months long. Calculating on an estimated 50 kg/m2/a of tomatoes within the nine month period and adding 25% for a year-round crop we reach 62,5 kg/m2/a under controlled conditions. For Vertical Harvest this would mean that a yield of 13.55 t throughout the year can be expected. When compared with the average data for central Europe once again, this would mean an area equivalent of 595.60 m2 would be needed.

---

1   http://www.verticalharvestjackson.com/faq-2/, retrieved 01.04.2015
2   ibid.
3   ibid.

VERTICAL FARM - REFERENCE MODELS

For leafy vegetables like lettuce and other microgreens Vertical Harvest with its growing carousel of 5,486.4 m2 (=18.000 sq.ft.) with an estimated 6kg/a output the annual yield is about 32,918.4 kg. The equivalent area needed for soil based agriculture with the same output, calculated for central Europe would be 15.382,42 m2. Adding the tomato and leafy yield the edible biomass output of Vertical Harvest is 46.46 t/a.

GREENHOUSE VOLUME: 1.968.92 m³
Building Footprint: 488,44 m2
Cultivation Area: 5,486,40 m2

3,17 %

Soil Based Eq.: 15.382,42 m2

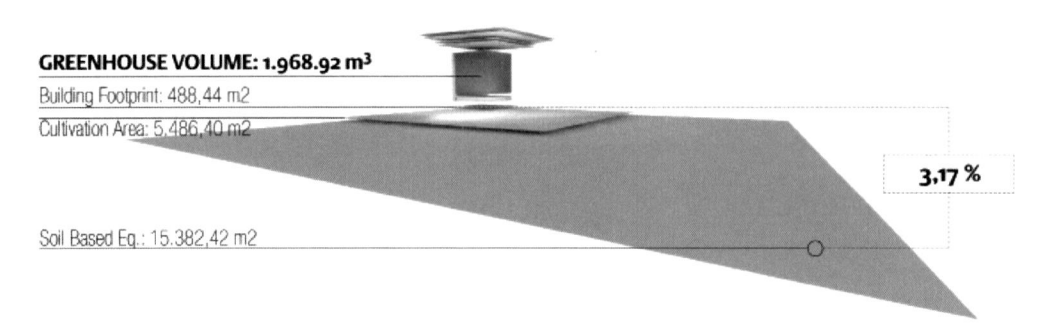

Fig.42.   Vertical Harvest: land use reduction, production volume and A/V-ratio

Comparing these data to the building footprint the resulting ratio is 468,44 m2 GFA building footprint to 6.082,00 m2 or 1/13. Compared to the actual greenhouse footprint of Vertical Harvest[1] we reach a ratio of 229,48 m2 to 6.082,00 m2 or 1/26. This value comes close to the estimated values from Vertical Harvest, which is: „Although Vertical Harvest is situated on a site that is 1/10 of an acre, the greenhouse will be able to produce the equivalent of 5 acres of traditional agriculture."[2]

---

1   The area was calculated as follows: greenhouse area groundfloor + greenhouse area 1st floor + greenhouse area 2nd floor/3 including horizontal circulation area within the greenhouse zones
2   http://www.verticalharvestjackson.com/faq-2/, retrieved 01.04.2015

SkyGreens
200 Lim Chu Kang Lane 2, Singapore 718804

http://www.skygreens.com

SkyGreens Pte. Ltd.
42 Kallang Place
Singapore 339170
AZ 85719
info@skygreens.com

Fig.43.   Skygreens, Singapore, Vertical production

Fig.44.   Skygreens, Singapore

CLIMATE DATA, Singapore, 1°17'N 103°50'E

hours/year	8,760
dayhours	4,417
sunshine hours	(44,82%) 1,980
kWh/m2/a	1,631.92
diffuse sunlight (kWh/m2/a)	1,127.05
direct sunlight (kWh/m2/a)	504.87
GJ/m2/a	5.87
kCal/m2/a	1,403,198
l Oil Eq./m2/a	158.30
av. kWh/m2/a	0.37
av. MJ/m2/a	1.33
av. kCal/m2/a	317.68
av. l Oil Eq./m2/a	0.04

Singapore has a tropical rainforest climate. Temperature, pressure and humidity are relatively constant, moving between 22°C and 35°C and 73% (morning) to 79% (evening) of relative humidity. More than two thirds of the total solar radiation are diffuse light. Roughly the half of dayhours are sunshine hours.

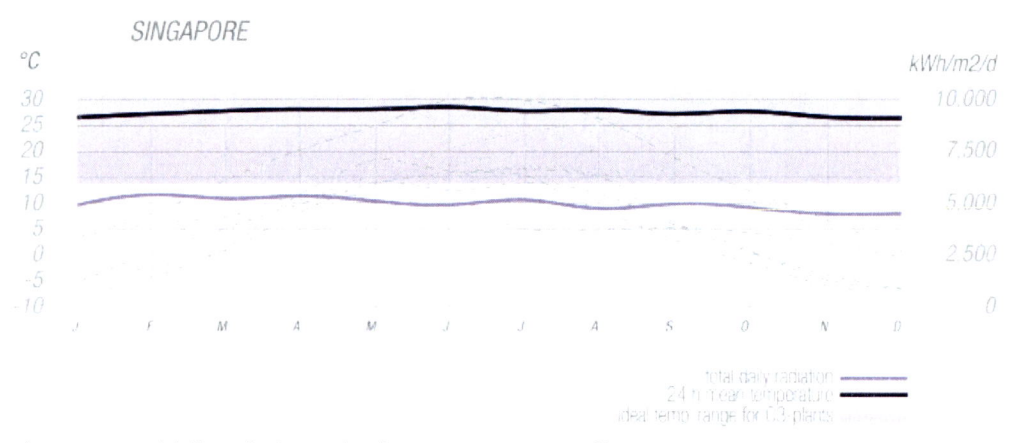

Fig.45.   Total daily radiation and 24 h mean temperature, Singapore

SKYGREENS, SINGAPORE

The first prototype was installed in 2009 and a research collaborative agreement was signed between Sky Greens and the Agri-Food and Veterinary Authority of Singapore (AVA) in April 2010. The multi-layer troughs in a rotating A-frame vertical structure, referred to as "A-Go-Gro" was then commercialized in 2012. Since then Sky Greens has been expanding and continuously installing additional Vertical Farms in Singapore, all of which are GPRO .in our classification.

GFA Building: ca. 196 m²

GFA Greenhouse: 196 m²

Cultivation Area: ca. 126 m²

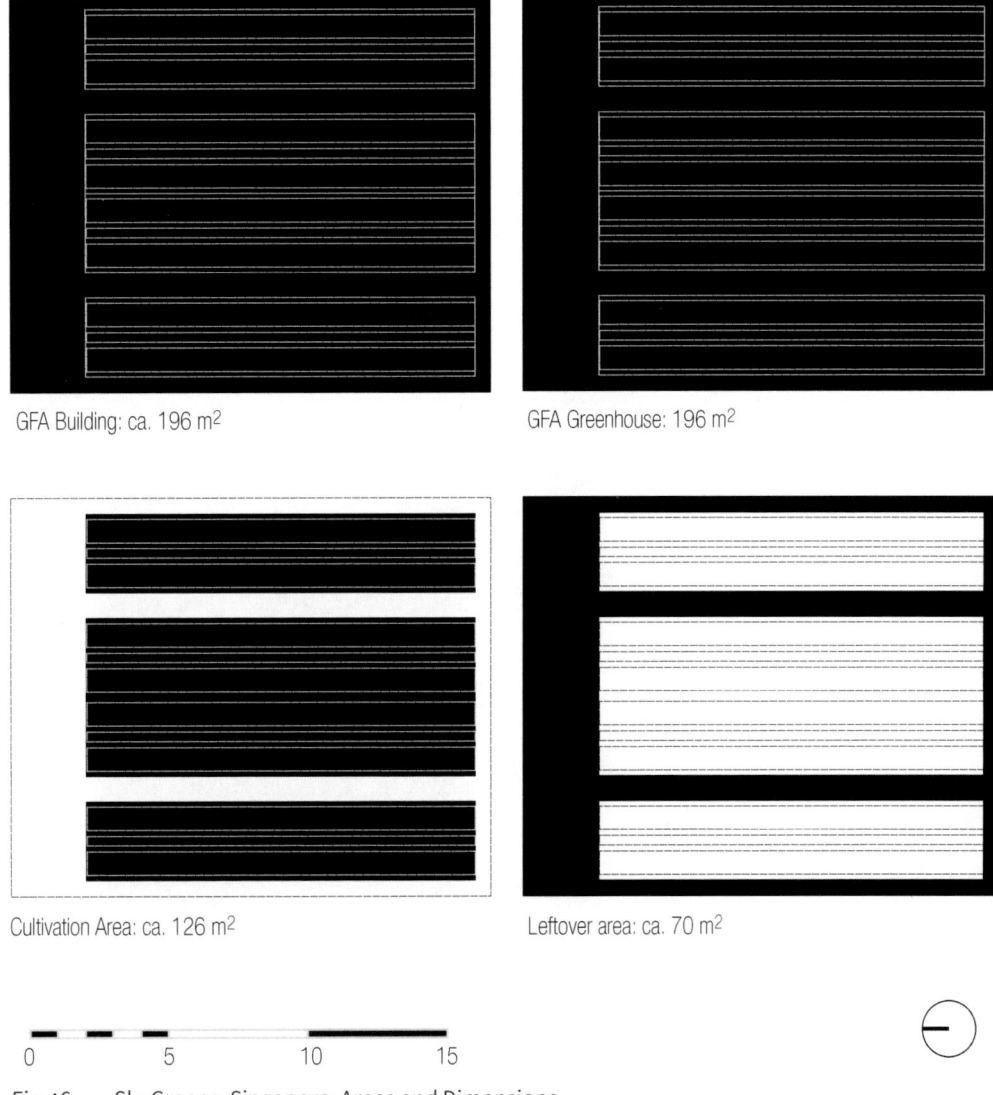

Leftover area: ca. 70 m²

0   5   10   15

Fig.46.   SkyGreens, Singapore: Areas and Dimensions

The cultivation method used is hydroponics. The A-frame within the greenhouse can reach up to 9 meters with 38 growing troughs. Similar to the idea of VertiCrop for Paignton Zoo, a conveyor belt was installed to ensure uniform light exposure for the

plants and this not in a horizontal circuit, but in a vertical one. The greenhouse receives sunlight also from above. This is made necessary by the production method.
Rainwater and recycled water are collected in overhead tanks and used not only for watering the plants via micro-sprinklers three times a day, but are also to support the patented Water Pulley System, which relies on flowing water and gravity to rotate the racks.

Considering an average crop cycle of 8 weeks[1] and considering that on one A-Go-Gro-System 20x38 lettuces (=780) are growing, and using the same average yielded weight of one lettuce, namely 250 g, the output per year (considering a 365 days-production) would be around 1.267,5 kg. The prototypical greenhouse with four A-shaped production systems then would produce around 5.070,0 kg or ca. 5 t/a on a building footprint of 196.16 m2.

The author was unable to obtain reliable data on food production output in Singapore. One reason might be that only 0.5% of the area of Singapore is currently in use for agricultural production.[2][3]

Fig.47.   Skygreens Singapore: land use reduction, production volume and A/V-ratio

---

1   SÄCHSISCHE LANDESANSTALT FÜR LANDWIRTSCHAFT, FACHBEREICH GARTENBAU 2004. Salate im Gewächshaus. Hinweise zum umweltgerechten Anbau. Managementunterlage. [Dresden]: Sächsische Landesanstalt für Landwirtschaft, Fachbereich Gartenbau. p.3
2   http://www.commonwealthofnations.org/sectors-singapore/business/agriculture/, retrieved 02.04.2015
3   This means that the population of Singapore (531 million) is completely dependent on food imports.

To keep up with comparisons for output data in central Europe an area equivalent of 2.369,15 m2 would be needed. The building footprint in this case is the same as the area of the greenhouse. The resultant ratio between the footprint of the built envelope and the potentially needed soil based agricultural land then is 1/12. This value is close to the official comparison published by Sky Greens on its website. "When compared with traditional monolayer farms, the Sky Greens patented Vertical Farming system intensifies land use and can result in at least 10 times more yield per unit land area."[1]

---

1   http://www.skygreens.com/technology/, retrieved 03.04.2015

Plantagon
Gumpekullavägen, Linköping 58278, Sweden

http://www.plantagon.com

Plantagon Headquarter
Rålambsvägen 17, 22nd floor
11259 Stockholm, Sweden

Fig.48.   Plantagon, Vertical Farm, 3D conveyor belt

Fig.49.   Plantagon, Vertical Farm, Linköping, Sweden

CLIMATE DATA, Linköping, Sweden: 58°24'39"N 15°37'17"E

hours/year	8,760
dayhours	4,385
sunshine hours	(33,52%) 1,470
kWh/m2/a	916.94
diffuse sunlight (kWh/m2/a)	517.62
direct sunlight (kWh/m2/a)	399.32
GJ/m2/a	3.30
kCal/m2/a	788,426
l Oil Eq./m2/a	88.94
av. kWh/m2/a	0.21
av. MJ/m2/a	0.75
av. kCal/m2/a	179.80
av. l Oil Eq./m2/a	0.02

Cold winters and mild summers determine the climate in Linköping from monthly minimum temperatures of -8°C in January to maximum temperatures of 20°C in July. The precipitation is distributed equally throughout the year between 10 and 15 days/month. Global total horizontal radiation is below 1000 kWh/m2/d. Only a third of the day hours are sunny.

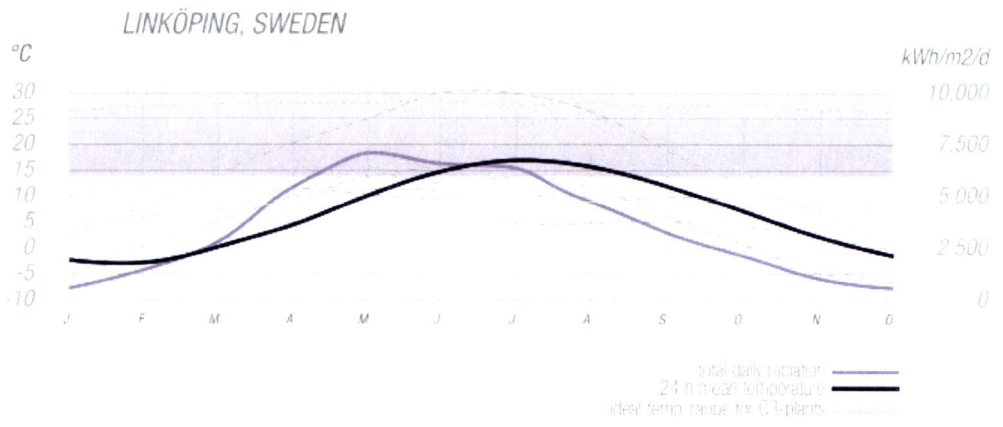

Fig.50.   Total daily radiation and 24 h mean temperature, Linköpin, Sweden

PLANTAGON, LINKÖPING, SWEDEN

Plantagon's proposal for a Vertical Farm in Linköping is the most holistic project of the greenhouses and projected Vertical Farms analyzed in this paper. In terms of the classification table it is a V+, a Vertical Farm with an additional program, which enables the use of synergy potentials on an energetic level. In addition the production entity embodies most of the necessary functions of the food supply chain, which is spatially divided here in sectors such as germination rooms, washing, preparation

GFA Building: ca. 1.005,70 m^2

GFA Greenhouse: ca. 425 m^2

Cultivation Area: ca. 1.115 m^2

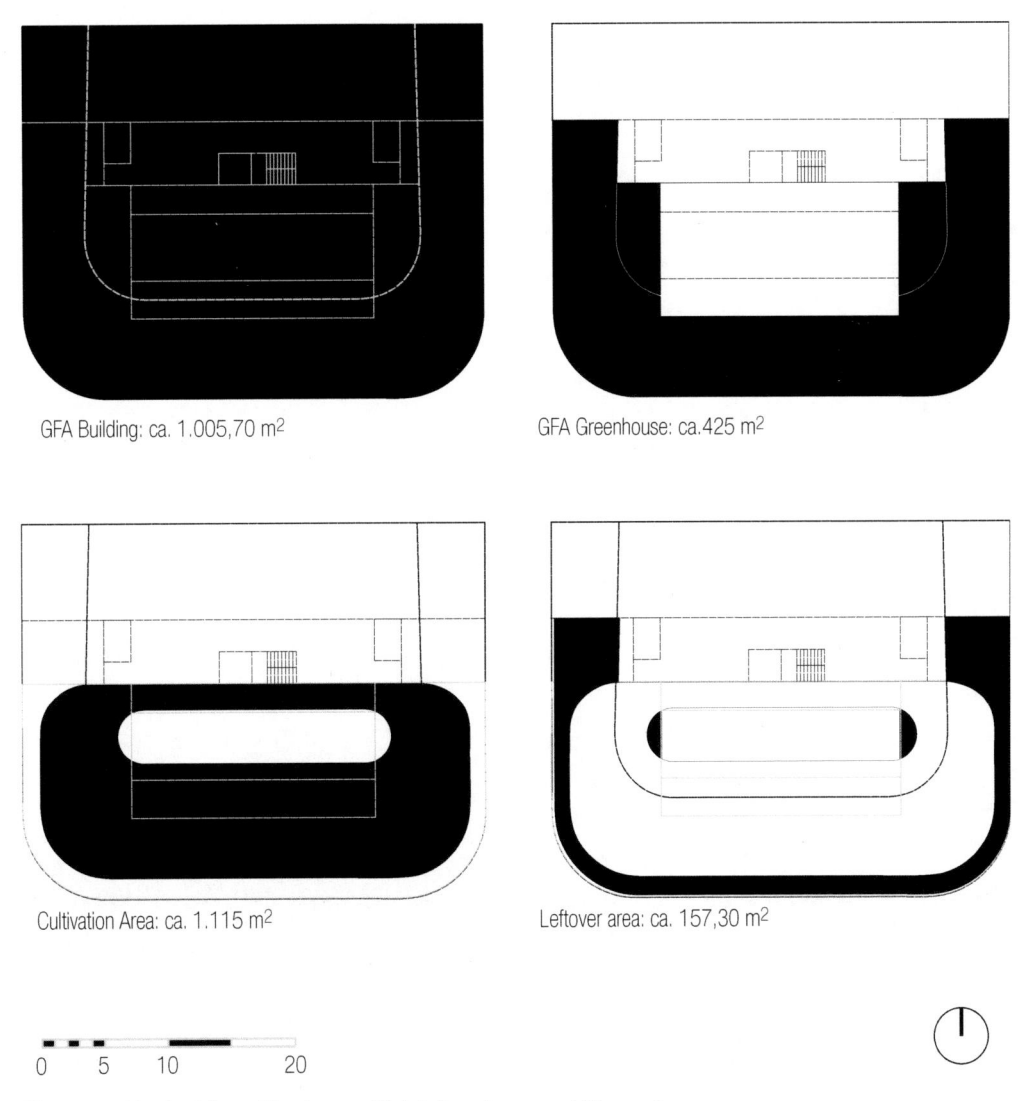

Leftover area: ca. 157,30 m^2

0 5 10 20

Fig.51. Vertical Farm Plantagon, Linköping, Areas and Dimensions

and storage.
The project was developed together with the city of Linköping, Tekniska Verken, the energy provider for Linköping and Sweco, a globally active sustainable engineering firm. An issue of primary interest in the project was to implement and optimize ener-

gy circuits between the Vertical Farm, the biodigester, run by Tekniska Verken and the district heating with an internal exchange of CO_2, excess heat and green waste.[1]

Typologically the building consists in a slender north-facing office building with a monovolumetric vertical greenhouse in the south, containing the 3D conveyor belt based on the form of a helix.

> „The different system designs basically all have the same production flow and location of equipment. The machinery is located in the basement on one or two floors and the trays are transported to the top of the helix by a special tray elevator. The crops grow during the slow transport down the helix and are ready for harvesting when they reach the end of the helix at the basement level. Food is harvested in batches using an automatic harvesting machine. After harvest, the trays and pots are disinfected, and the pots are separated and replanted with another seed for the next round in the cultivation loop. After germination, the pots are recombined with the trays and elevated to the top of the growing helix to repeat the process."[2]

The helix itself transports the trays from top to the basement level. The cultivation area it provides is 1,999.20 m2[3] when pak choi is planted. Calculated with the expected edible biomass of 1.300 kg/d (=4.000 plants/d), as reported at the Urban Agriculture Summit by Sweco-Horticulturist Susanna Hultin, and assuming the non-edible portion of this to be 10% of the total weight, the annual biomass output is 521.95 t. For reasons of comparison in this calculation an optimized assumption of a greenhouse pak choi yield is used of 5.25 kg[4]/ready of harvested tray[5]. the resulting effective all year-cultivation area within the Vertical Farm in Linköping would be seven times higher than the cultivation area of all the trays within this system, namely 14,592.40 m2 or 1.45 ha[6].

A tray moves along the helix in 50 days or 142.64m or 2.85m/d. By dividing all the trays by 50 (3.331 trays) we get the daily number of harvest trays ready for harvest. With 66.6 trays ready to harvest a day each with 15 plants of 350g each, the daily biomass output (marketable and non-marketable) is 349.65 kg or 127.62 t/a. Even by calculating with a shorter crop rotation (35 days) and a higher weight of the plant, this re-

1   http://plantagon.com/urban-agriculture/industrial-symbiosis, retrieved 03.04.2015
2   http://www.handelskammer.se/files/2011404_ps_en_plantagonsweco.pdf, retrieved 03.04.2015
3   own calculation: helix length = 999,48 m; distance from tray to tray = 30 cm; resulting trays on helix: 3.331,6 trays 300 cm x 20 cm = 0,6 m2.
4   350 g/Pak Choi; 90% (318,18g) is edible biomass
5   15 Pak Choi plants per 325g/head; http://www.lel-bw.de/pb/site/lel/get/documents/MLR.LEL/PB5Documents/lel/pdf/a/Alternative%20Herbstkulturen%20-%20Heike%20Sauer%20LVG%20Heidelberg.pdf, retrieved 07.04.2015
6   If every tray needs 50 days from seed to harvest, every tray it is used 7.3 times a year.

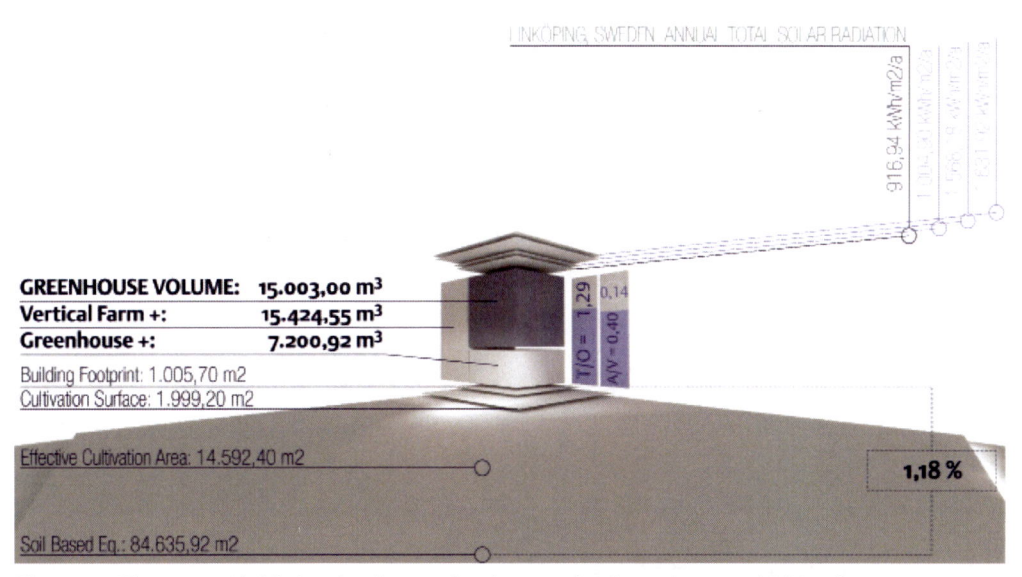

Fig.52.   Plantagon, Linköping: land use reduction, production volume and A/V-ratio

sult is far less than the calculated expected year-round harvest communicated by Plantagon. For this diagram the author takes the verifiable data from the above listed empirical data of greenhouse cultivation: On an effective area of 1.45 ha 127.62 t is produced. This is 88.01 t/ha or 5.8 times more than the average pak choi yield per hectare in central Europe.

14.592,40 m2 or 1,45 ha Pak Choi[1] soil based cultivation in a climate zone like Linköping could take place for maximum 8 month/year. The expected biomass output can reach up to 15t/ha[2]. The ratio between the building footprint (1,005.70 m2) and the agricultural land needed is 1/29, the ratio between the greenhouse footprint (376.02 m2) and the agricultural land needed is 1/77.

Even though the numbers communicated on the Urban Agriculture Summit seem way too optimistic, it can still be pointed out here that this production method with the robotic 3D-conveyor belt changes the basic cultivation surface available here into one with an effective cultivation area throughout the year which is seven times higher.

---

[1] Pak choy is a cool season crop that prefers moist and uniform conditions in full sunlight. High temperatures with long days will induce bolting, especially in the white-stemmed varieties. The ideal temperature during growth is 15-20°C and, while best grown in spring and autumn, it can be grown all year round. Most varieties of Pak choy can tolerate light frosts., https://www.daf.qld.gov.au/plants/fruit-and-vegetables/vegetables/asian-vegetables/Pak-choy, retrieved 08.04.2015

[2] Roughly 50% of greenhouse yields. https://researcharchive.lincoln.ac.nz/bitstream/10182/670/3/Fu_Magrsc.pdf, retrieved 08.04.2015

VERTICAL FARM - REFERENCE MODELS

This Vertical Farm, with a building footprint of 1,005.70 m2 a cultivation surface of 1,999.20 m2 was implemented to illustrate the benefits this patented production method brings for the reduction of arable land use while maintaining the same edible biomass output. Due to its continuous movement and the seed- and cultivation intervals this cultivation surface is effectively seven times higher, throughout the year at 14,592.40 m2. To obtain the total (edible and non-edible) biomass with pak choi as calculated above, based on an average central European yield, an area of 84,635.92 m2 of soil based agriculture was compensated by this Vertical Farm.

GREENHOUSE VOLUME: 15.003,00 m³
Building Footprint: 1.005,70 m2

1,18 %

Soil Based Eq.: 84.635,92 m2

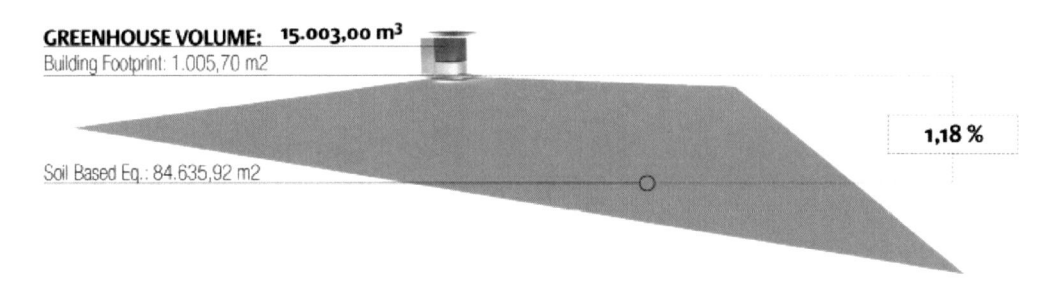

Fig.53.   Plantagon, Linköping: land use reduction, production volume and A/V-ratio, large scale

4. Vertical Farm - Substitution of Natural Growth Factors

4.1. Goals and methods

The amount of light plants need for growth in closed conditions is examined in this chapter. A brief review of the basic physics of light will be useful here to distinguish between light for humans and light for plants, or light in photosynthesis and in architecture. Differentiation between the quality and the quantity of light, between photometry and radiometry are necessary to build up the Simulation Model in Chapter 5.

Assumptions on the overall energy value the sun offers for plant growth already tell us that lighting power for Vertical Farms will be the significant energy consuming system within the building. This assumption is also based on the results of two Master theses by Chirantan Banerjee[1] and Gordon Graff [2], although both these works only touch on the complexities of light use in plant growth, nevertheless both of these works highlight the relevance of lighting within the energy performance of the Vertical Farm.

This work, however, aims to go beyond interpolating estimations of light needs of plants to an overall energy consumption for artificial lighting, and thus deepened the scope for the question of which quality and quantity of light is necessary for the production of a specific vegetable or fruit plant.
Lycopersicon esculentum (Mill.) (solanaceae), commonly known as tomato, stands in the spotlight of this research.

Firstly an introduction in the physics of light will be given, the difference of perception of light between humans and plants will be explained. Subsequently a picture of the state of the art in research about PPFD (Photosynthetic Photo Flux Density), DLI (Daylight Integral) and PAR (Photosynthetically Active Radiation) will be given.

---

[1] BANERNJE, C. 2012. Market Analysis for Terrestrial Application of Advanced Bio-Regenerative Modules: Prospects for Vertical Farming. Master of Science Masterarbeit, Rheinische Friederichs-Wilhelms-Universität, Hohe Landwirtschaftliche Fakultät.
[2] GRAFF, G. 2011. Skyfarming. Master of Architecture, Waterloo, Ontario, Canada.

In addition the different growth stadiums of the *L. esculentum* will be examined and specified for its lighting-, temperature and water requirement. Lastly the edible biomass-output per year can be calculated, related to the energy-consumption/kg and compared to soil-based *L. esculentum* and greenhouse *L. esculentum*.

Research and statistics on horticulture and plant physiology necessary for this thesis, in addition, got supported by the Institute for Plant Sciences at the University for Natural Resources and Life Sciences, Vienna. To evaluate statistics and numbers they got discussed and compared with empirical data provided by „Zeiler"-greenhouses in Vienna across excursions and interviews. It is one of the largest companies in Eastern Austria and, to this day, the only year-round producer of tomatos.

4.2. Introduction and basics

Translation of light to sugar is crucial factor in food production. Not only the available light diminishes on Earth as one moves from the equator to the poles, but also the length of temperate seasons. The practice of greenhouse production has been increasing in most countries ever since the 17th century in an effort to prolong the food production seasons. But greenhouses reach their limits as they encounter power demands for temperature heating and light at the feasibility threshold for food production in energy and economic terms. The greenhouse skin must generally be optimized for maximum light transmittance and to minimize heat transmission.
Enlarging the photoperiod and the length of crop rotation calls for supplementing or substituting natural growth factors, such as photosynthetically active radiation. This makes it necessary to investigate in more detail which part of the total solar radiation actually is crucial for sugar production and plant growth.

4.2.1. Photometry and Radiometry

Radiation from the sun can be distinguished by its quality or quantity. Quality, the waveband of the light or the distribution of the wavelength within the waveband, is central to distinguish the energy content of photons by measuring wavelengths and frequency and for defining whether it is visible or invisible radiation. The quantity, or intensity or the amount of energy of specific wavebands enables the differentiation between useful and non-useful, essential and harmful energy, both in photometry and radiometry.

Photometry is the science of measuring light. The reference is the sensitivity for brightness of the human eye. Radiometry is concerned with the measurement of radiant energy in terms of power. "In modern photometry, the radiant power at each wavelength is weighted by a luminosity function that models human brightness sensitivity."[1]
A difference in radiometry today, especially for plant physiology, is that the perceived brightness of light is no longer relevant, but the focus is now on the energy content of photons within PAR (Photosynthetic Active Radiation), and more specifically the quantity of photons between 400 nm and 500 nm (blue light) as well as photons between 600 nm and 700 nm (red light) and their energy content. "This quantity can be measured as the number of photons, or as a total energy value. Whenever a num-

---

1 http://en.wikipedia.org/wiki/Photometry_%28optics%29, retreived 03.08.2014

ber of radiation quantity (intensity) is given, the wavelength(s) involved must also be given (quality), or else the number has little useful value"[1]

Photometric measurements are of central significance for architecture. The wavelength 555 nm is that most relevant for luminance and illumination. The average human eye is most sensitive at 555 nm, or a frequency of 540 THz. "Photometry describes lighting conditions with the human eye as primary sensor." "The spectral responsivity curve of the standard human eye at typical light levels is referred to as the CIE Standard Observer Curve (photopic curve), and covers the waveband of 380 - 780 nm. The human eye responds differently to light of different colors and has maximum sensitivity between yellow and green. In order to make accurate photometric measurements of various colors of light, or from differing types of light sources,

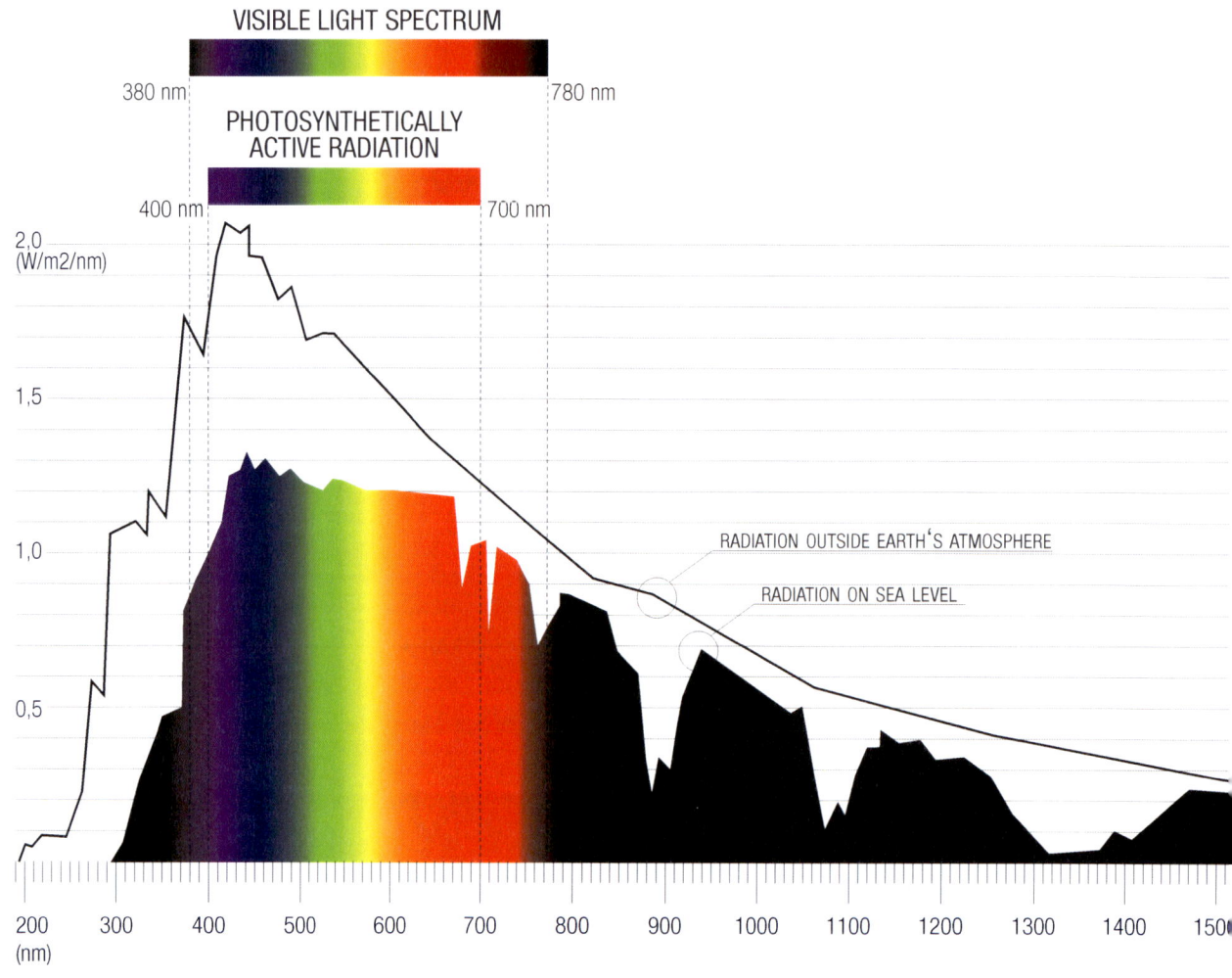

Fig.54.   Electromagnetic radiation with ratio of light and PAR

---

1   GIACOMELLI, G. 1998. Components of Radiation Defined: Definition of Units, Measuring
     Radiation Transmission, Sensors. CCEA, Center for Controlled Environment Agriculture,
     rutgers university, Cook College.

a spectral responsivity curve for a photometric sensor must match the CIE photopic curve very closely."[1]

Photometry is therefore concerned with visible light (from the perspective of the human eye); the corresponding quantities are as follows:

- *Luminous flux: radiation coming from a source per unit time (cd x sr), where sr is the solid angle, expressed in lumen (lm).*
- *Luminous energy: or quantity of light is not the same as radiant energy. The quantity of light referrs only to the amount of visible light (from 380 to 780 nm), expressed in lumen per second (lm s). This entity sometimes is expressed also in talbots.*
- *Luminous intensity: is the luminous flux in a particular direction per unit solid angle. The SI unit of luminous intensity is candela (cd).*

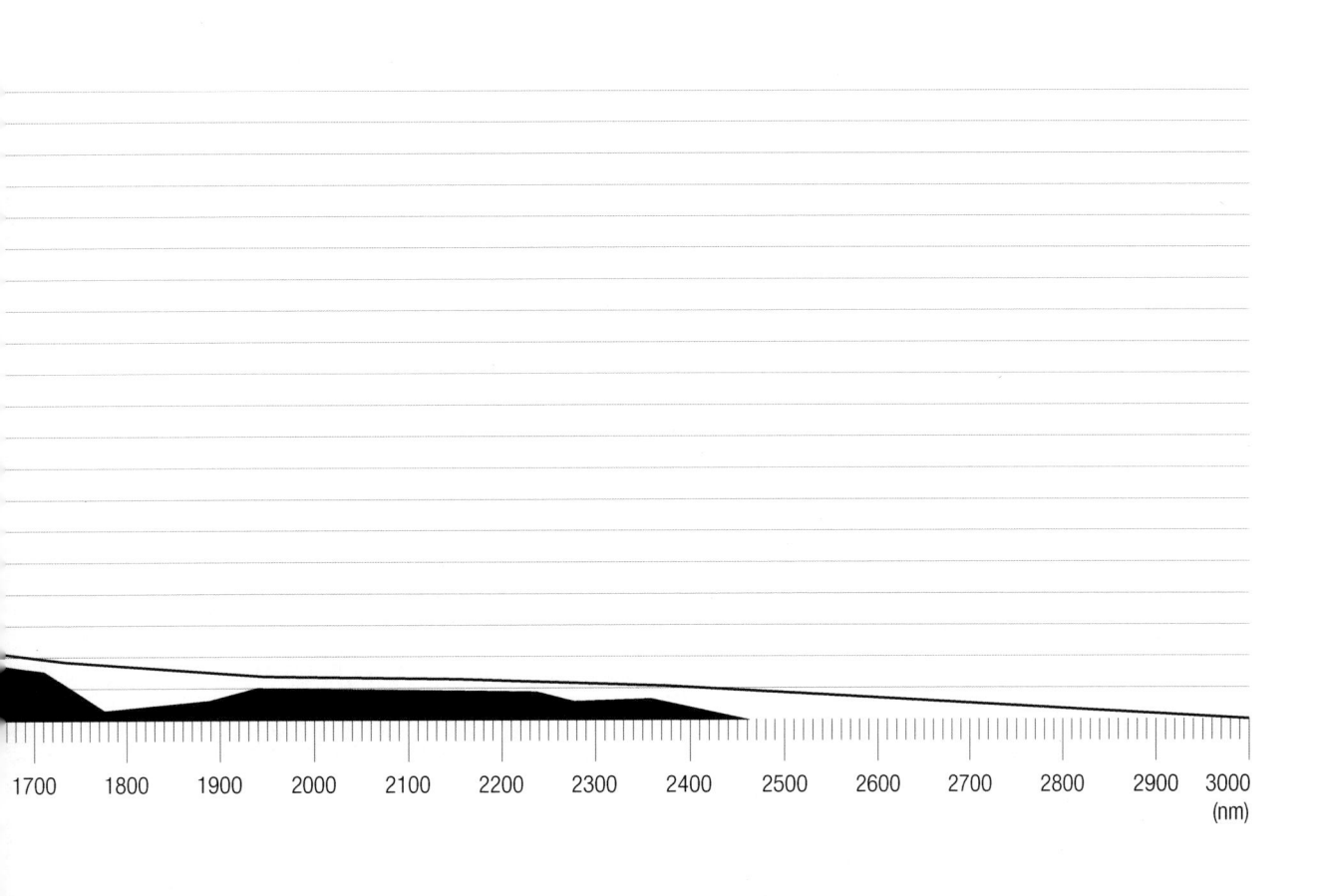

---

[1] http://www.licor.com/env/pdf/light/Rad_Meas.pdf, p. 3. Retrieved 03.11.2015

• *Illuminance is the density of the luminous flux, incident at a point on a surface. In architecture the density of the luminous flux is one of the prior entities. Illuminance gets measured by Luxmeters which have the highest spectral response on a wavelength of 555 nm or a frequency of 540 THz.*

The visible spectrum for human eye sensitivity ranges from approx. 380 nm (violet) to approx. 780 nm (dark red). Luxmeters to measure illuminance are calibrated to exactly this wavelength. Illuminance is thus the key value for light in architecture.

By contrast with photometry, radiometry focusses on the totality of electromagnetic light, including the range from 3^{11} and 3^{16} Hz corresponding to wavelengths from 0.01 and 1000 micrometers and therefore includes UV light, visible and infrared light.

It is necessary to divide electromagnetic radiation into the visible and invisible range. The differentiation between radiometry and photometry is essential for sensitizing to different qualities of light measurements in architecture intended for human use and also in architecture provided for plants.

The electromagnetic spectrum summarizes total solar radiation. In contrast to photometry, where only the spectrum visible to the human eye is taken into account.

Radiometry measures radiant energy (SI unit is J [Joule]), an interchangeable form of energy. The radiant energy flow rate in form of specific electromagnetic waves is called radiant flux (W). „Radiant flux can be measured as it flows from the source (sun, natural conditions), through one or more reflecting, absorbing, scattering and transmitting media (the Earth's atmosphere, a plant canopy) to the receiving surface of interest (a photosynthetizing leaf)."[1] Total solar radiation is measured by the solarimeter between the wavelengths from 300 nm and 3000 nm.

Plants need a very specific range of solar radiation from 400 nm to 700 nm; this is the so called Photosynthetic Active Radiation (PAR)[2]. Although plants also use a very small percentage beyond 400 nm and 700 nm, the essential radiation for photosynthesis lies in the range between 400 nm and 500 nm (blue light) and between 600 nm and 700 nm (red light). Chlorophyll A and B and carotenoids are most sensitive to these very specific ranges. This means that the sensitivity of the human eye and the sensitivity of plants for photosynthesis are not congruent.[3]

---

1 http://www.licor.com/env/pdf/light/Rad_Meas.pdf. p.1, retrieved 13.08.2014 - This article is referring to LANG, O.L. et al. 1981. Physiological plant ecology. Chapter: Photosynthetically active radiation. Springer-Verlag. Berlin, Heidelberg, New York
2 SCHOPFER, P. & BRENNICKE, A. 2010. Pflanzenphysiologie, Heidelberg, Spektrum Akademischer Verlag, p.167
3 ibid. p.445

„Although photometric measurements have been used in the past in plant science[1], PPFD and irradiance provides the preferred measurements in advanced plant and greenhouse research. The use of the word ‚light' is inappropriate for plant re-search. The terms "ultraviolet light" and "infrared light" clearly are contradictory."[2]

The quantities corresponding to photometric units are as follow:

- Radiant flux: the amount of radiation coming from a source per unit of time (W [Watt]).
- Radiant Energy is the radiant flux leaving a point on the source per unit of time. Like all forms of energy, this SI unit is the joule (J). This term is usually used if emitted radiation is measured in the surrounding environment. This entity is also interchangeable with Watts, because a Joule per second equals one Watt (Js = W).
- Radiance is the radiant flux emitted by a unit area of a source or scattered by a unit area of a surface (W m-2 sr-1 [Watt per m2 steridian]).
- Irradiance is the radiant flux incident on a receiving surface from all directions (W m-2 [Watt per m2]).

Within the scope of this fact it also emerges that lux (lumen/m2) as the SI unit for illuminance and luminous emittance to measure luminous flux/m2, a basic measuring procedure in architecture can no longer be used in lighting analysis of the production facility for Vertical Farms. The Lux meter must be substituted with the quantum sensor, which is limited to the PAR (photosynthetic active radiation from 400 nm to 700 nm) with its output value of ($\mu mol/m^{-2}/s^{-1}$) and the spectroradiometer, an instrument which splits the incoming light into separate wavelengths or wavebands and then measures the irradiance of the photons in these wavelengths. It measures the spectral irradiance (SI) in the units $\mu mol/m^{-2}/s^{-1}$ or W/ m2.

---

1 ATANAS, G.D. 2005. Integrierte Produktion von Tomaten (*Lycopersicon esculentum* Mill.) im Gewächshaus unter besonderer Berücksichtigung der integrierten Bekämpfung der Weissen Fliege), Dissertation, Humboldt-Universität Berlin., p.889
2 ibid.

4.2.2. Light for plants - Light for humans

THE SOLAR CONSTANT

"(...) [is] the total radiation energy received from the Sun per unit of time per unit of area on a theoretical surface perpendicular to the Sun's rays and at Earth's mean distance from the Sun. It is most accurately measured from satellites where atmospheric effects are absent. The value of the constant is approximately 1,366 kilowatts per square meter.[1] The "constant" is fairly constant, increasing by only 0.2 percent at the peak of each 11-year solar cycle. Sunspots block out the light and reduce the emission by a few tenths of a percent, but bright spots, called plages, that are associated with solar activity are more extensive and longer lived, so their brightness compensates for the darkness of the sunspots. Moreover, as the Sun burns up its hydrogen, the solar constant increases by about 10 percent every billion years."[2]

„The solar irradiance is measured by satellite near the outer surface of Earth's atmosphere."[3]

The energy content of the solar constant, calculated per year and m2 of solar irradiation, unfiltered by the atmosphere, is equivalent to more than 43 GJ.

---

[1] 1361 watts per squaremeter per second
[2] http://www.britannica.com/EBchecked/topic/552889/solar-constant, retrieved 10.09.2014
[3] Kopp, G.; Lean, J. L. (2011). „A new, lower value of total solar irradiance: Evidence and climate significance" (PDF). Geophysical Research Letters 38: n/a. Bibcode:2011GeoRL..38.1706K. doi:10.1029/2010GL045777.

SOLAR OR GLOBAL RADIATION - DIFFUSE AND DIRECT RADIATION

"Solar radiation is also known as global radiation, meaning that it is the sum of direct shortwave radiation from the sun and diffuse sky radiation from all upward angles."[1]
"(...)[R]adiation has two distinct directional properties when it reaches the ground. Direct radiation arrives from direction of the solar disk and includes a small component scattered directly forward. The term diffuse describes all other scattered radiation received from the blue sky (including the very bright aureole surrounding the sun) and from clouds, either by reflection or by transmission."[2]

It is also necessary to state that light transmission is the same in both direct and diffuse radiation. "Direct or diffuse light does not have different PAR values. This means that our eyes perceive differences in lumens between direct and diffuse light. Diffuse light appears dimmer to us even though the total light transmission is not decreased."[3]
Recent findings in the Netherlands even prove that diffuse light increases photosynthesis up to 25 %. This point will be highlighted below: light is not uniformly distributed in greenhouses, but this can be improved if the light is diffuse. To determine the effect of diffuse light on crop growth and development, an experiment with *L. esculentum* crop was conducted from December 2010 to November 2011 under commercial crop management. Three kinds of glass were used as greenhouse covering: standard glass (no diffuse light, 0% haze) and two types of diffuse glass which transformed an increasing fraction of the direct irradiation into diffuse irradiation (45% and 71% haze).
„As presented by Dueck et al.[4] yield increased by 7.8% under 45% haze and by 9.4% under 71% haze, compared to the reference. During the experiment we performed measurements in order to understand these effects. Diffuse light penetrated deeper and more homogeneously into the canopy, which led to higher photosynthesis rates in the middle and bottom canopy layers. Furthermore, less photoinhibition was measured under diffuse light treatment when the outdoor irradiation was high. Under sunny conditions the temperature of upper leaves in the canopy was 3 to 5 °C lower in the greenhouses with diffuse glass compared to the control, while greenhouse air temperatures were comparable. The leaf anatomy, canopy structure, total nitrogen and chlorophyll contents of top, middle and bottom canopy layers were also studied in order to further explain the increased production under diffuse light. The results showed that diffuse glass on greenhouses is one way to improve the light use efficiency of greenhouse crops."[5]

---

1 http://www.fao.org/docrep/x0490e/x0490e07.htm, retrieved 10.09.2014
2 John L. Monteith et.al „Principles of Environmental Physics: Plants, Animals and the Atmosphere", AP, fourth edition, p. 58 - 59
3 http://www.greenhousecatalog.com/greenhouse-light, retrieved 11.09.2014
4 DUECK, T., JANSE, J., LI, T., KEMPKES, F. & EVELEENS, B. 2012. Influence of Diffuse Glass on the Growth and Production of Tomato. VII International Symposium on Light in Horticultural Systems, 75-82.
5 https://www.wageningenur.nl/en/Publication-details.htm?publicationId=publication-way-343239373835, retrieved 14.09.2014

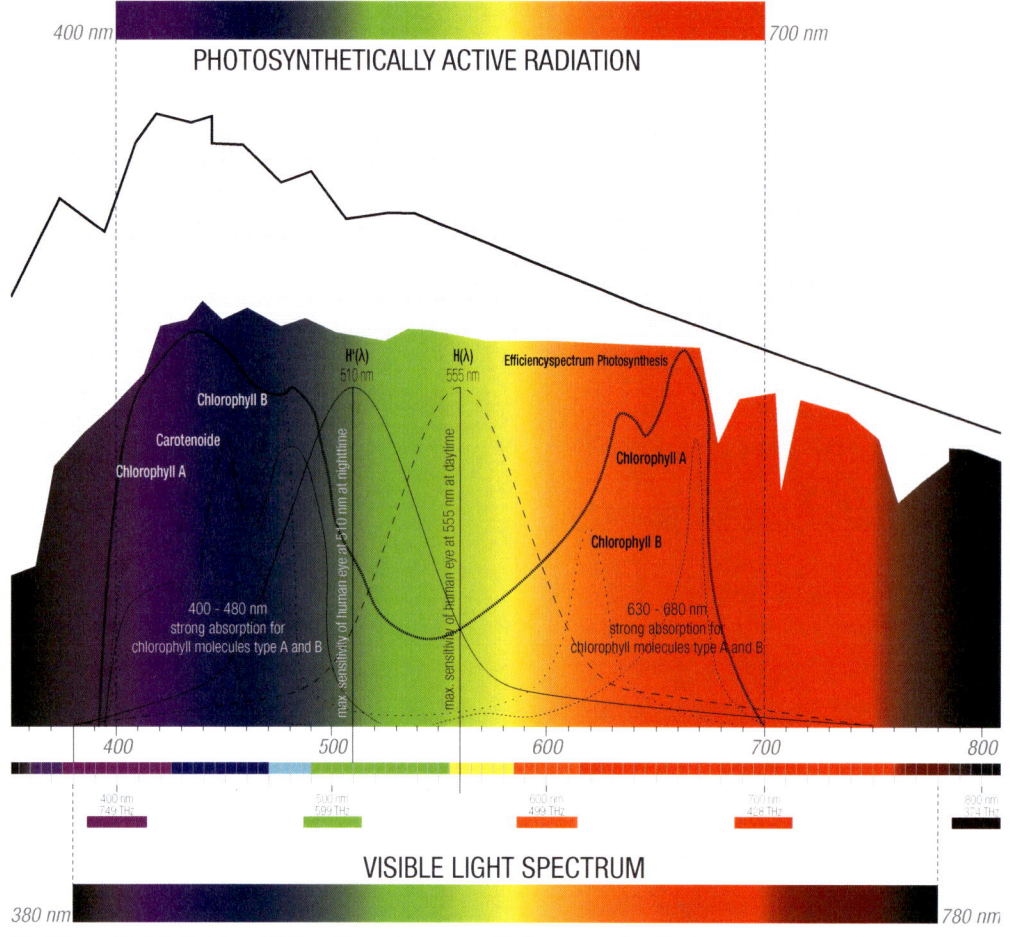

Fig.55. Visible light vs. PAR and its related sensitivity curves for the human eye and plants

PHOTOSYNTHETICALLY ACTIVE RADIATION - PAR

"The PAR, Photosynthetically Active Radiation, comprises the waveband 400 to 700 nm, which are the limits of wavelengths that are of primary importance for plant photosynthesis. The PPFD, Photosynthetic Photon Flux Density is the number of photons in the PAR waveband that are incident on a surface in a given time period ($\mu mol/m^{-2}/s^{-1}$). The quantum sensor will measure this value. A very clear sky value will approach approx. 2000 $\mu mol/m^{-2}/s^{-1}$ PAR."[1]

The PPFD number for a clear sunny sky differs by up to 15% in different studies, from 1700 $\mu mol/m^{-2}/s^{-1}$ (also used by Gene Giacomelli) to 2000 $\mu\ mol/m2/d$. Most conversion calculators online use the factor 0.018 to convert lux to $\mu mol/m^{-2}/s^{-1}$ and the factor 0.219 from Photons to W (sunlight) or 4.57 from W$_{PAR}$ to Photons.[2]

„Radiation can either be reflected, absorbed or transmitted once it impacts a surface. The properties of the material will determine what proportion of the three will be; however, the sum of the energy reflected, absorbed and transmitted must be 100%. The properties are often abbreviated by the Greek symbols α, ρ and τ, which represents reflectance, absorbtance and transmittance. There are standard test procedures for determining each. The leaf will typically absorb nearly 95% of wavelengths between 400 - 700 nm, while only 5% of the 700 - 800 nm waveband is absorbed. Of the remaining 95% of the 700-850 nm waveband, approximately 45% is reflected, and 45% is transmitted."[3]

---

1   GIACOMELLI, G. 1998. Components of Radiation Defined: Definition of Units, Measuring Radiation Transmission, Sensors. CCEA, Center for Controlled Environment Agriculture, rutgers university, Cook College. p.5 ff

2   1. GIACOMELLI, G. 1998. Components of Radiation Defined: Definition of Units, Measuring Radiation Transmission, Sensors. CCEA, Center for Controlled Environment Agriculture, rutgers university, Cook College. p. 5 ff
2. http://www.skyeinstruments.com/wp-content/uploads/LightGuidanceNotes.pdf, retrieved 12.11.2014
3. http://www.licor.com/env/pdf/light/Rad_Meas.pdf
4. http://www.controlledenvironments.org/Growth_Chamber_Handbook/Plant_Growth_Chamber_Handbook.htm, retrieved 10.10.2013
5. http://www.egc.com /useful_info_lighting.ph, retrieved 10.10.2013

3   GIACOMELLI, G. 1998. Components of Radiation Defined: Definition of Units, Measuring Radiation Transmission, Sensors. CCEA, Center for Controlled Environment Agriculture, rutgers university, Cook College. p.6

The ratio of PAR [W] within the total solar radiation [W] changes during the day and the year. In horticulture and agriculture the quantity of PAR [W], is usually calculated, depending on the location, by using factors from 0.44[1] to 0,50[2][3].

Based on the definition of units and measurements of Giacomelli's paper[4] gives us the possibility to calculate the photons and therefor the energy content of the color within the light spectrum which is most effective for photosynthesis.

	[nm]	[THz]	[m]	[Hz]	[nm]	[THz]
red	700-630	420-480	0.000000700	428,532,082,857,143	700	428
orange	630-590	480-510	0.000000610	491,758,127,868,852	610	491
yellow	590-560	510-540	0.000000575	521,691,231,304,348	575	521
green/yellow	555	540	0.000000555	540,490,915,315,315	555	540
green	560-490	540-610	0.000000525	571,376,110,476,190	525	571
blue	490-450	610-670	0.000000470	638,239,272,340,425	470	638
purple	450-400	670-750	0.000000400	749,931,145,000,000	400	749

Tab.10.   Light Spectrum - Wavelength, Frequency and Photon Energy

---

1   GIACOMELLI, G. 1998. Components of Radiation Defined: Definition of Units, Measuring Radiation Transmission, Sensors. CCEA, Center for Controlled Environment Agriculture, rutgers university, Cook College.
2   http://www.landwirtschaftskammer.de/gartenbau/beratung/technik/artikel/lichtwerte-umrechnen.htm, retrieved 12.05.2015
3   „Tomaten Zeiler" for its greenhouses use this factor to get PAR-values from its solarimeter-sensors placed on the rooftops. S. appendix „Excursion Tomaten Zeiler.
4   ibid, referring to G.H.M. Kronenberg and R.E. Kendrick, in: 1986: R.E. Kendrick and G.H.M. Kronenberg (Eds.), „Photomorphogenesis in Plants, Nijhoff, Dordrecht, pp.99-114

PHOTOSYNTHETIC PHOTON FLUX DENSITY - PPFD

The PPFD, Photosynthetic Photon Flux Density is the number of photons within the PAR -waveband that is incident on a surface in a give[n] time period (µmol/m^{-2}/s^{-1}). The quantum sensor will measure this value.

When considered as a photon it may be expressed in energy terms, Watts per square meter (W/m2), or as the number of photons (moles of photons) µmol/m^{-2}/s^{-1} . Wavelength as units of meters, typically nanometers (nm) [...] or micrometers (µm). Fre-

[nm]	[THz]	Js [W] / photon	Js [W] / mol	kJ [kW] / mol	kcal / mol
700	428	0.000000000000000000283948339399	170,997.70	171	40.891
610	491	0.000000000000000000325842356687	196,226.87	196	46.869
575	521	0.000000000000000000345676239268	208,171.12	208	49.738
555	540	0.000000000000000000358133040683	215,672.78	216	51.651
525	571	0.000000000000000000378597785865	227,996.94	228	54.521
470	638	0.000000000000000000422901782083	254,677.43	255	60.977
400	749	0.000000000000000000496909593948	299,245.98	300	71.738

quency (f, A/N) has units of cycle per second. Together they are related as parameters of a photon of light by the constant c, the speed of light (299.792.458 m/s, A/N). The frequency of the photon is equal to the speed of light divided by wavelength of the photon. The energy of a wavelength of light is equal to Planck's constant (h = 6,626·10-34 Js, A/N) multiplied by the speed of light and divided by the wavelength. From this relationship, an important fact is determined. For radiation (light), as its wavelength increases, its energy decreases, and as the wavelength decreases, the energy increases. Thus short wave blue light has more energy than longer wave red light."[1]

„Mol is a unit of measurement used in physics and chemistry to express amounts of elements, defined as the amount of any substance that contains as many elementary entities (e.g. atoms, molecules, ions, electrons) as there are atoms in 12 grams of pure carbon-12 (12C), the isotope of carbon with relative atomic mass of exactly 12 by definition. This corresponds to the Avogadro constant, which has a value of 6.02214129(27)·10^{23} elementary entities of the substance."[2]

A mole of photons, therefore consists in 602 trillion light particles. This entity is used to define the Daylight Integral (DLI) and is described in more detail on the next page.

---

1 GIACOMELLI, G. 1998. Components of Radiation Defined: Definition of Units, Measuring Radiation Transmission, Sensors. CCEA, Center for Controlled Environment Agriculture, rutgers university, Cook College.
2 International Bureau of Weights and Measures (2006), The International System of Units (SI) (8th ed.), pp. 114–15, ISBN 92-822-2213-6

DAYLIGHT INTEGRAL - DLI

„ (…) DLI, the daylight integral, is the cumulative amount of photosynthetic light that is received each day. The DLI is measured as the number of moles of light (mol) per square meter (m2) per day (d1), or mol/m2/d. The DLI can have a profound effect on root and shoot growth of seedling plugs, root development of cutting and finish plant quality attributes such as stem thickness, plant branching and flower number."[1]

DLI is measured by the cumulative amount of rain or light received during a 24-h-period. It is dependent on the time of the year (sun's angle), location, latitude and cloud cover and the daylength (photoperiod).

In the context of greenhouses or Vertical Farms this is additionally nfluenced by the glazing type, the structure and all obstructions, hanging baskets, etc. Generally we can assume that on earth DLI varies from 5 to 60 mol/m2/d. In greenhouses DLI rarely exceeds 30 mol/m2/d, because of shading applied to prevent excessive temperatures. Target minimum DLI inside a greenhouse should be from 10 to 12 mol/m2/d. On this point it is necessary to highlight the difference of minimum DLI found in the literature. Cultivars used for outdoor (soil based agriculture) tendentially need a higher number in the context of DLI supply. Most greenhouse plants, especially F1-hybrids of *L. esculentum* are optimized to germinate, grow and ripen best with lower temperatures and lower light measurements. This explains a DLI-range, e.g. for *L. esculentum* from 10 mol/m2/d to 20-30 mol/m2/d.[2]

---

1 RUNKLE, E. 2006. Do you know what your DLI is? Available: http://www.hrt.msu.edu/energy/Notebook/pdf/Sec1/Do_you_know_what_your_DLI_is_by_Runkle.pdf.
2 JONES J. Benton, 2007. Tomato Plant Culture : In the Field, Greenhouse, and Home Garden, Second Edition, Edition 2, CRC Press, p.58

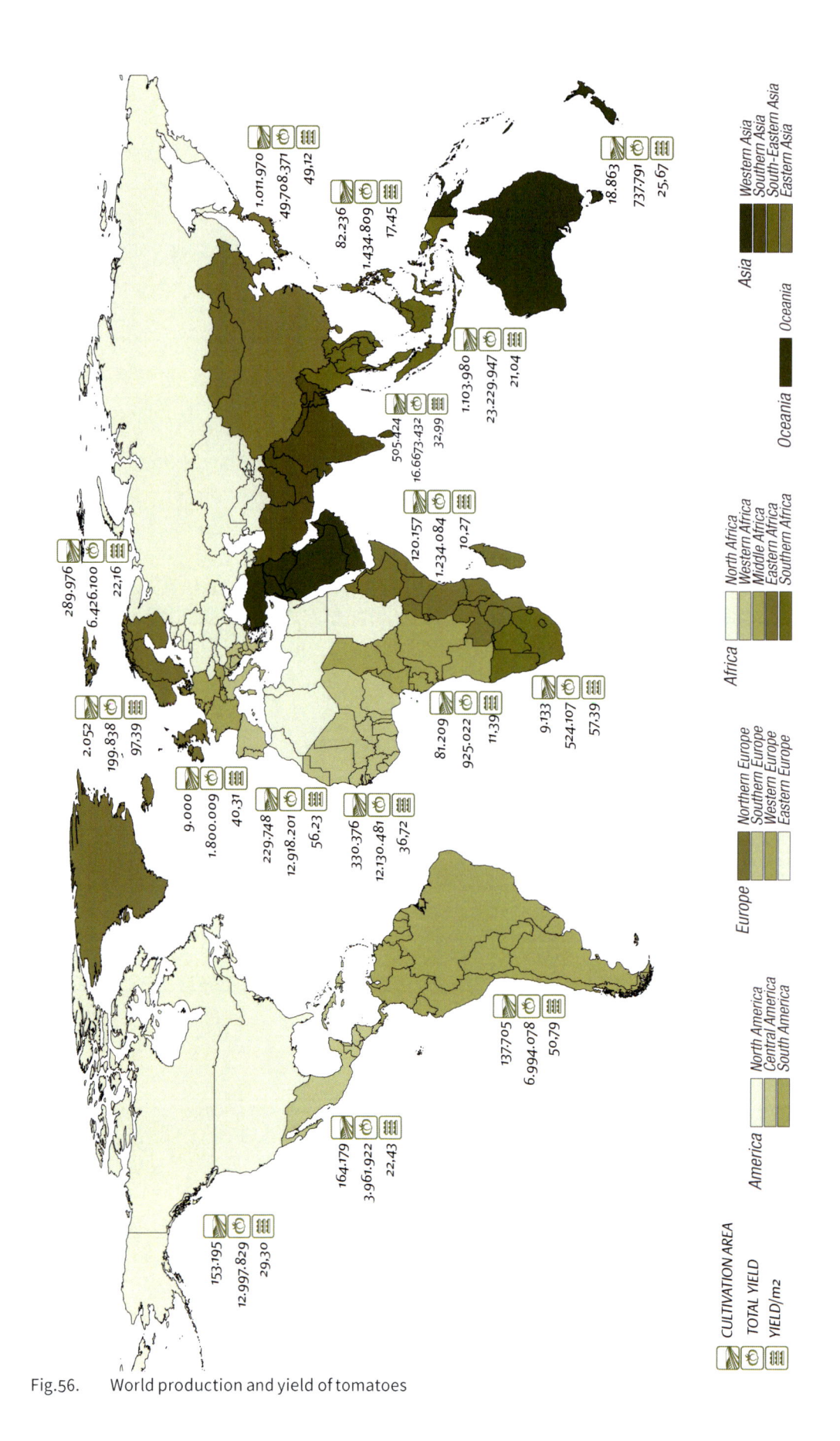

VERTICAL FARM - SUBSTITUTION OF NATURAL GROWTH FACTORS

Fig.56.   World production and yield of tomatoes

4.3. The Tomato - *Lycopersicon esculentum* (Mill.)

The simulated Vertical Farm will use *L. esculentum* for the following primary reasons:

- *L. esculentum* is the most widely produced vegetable in the world.[1]
- *L. esculentum* is a vegetable with a strong growth in production quantity worldwide.[2]
- Huge landfills consumed for *L. esculentum* Production, for soil based agriculture and greenhouses. (Fig.59 with numbers retrieved from FAO)
- *L. esculentum* is probably one of the best researched vegetable, ideal for data availability for growing conditions.
- *L. esculentum* is one of the plants with the highest daylight needs for photosynthesis.

"While *L. esculentum* continues to be one of the most widely grown plants, the production and distribution of *L. esculentum* fruits have been changing worldwide. Smaller, flavorful *L. esculentum* are becoming more popular than beefsteak *L. esculentum*, greenhouse-grown *L. esculentum* cultivars are one of the most researched and developed vegetables, optimized for greenhouse production and, as a consequence a potential product for Vertical FarmVertical Farming. Its high daylight needs for photosynthesis will sharpen the potential limits of a stacked greenhouse type in plant production. Its high daylight needs for photosyntesis will sharpen the potential limits of a stacked greenhouse type in plant production.

4.3.1. General Data - Origin and Distribution

L. esculentum is one of the major crops and main vegetables consumed in many countries. It has its origins in South America, where the "xitomatl" was cultivated by the Aztecs.[3] The earliest known reference in Europe is a description by Pietro Andrea Matthioli[4] who classified the „golden apple"[5] as a nightshade plant and a mandrake, a category of food known as an aphrodisiac. This may be one of the reasons why the Catholic Church defamed the fruit and called it as the "fruit of the devil"; it was forbidden and the successful distribution of the fruit was stopped for at least 200 years.[6]

---

[1] http://faostat3.fao.org/browse/Q/QC/E, retrieved 31.10.2015
[2] ibid.
[3] http://www.epicureantable.com/articles/atomatohis.htm, retrieved 19.09.2014
[4] McCue, George Allen. „The History of the Use of the tomato: An Annotated Bibliography." Annals of the Missouri Botanical Garden (Missouri Botanical Garden Press) 39, no. 4 (November 1952): p.291
[5] In Italy the *L. esculentum* is called pomodoro, literally translated the golden apple.
[6] LUNDY, R. 2006. In Praise of tomatoes, New York, Lark. p.42

VERTICAL FARM - SUBSTITUTION OF NATURAL GROWTH FACTORS

WORLD TOMATO PRODUCTION 2011

4.731.999 ha   37,17 t/ha   158.019,580.71 t
                                23,07 g/cap/a

AREA HARVESTED   world food and food manufacture production 6,824,143,000 t
YIELD PER AREA   996,53 kg/cap/a
TOMATO PRODUCTION

Fig.57.   World production of tomatoes

Since the fifties of the last century, the *L. esculentum* increased in its popularity in Europe at great speed until it became the one of the most widely produced and consumed vegetables all over the world.[1]

The world primary production of food and food manufacturing is currently 6,824,143,000 t or 996 kg/cap/a. Tomatoes are cultivated on a surface area of 4.731.999 ha. The world average *L. esculentum* production is 158,019,580.71 tonnes or 23 kg/cap/a or 2.3% of the world's total food and food manufacturing production.

L. ESCULENTUM - WORLD PRODUCTION

WORLD BIGGEST TOMATO PRODUCERS 2011

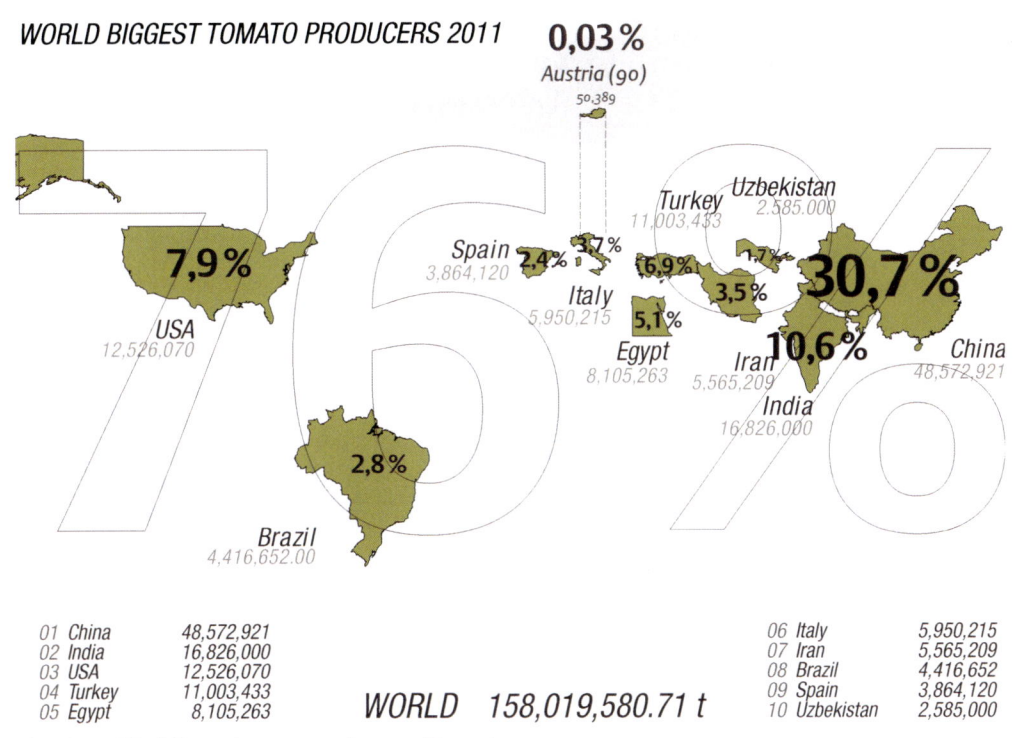

01 China	48,572,921
02 India	16,826,000
03 USA	12,526,070
04 Turkey	11,003,433
05 Egypt	8,105,263
06 Italy	5,950,215
07 Iran	5,565,209
08 Brazil	4,416,652
09 Spain	3,864,120
10 Uzbekistan	2,585,000

WORLD 158,019,580.71 t

Fig.58.   World's ten largest producers of tomatoes

From all vegetable produced worldwide the *L. esculentum* is on top of the list. 16,58% or 158.019.580,71 t of all vegetable produced (953.272.659 t) are *L. esculentum*.[2]

Comparing the different food zones of Kastner (Fig. 59) we see a huge difference in yield/ha. Depending on energy input, level of technology, mechanization and climate *L. esculentum* crop yield varies from 6.57 t/ha in Western Africa to 97.39 t/ha in Northern Europe. The ten biggest *L. esculentum* producing countries of the world produce more than 76% of the total global *L. esculentum* crop. 30,7% of it are produced in China, followed by India (10,6%), USA (7,9%), Turkey (6,9%), Egypt (5,1%), Iran (3,5%), Brazil (2,8%), Italy (3,7%), Iran (3,5%), Spain (2,4%) and Uzbekistan (1,7%) as seen in the diagram the page before.

---

1   FAOSTAT, world tomato production, area harvested and yield, retrieved 26.09.2014
2   ibid.

ITALY AND AUSTRIA

Some of the production and consumption data for Austria is related to data from Italy for reasons of comparison. Italy is the biggest *L. esculentum*-producer in Europe and also the biggest per capita consumer of this vegetable. Italy produces 5,950,215 tons of fresh tomatoes a year on an area three times that of Vienna, on 103,858 ha. The yield calculated by FAO is 57.29 t/ha/a. The per capita consumption is about 60.5 kg/a or 30 kcal/cap/day are covered by *L. esculentum* consumption.

In Austria *L. esculentum* cultivation started intensively after the Second World War and since then it is has been increasing continuously. Austria is at place 90 among world *L. esculentum* producers. The per capita consumption ranges between 16 kg/cap/a (FAO-data) to 27,7 kg/cap/a (Statistik Austria)[1]. The per capita consumption ranges between 16 kg/cap/a (FAO-data) to 27,7 kg/cap/a (Statistik Austria) The per capita consumption is much lower than in Italy. On 185 ha in Austria in 2012 52,032 tons were produced. This is explained by the cultivation method. More than 94 % of all crop yield is cultivated in greenhouses or plastic tunnels.[2]

AUSTRIA TOMATO PRODUCTION 2011

	t	ha	%
Tomatos total	52,032	183	100%
Tomatos Glass/Foil	**51,606**	**173**	**99.18**
Tomatos Soilbased	427	10	0.82

Austria land use:
- forests 40,50 %
- agr. area 34,30 %
- soil based agriculture 17,00 %
- other 8,74 %

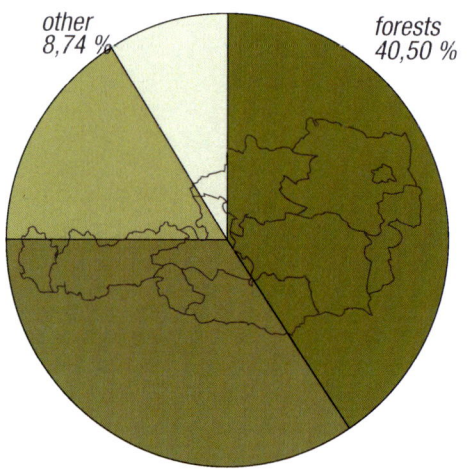

AUSTRIA
83.878,99 km2

50.389 t/a   185 ha   272 t/ha
                         27,2 kg/m2

Fig.59. Land use of Austria and Vieanna and Tomato-production

---

1 STATISTIK AUSTRIA, Ernteerhebung. Erstellt am 29.11.2012. - 1) Anbaufläche lt. Auskunft der Landwirtschaftskammern und Erzeugergenossenschaften; retrieved 26.10.2014

2 http://www.wien.gv.at/statistik/wirtschaft/tabellen/gemueseernte-anbauflaeche.html; Koordinationsstelle der Landwirtschaftskammer Wien, Mag. Doris Reinthaler et al. retrieved 26.10.2014,

PRODUCTION IN AUSTRIA AND VIENNA

L. esculentum production in Austria in 2012 was 52,032 tons/a. *L. esculentum* production in Austria is steadily increasing. Official data from Statistik Austria show an increase from 2011 to 2014 of more than 8 % and reached a production weight of 54.469 tons.[1] An interesting point in this context is, that even Austria is strongly shaped by intense agriculture, 37,5% of the overall *L. esculentum* production comes from the agricultural area within the city border of Vienna. Some 15 % of the city's surface is used for agricultural production. In 2012 Vienna has produced 19.385 tons of *L. esculentum* on a surface of 45 ha. There is no soil based *L. esculentum* production outside of greenhouses or foil tunnels in Vienna.

Crop yield in Vienna's greenhouses per year is 430 t/ha or 43 kg/m2 compared to Austria with 272 t/ha or 27,2 kg/m2.[2] Vienna could deliver to its inhabitants 50% of per capita *L. esculentum* consumption - during the the period of ideal climate conditions. Actually Austria is producing 16 % of its *L. esculentum* consumption. Most of its imports come from Italy, Spain and the Netherlands.[3]

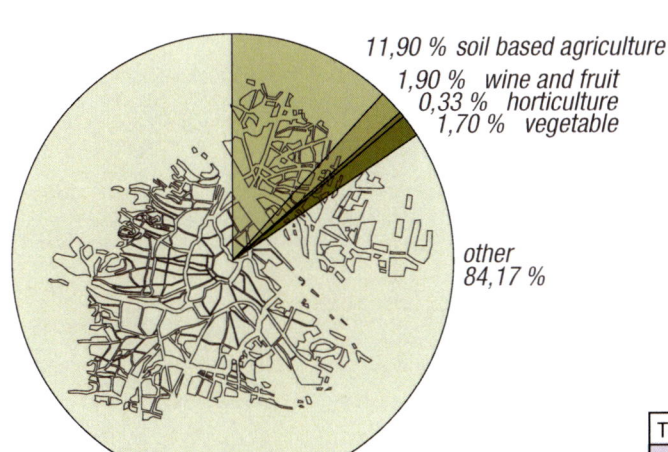

11,90 % soil based agriculture
1,90 % wine and fruit
0,33 % horticulture
1,70 % vegetable
other 84,17 %

VIENNA

	t	ha	%
Tomatos total	19,385	45	100%
Tomatos Glass/Foil	**19,385**	**45**	**100.00**
Tomatos Soilbased	0	0	0.00

VIENNA
414,87 km2

19.385 t/a   45 ha   430 t/ha
                      43 kg/m2

---

1 http://www.wien.gv.at/statistik/wirtschaft/tabellen/gemueseernte-anbauflaeche.html; Koordinationsstelle der Landwirtschaftskammer Wien, Mag. Doris Reinthaler et al. retrieved 26.10.2014,
2 ibid. retrieved 27.10.2014,
3 ibid.

4.3.2. *L. esculentum* and Light

Lighting will be the significant energy consuming system within the building, this work decided to concentrate on a product which is intensively PAR-light-dependent throughout the whole crop rotation. The results should thus take into account and present a certain worst case scenario.

Different crops differ from each other enormously in regard to the needed PPFD they need for photosynthesis and do so up to a factor 150, e.g. *L. esculentum* largely needing around 300 µmol/m^{-2}/s^{-1} , while strawberries make do with a mere 2 µmol/m^{-2}/s^{-1} [1] [2]. The specific cultivar the author will examine is *L. esculentum*, a F1-hybrid[3] Furthermore it is an economically important cultivar.[4]

Certain requirements are necessary for optimal *L. esculentum* growth. "The key requirements are light, carbon dioxide (CO2), water, adequate temperature, and sufficient and proper nutrients."[5]

Plant physiology within controlled environments is a complex matter. Integrating all of the components which affect the photosynthesis and morphogenesis of plants would go beyond the scope of this dissertation. It is thus necessary to concentrate on the most important parameters for plant growth which directly is interlinked with space and energy.

This chapter outlines space and energy influencing parameters for *L. esculentum*, namely lighting conditions, temperature, water supply, cultivation method and plant heights.

Lighting conditions, the ratio between available daylight and best PPFD-density for photosynthesis directly influence the production method and the building shape. These two parameters in turn directly influence the overall energy need for lighting. The optimum temperature for plant growth and maximum crop yield is directly in-

---

1   Künstliche Beleuchtung im Gartenbau, Philipps, AEG
2   http://www2.produktinfo.conrad.de/datenblaetter/100000-124999/101861-an-01-ml-PRONOVA_LUX_QUANTUM_METER_de_en.pdf, retrieved 12.11.2014
3   Crossing two genetically different plants produces a hybrid seed. This can happen naturally, and includes hybrids between species (for example, peppermint is a sterile F1 hybrid of watermint and spearmint). In agronomy, the term "F1 hybrid" is usually reserved for agricultural cultivars derived from two parent cultivars. These F1 hybrids are usually created by means of controlled pollination, sometimes by hand-pollination. For annual plants such as *L. esculentum* and maize, F1 hybrids must be produced each season. http://en.wikipedia.org/wiki/F1_hybrid, retrieved 04.08.2014
4   HENDRICKS Patrick, 2012. Life Cycle Assessment of Greenhouse Tomato (Solanum lycopersicum L.), University of Guelph, Thesis
5   HENDRICKS Patrick, 2012. Life Cycle Assessment of Greenhouse Tomato (Solanum lycopersicum L.), University of Guelph, Thesis

fluenced by the orientation of the plants, the shape of the facade and the A/V-ratio of the Vertical Farm. Water supply and bed sizes are needed for a schematic arrangement of the production method in line with the expected plant height through the cultivation period.

Photosynthesis is a complex physical and chemical process and a full presentation of this subject would extend beyond the constraints of this dissertation. In order to establish estimates for energy consumption or additional lighting needed by plants to a useful standard of accuracy, it is essential to explain the basic process of photosynthesis.

Also noted on this point should be that morphogenesis, "the shaping of an organism by embryological processes of differentiation of cells, tissues, and organs and the development of organ systems according to the genetic ‚blueprint' of the potential organism and environmental conditions"[1], will not be treated in this work. Of course specific light qualities lead to differences in plant growth, e.g. the blue spectral range leads to compact, red spectral range to elongated plant bodies.[2] But several other parameters influencing morphogenesis of plants cannot be identified as space-influencing factors.

---

1   http://www.britannica.com/EBchecked/topic/392779/morphogenesis, retrieved 11.09.2014
2   Künstliche Beleuchtung im Gartenbau, Philipps, AEG

4.3.3. *L. esculentum* and Photosynthesis

Light, or more precisely PAR wavelengths, are absorbed by the pigment chlorophyll. In more developed plant species we find Chlorophyll A and Chlorophyll B. The ratio between Chlorophyll A and Chlorophyll B is 3:1. Carotene and xanthophyll are also pigments and play an important role in photosynthesis (Fig. 58 on page 168). But in contrast to Chlorophyll a and b, carotene and xanthophyll act like transporters after absorbing the energy of photons of a specific wavelength, and they send the chemical energy to the chlorophyll. The task carotene and xanthophyll perform is to enlarge the assimilation spectrum or in other words the plant-sensitivity-curve for photosynthesis.

$$\text{Photosynthesis: } 6CO_2 + 6H_2O = C_6H_{12}O_6 + 6O_2$$

The spectrum where photosynthesis occurs was first recorded and published in 1973 by K.J. McCree. The sensitivity curve of 22 different plants was observed (e.g. barley, soya and *L. esculentum*). These results show that there is a marked decrease of photosynthetic reaction the closer the wavelength comes to the blue light range.
Recent publications in the Netherlands from the Institute for Horticulture in Wageningen show that there is evidence that the role of blue light for the photosynthesis was underestimated by McCree. The results suggest that light within the waveband from 530 to 670 nm is the most effective, light waves shorter than 400 nm and longer than 700 nm is virtually insignificant.

$$\text{Respiration: } C_6H_{12}O_6 + 6O_2 = 6CO_2 + 6H_2O$$

4.3.4. Factors affecting the rate of Photosynthesis

Light Compensation Point

There is a break-even point when the plant is producing as much sugar as it needs for respiration. This point is defined as the light compensation point. As light increases (and water is available) carbon production also increases. The plant thus exceeds its carbon production, the surplus is transformed into glucose. Exceeding the light compensation point, is the main goal in food production as a fundamental principle.
By increasing brightness and intensity within PAR the photosynthesis-rate also increases, "but only up to a certain point, beyond which increasing the brightness of light has little or no effect on the rate of photosynthesis. (...) The light intensity at which the net amount of oxygen produced is exactly zero, is called the compensation point for light."[1] At this point the consumption of oxygen by the plant due to cellular respiration is equal to the rate at which oxygen is produced by photosynthesis.
"The compensation point for light intensity varies according to the type of plant, but it is typically 40 - 60 W/m2 for sunlight. The compensation point for light can be reduced (somewhat) by increasing the amount of carbon dioxide available to the plant, allowing the plant to grow under conditions of lower illumination."[2]

Light Saturation Point

On the other side of the "photosynthesis activating point" of what we could call the light compensation point, is another point essential in plant cultivation: the light saturation point.

Photosynthesis continues to produce sugar from CO2 and H2O until it reaches the saturation point. At this point carbon production can no longer occur.[3]
"The saturation point describes the amount of light that is beyond the capability of the chloroplast to absorb. Photosynthesis still occurs, but the amount of light has exceeded the amount of pigments that are available for absorption."[4] This saturation point is different for every plant. "Different plants have different saturation points, determined by the number of pigments in their chlorophyll cells. Plants that typi-

---

1 http://Tomatosphere.org/teachers/guide/grades-8-10/plants-and-light, retrieved 12.09.2014
2 ibid.
3 SCHOPFER, P. & BRENNICKE, A. 2010. Pflanzenphysiologie, Heidelberg, Spektrum Akademischer Verlag. p.445 ff.
4 http://www.ehow.com/about_6535863_definition-plant-light-saturation.html, retrieved 12.09.2014

cally grow in shaded areas have lower saturation points, while those that grow in areas more exposed to light have higher saturation points. The integrated photon flux, CO2 concentration, and atmospheric humidity are critical parameters, with a photon flux, of 20 to 30 mol/m2/d being optimum for most plants, including *L. esculentum*."[1] This high value is referred to *L. esculentum* growing on soil under a free sky.

Evaluating the light compensation point and the light saturation point for a specific plant type is crucial for all plants grown under artificial conditions in greenhouses or Vertical Farms. Due to the fact that solar radiation and therefore the amount of PAR also decreases within a built environment, electric lighting is essential for plants which need high amounts of light, both in greenhouses and in Vertical Farms.

As already noticed in the beginning on this chapter, every plant is different, has individual needs of PPFD, CO2 and water. On this point of the work it is necessary to limit research results on lighting on cultivars of *L. esculentum* argued as follows:t

- *L. esculentum* is a fruity vegetable with one of the highest light density needs for photosynthesis, morphogenesis and fruit development.
- *L. esculentum,* as a developed F1-hybrid , has already been optimized and adapted to conditions with less light and lower temperature compared to soil-based *L. esculentum*
- *L. esculentum* needs a remarkable amount of water for its growth and fruit development in soil based agriculture.

L. esculentum is one of the most extensively researched vegetables, especially in the Netherlands, where considerable data are available.[2]

---

[1] JONES J. Benton, 2007. Tomato Plant Culture : In the Field, Greenhouse, and Home Garden, Second Edition, Edition 2, CRC Press, p.58

[2] On this point although it is necessary to claim that data about PPFD, DLI, photoperiodism etc. are not coherend, but can be used as a fan of data which enable assumtions for the simulation model, starting below.

LIGHT FOR L. ESCULENTUM - A RESUMÉE

We have seen that data about PPFD for *L. esculentum* in greenhouse production differ greatly from study to study. There are more than 5000 cultivars of *L. esculentum*. Most greenhouse *L. esculentum*, however, are F1-hybrids already optimized for greenhouse production, where sunlight is the determinant. Every cultivar is different and therefore has its own individual "ideal" PPFD-curve during crop rotation. The author also has seen that sometimes there is still confusion even in horticulture "about behavior and terminology dealing with radiation more than almost any other factor." (...) In the control of plant growth (...) there are at least five types of information that may be derived from the radiation en-vironment:

1) radiation quantity [W/m2, Ed.]
2) radiation quality [spectral distribution, PAR, Ed.]
3) direction of radiation
4) duration of radiation (timing of light-dark transition) [time or DLI, Ed.] and
5) polarization.

Of these five groups, industry has utilized only 1) and 4) to any significant extent in design and management decision."[1] But numerous published papers and the lively scientific activity focused on this area carried out over the past few years has put us in a position to apply approved data, e.g. the ideal PPFD for F1 hybrids of *L. esculentum* optimized for greenhouses during the phase of fruit development. PPFD during the establishment, vegetative growth, flowering and fruit-set from different studies are averaged out. The diminishing PPFD-factor through the crop rotation period is an assumption, based on discussions the with "Zeiler". These values subsequently were evaluated at the Department of Crop Sciences at the University for Natural Resources and Life Sciences. The result is an ideal lighting demand curve throughout the whole crop rotation of *L. esculentum*, representing sigmoid growth curve visible in diagramatic form on the next page. The establishment period (seeding and transplanting) is excluded within the simulation model. Low light and space requirement in the first weeks of the crop rotation led to this decision. Crop rotation and lighting analysis within the simulation model therefor starts after transplanting with the start of the vegetative growth.

For the sake of completeness it must be said that the needed PPFD can be maximized or minimized by reducing or rising the CO_2-level within the Vertical Farm. Water supply and nutrient composition are also directly interconnected with the ideal PPFD-curve. On this field intensive research is going on and more need of research can be assumed. Considering all these factors would go beyond this doctoral thesis.

---

1   HANAN Joe J. et al.1998. Greenhouses, Advanced Technology for Protected Horticulture, Boca Raton, Kondon, New York, Washington, D.C., CRC Press, p.91

Four PPFD-amounts, following a sigmoid growth curve, now get defined for *L. esculentum* production within the Vertical Farm:

- 50 µmol/m^{-2}/s^{-1} during the establishment (seeding, (trans-)planting)[1]
- 150 µmol/m^{-2}/s^{-1} during the vegetative growth (development and photomorphogenesis)
- 150 - 300 µmol/m^{-2}/s^{-1} during the flowering period to the first fruit-set (blossoms, pollination, first fruits and fruit growth)
- 300 µmol/m^{-2}/s^{-1} during ripening to first harvest (fruit growth, lycopene production)[2][3]
- 300 - 100 µmol/m^{-2}/s^{-1} along the rest of the crop rotation

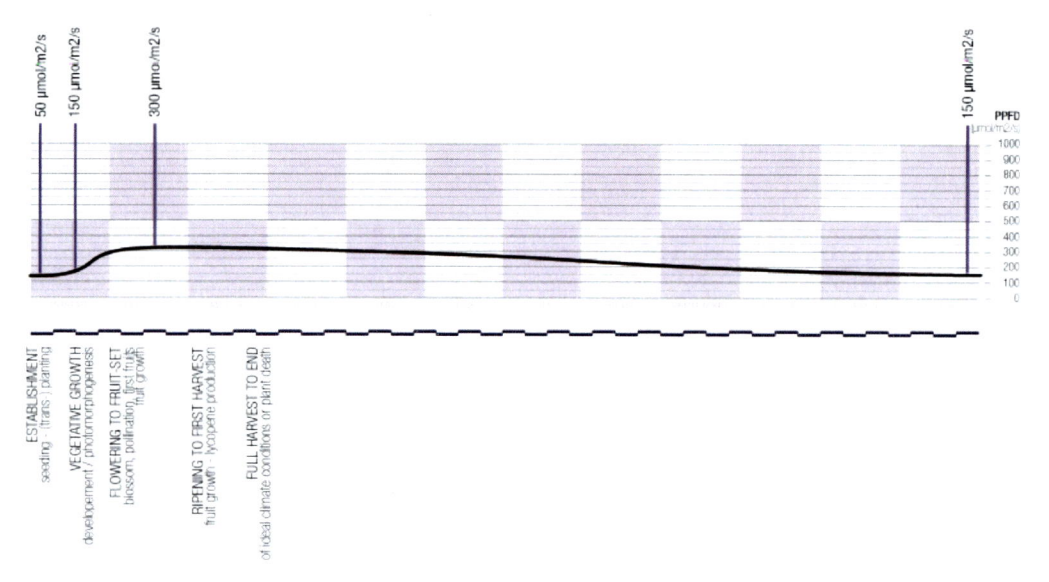

Fig.60. *L. esculentum*: PPFD need change along the crop rotation

---

1 http://www2.produktinfo.conrad.de/datenblaetter/100000-124999/101861-an-01-ml-PRONOVA_LUX_QUANTUM_METER_de_en.pdf, retrieved 29.04.2014; A range from 45 to 55 µmol/m-2/s-1 is recommended.
2 HANFORD, A. J. 2004. Advanced Life Support. Baseline Values and Assumptions Document. Houston: National Aeronautics and Space Administration. p.59
3 Researches from LumiGrow (http://www.lumigrow.com/) who supported calculations of this work, calculate with a range of 300 - 380 µmol/m-2/s-1

4.3.5. Temperature and other Growing Conditions

TEMPERATURE

All phases within the lifetime of the *L. esculentum* plants, from germination, to plant growth, flowering and pollination, fruit-set, photosynthesis and yield - are all influenced by temperature.

"A day temperature from 70 to 82°F [approx. 21°C to approx. 28°C] is optimum, while night temperature from 62 to 64°F [approx. 16,5°C to approx. 17.5°C] is optimum for greenhouse *L. esculentum*. During cloudy weather, a temperature closer to the lower end of these ranges is preferred, while in sunny weather, temperatures closer to the higher end are better." "[1]

These ranges were compared to differend studies and experiments with greenhouse *L. esculentum* and, by the end of the chapter, compared with the production practices of „Zeiler".

MINIMUM AND MAXIMUM TEMPERATURES

The ideal temperature is dependent of direct and diffuse solar radiation, the relative air humidity, water and CO_2 concentration in the air. An overview of different findings will be given below. But there are physiological limits beyond the ideal temperature for the best and biggest yield.

L. esculentum plants prefer warm weather. Temperatures below 10°C or below delay seed germination and vegetative development are inhibited. The consequences are a reduction of fruit-set and an impairment of fruit ripening.[2] *L. esculentum* can scarcely absorb nutrients at all when temperatures sink below 12°C. Below 10.5°C the degree of growth is negligible. The reduction of nutrient uptake starts below 14°C. Beyond 32.5°C the evaporation cooling through leaf transpiration starts to diminish and so called water stress begins. Beyond 35°C lycopene, a carotenoid which gives the fruit its characteristic red color no longer develops. This situation thus inhibits the development of normal fruit color and it also reduces fruit-set.

Beyond 36°C blossom drop (flower abortion) starts, especially if these temperatures already occur in the early morning period and last for a number of consecutive

---

1 SNYDER, R. G. [2010]. Greenhouse Tomato Handbook. Mississippi: Mississippi State University Extension Service.
2 JONES J. Benton, 2007. Tomato Plant Culture : In the Field, Greenhouse, and Home Garden, Second Edition, Edition 2, CRC Press p.18

hours.[1] Temperature differences of between 4°C to 8°C in daytime and nighttime improves germination, growth and development, and also flowering and yield.[2]

The range within the minimum and maximum temperature, where *L. esculentum* production should take place for high quality yield is between 20 and 24°C.[3]

Through the crop rotation of the *L. esculentum* plants the following approximate temperature values can be defined:

Establishment, from seedlings to (trans-)planting:
    19,5 - 21,5°C[4]

Vegetative growth - development, photomorphogenesis
    19,5 - 21,5°C[5]

Flowering to fruit set - blossoms, pollination and first fruit set
    18,5-20°C[6]

Fruit growth
    18-22°C[7]

Ripening to first harvest - fruit growth, lycopene production, sugar production
    22-24°C[8]

Full harvest to the end of the crop rotation
    22-24°C[9]

Taking into consideration the rate of truss production, the opening rate during flowering, the fruit development /time, the number of flowers and set-fruits, the best values lie at 22°C.[10] The only higher value compared to other temperatures we find at the mean fruit size (g) at 18°C.

---

1   JONES J. Benton, 2007. Tomato Plant Culture : In the Field, Greenhouse, and Home Garden, Second Edition, Edition 2, CRC Press. p.18

2   Voican, V., Lăcătus, V. and Tănăsescu, M. 1995. GROWTH AND DEVELOPMENT OF TOMATO PLANTS RELATED TO CLIMATIC CONDITIONS FROM SOME AREAS OF ROMANIA. Acta Hort. (ISHS) 412:355-365

3   ibid. p.18

4   LANDESANSTALT FÜR LANDWIRTSCHAFT, FACHBEREICH GARTENBAU 2004. Gewächshaustomaten. Hinweise zum umweltgerechten Anbau. Managementunterlage. [Dresden]: Sächsische Landesanstalt für Landwirtschaft, Fachbereich Gartenbau.

5   ibid.

6   ibid.

7   2001, S.R.Adams et al. „Effect of Temperature on the Growth and Development of Tomato Fruits", Horticulture Research International, Wellesboune, Warwick CV35 9EF, UK, Annals of Botany 88: 869-877, 2001, p.869-877

8   Excursion Report, see „Appendix"

9   ibid.

10   ADAMS, S. R., COCKSHULL, K. E. & CAVE, C. R. J. 2001. Effect of Temperature on the Growth and Development of Tomato Fruits. Annals of Botany ; 88, 869-877. The experiment was made with the Tomato „Liberto" with constant temperatures at 14, 18, 22 and 26°C.

„Air temperature can have a marked affect on the atmospheric demand (moisture requirement) of the *L. esculentum* plant, increasing with increasing air temperature."[1] Water requirements for soil based *L. esculentum* can increase fivefold between ideal and extreme temperatures. „However, the relationship between air temperature and relative humidity can moderate the transpiration rate, reducing the atmospheric demand with increasing humidity."[2]

Comparing these data with the interview at "Zeiler" we see the following analogies. Zeiler maintains a maximum temperature between 20 and 24°C during the heating period in a greenhouse in the south of Vienna,. In summer the greenhouse is not cooled as temperatures increase. The roof of the greenhouse is covered with a special color on a chalk basis instead, to increase the reflection of direct solar irradiation. The greenhouse is naturally ventilated, air exchange takes place through openings in the glass roof.

During the heating period, again the temperature ranges from 24°C to a minimum of 14°C. The reduction from 24 to 14°C occurs rapidly after sunset, when all the ventilation dampers are immediately opened. The findings of this practice are an increase in blossom-production and fruit-set, a fact which is also confirmed by studies in greenhouses from Sachsen, Germany, where the night-temperature rated value of approx. 15°C for approx. 4 hours is reduced for 1 to 2 K. This supports the generative plant growth and also has positive effects on fruit-set.[3] At the Zeiler greenhouse, the temperature increase from 14 to 24°C takes place slowly supported by the daylight after the respiration-phase.

WATER

Water is a scarce resource in most parts of the world. The amount of precipitation falling on the world land surface of 150.000.000 km2 is about 110.000 km3/a. Two thirds of this is evapotranspirated by vegetation on the land surface. The remaining volume "feeds" rivers and lakes and aquifers. These are the renewable freshwater resource of the world. The water withdrawal for municipal, industrial and agricultural purposes is returned to the environment after a certain time period.[4]
In this context the percentage of water withdrawal by traditional agriculture, how much water a *L. esculentum* plant needs to produce high quality and high quantity yield are important questions for the for the Vertical Farm and so too is the issue of

---

1 JONES J. Benton, 2007. Tomato Plant Culture : In the Field, Greenhouse, and Home Garden, Second Edition, Edition 2, CRC Press, p. 18
2 ibid. p.18
3 SÄCHSISCHE LANDESANSTALT FÜR LANDWIRTSCHAFT, FACHBEREICH GARTENBAU 2004. Gewächshaustomaten. Hinweise zum umweltgerechten Anbau. Managementunterlage. [Dresden]: Sächsische Landesanstalt für Landwirtschaft, Fachbereich Gartenbau. Chapter 1.7.2
4 http://www.fao.org/nr/water/aquastat/water_use/index.stm, retrieved 24.09.2014

estimating to what extent Vertical Farms can reduce the water consumption in agricultural production.

On a global average use 70% of fresh drinking water is used worldwide for soil-based agriculture[1] This value, of course varies greatly between different countries where the values of water withdrawal for agricultural use range from 91% to 2%.[2]

Especially *L. esculentum* have a vaste water content, up to 96%[3]. The plant absorbs water by its roots and through irrigation and fertigation.

Irrigation is the technical term for artificial application of water in non or low rain-fed agricultural areas. In greenhouses and Vertical Farms irrigation is an intrinsic subsystem with a central advantage to soil-based outdoor agriculture: Through drip irrigation the plant gets its water and nutrients exactly where it absorbs it.

Moisture requirements on the field vary from 2.000 to 10.000m3/ha/a. „A mature *L. esculentum* plant may wilt during an extended period of high air temperature if the plant is not able to draw sufficient water through its roots, a condition that can occur if the rooting medium is cool or the rooting zine is partially anaerobic. Also the size of the root system may be a factor. Just how large the root system must be to ensure sufficient rooting surface for water absorpiton is not known."[4]

In this context we talk about evapotranspiration, which is the sum of evaporation of water from the soil and the transpiration from the plants. The amount of water required can thus be drastically minimized compared to conventional soil-based agriculture because of eliminating water losses through evaporation and water dissipation to root areas of the growing crop.[5]

The high-tech-greenhouse and the Vertical Farm are conceived as closed environmental systems where water recovery is possible. The aim of implementing closed water cycles in Vertical Farms thus makes sense and to a certain extent, has also been an actively pursued objective in a number of dif-ferent projects.[6 7 8]

"In greenhouse vegetable crops, the irrigation water-use efficiency (WUE), expressed as the ratio between marketable crop production and total crop irrigation supply, is much higher than in open field crops due to the low evaporative demand inside the greenhouse that reduces water requirements and the higher productivity of greenhouse-grown crops. (...) In unheated plastic greenhouses in the Mediterranean Basin,

---

1 http://www.fao.org/nr/water/aquastat/water_use/index.stm, retrieved 24.09.2014
2 ibid.
3 „Nutrient composition and antioxidant activity of eight Tomato (*Lycopersicon esculentum*) varieties, J.L. Guil-Guerrero, M.M. Rebolloso-Fuentes Article
4 JONES J. Benton, 2007. Tomato Plant Culture : In the Field, Greenhouse, and Home Garden, Second Edition, Edition 2, CRC Press, p.18
5 ibid.
6 http://plantagon.com/urban-agriculture/industrial-symbiosis, retrieved 24.09.2014
7 http://urbanfarmers.com/projects/basel/, retrieved 20.09.2014
8 http://www.plantchicago.com/non-profit/farms/plantaquaponics/, retrieved 20.09.2014

WUE was similar between crops grown in soil or substrate, and increased under the following conditions:

improved greenhouse structure
increased length of growing season
recirculation of nutrients in substrate-grown crops

The highest WUE values of 45 (substrate-open system) and 66 kg m3 (sub-strate-closed system) were for *L. esculentum* grown in the Netherlands with glass-houses."[1] An enormous improvement in efficient water consumption results in the change-over from soil based agriculture to closed systems such as greenhouses or Vertical Farms.

Water requirements for *L. esculentum* are normally the sum of the water lost for eva-poration of the soil, transpiration of the plant and the "incorporated" water of the plant itself. A rule of thumb exists to calculate the amount of the daily water supply needed for greenhouses [ml/m2]: when the daily radiation [J/cm2] is multiplied by the factor 3, e.g. with a radiation of 1,000 J/cm2 three liters of water are needed. This approximate value is only valid for mature indeterministic *L. esculentum* plants[2]. The following calculations are for media and calculating 2.5 plants per m2.

During the crop cycle there is a steady increase in water requirement until a peak at the harvest, when the plant reaches its maximum fruit output and after which the water requirement shrinks again.

During the establishment and vegetative growth phases for *L. esculentum* plants 0.75 l/m2 are required. During the flowering to the first fruit-set through the pe-riod of fruit growth the plant water needs are doubled from 1.65 l/m2. 3.15 l/m2 of water from the weeks of ripening to the first harvest while 7 l/m2 are needed during the period of full harvest. After the plant reduces its fruit development, the water requirement shrinks from 7 to 5 l/m2 within four weeks and diminishes water uptake until the end of the cropping season with 2 l/m2 of water.[3]

---

1 W.BAUDOIN et al. 2013. Food and Agriculture Organization of the United Nations, Plant Production and Protection Division. „Good Agricultural Practices for greenhouse vegetable crops - Principles for Mediterranean climate areas", p.130
2 Indeterministic *L. esculentum* plants normally are used in greenhouses, while deterministic plants are used for soil based agriculture outside of protected environments. The main difference between these two types are that indeterministic plants can reach a height up to 3m while deterministic plants from their morphology are bushy and reach a maximum height of approximately 80 to 90 cm.
3 SÄCHSISCHE LANDESANSTALT FÜR LANDWIRTSCHAFT, FACHBEREICH GARTENBAU 2004. Gewächshaustomaten. Hinweise zum umweltgerechten Anbau. Managementunterlage. [Dresden]: Sächsische Landesanstalt für Landwirtschaft, Fachbereich Gartenbau. Chapter 15

Similar can be found in the Greenhouse Tomato Handbook[1], where 50 ml per plant are given for new transplants and reach an amount of 3 quarts (2.7l) for a mature plant on a sunny day. "Generally, 2 quarts per plant per day are adequate for fully grown or almost fully grown plants. Monitor plants closely, especially for the first couple of weeks following transplantation, so that the volume of water can be increased as needed. (…) Most growers use from 6 to 12 waterings per day once plants are established."[2]

These figures are also comparable with the practical experience of the "Zeiler" greenhouse in Vienna, with a daily water supply of 8l/m2. Per m2 on average 2.75 plants are produced, and every plant needs approx. 2,96 l of water/day.
We find a higher value of water quantity per plant in fertigation studies of the University of Arizona where an average of 4 l/plant/day is mentioned. On average of 2,5 plants/m2 (2.5 heads on one stem) the amount of water as nutrient solution reaches a value of 10 liters/m2/day. The "water needs may be doubled!" in desert areas if evaporative cooling is used.[3]

4.3.6. Growing media and plant density

Industrially grown greenhouse *L. esculentum* are on the increase around the world, especially in Europe, North America and China. The most extensively used growing medias today are rockwool, perlite, media containing peat moss and coconut coir.

Recent studies at the University of Arizona, focusing on the influence in crop yield of *L. esculentum* clearly show that differences in crop output between the different medias is negligible. Water consumption and nutrient distribution, however, have an altogether different priority pattern and their effectiveness depends directly on the substrate. In order to achieve comparable results regarding water consumption, evapotranspiration of plants, which are an important influencing factor of the indoor climate control, the author chose to use coconut coir for the following reasons:

Coconut coir is abundantly available at low cost and compared to rockwool, it is a renewable material. The water holding capacity (and the bond of macronutrients) are

---

1 SNYDER, R. G. 2010. Greenhouse Tomato Handbook. Mississippi: Mississippi State University Extension Service.
2 ibid.
3 http://ag.arizona.edu/ceac/sites/ag.arizona.edu.ceac/files/pls217nbCH10_0.pdf, retrieved 17.09.2014

also better compared with other media. Coconut coir is "far less costly than rockwool and any media containing peat moss."[1]

Recent findings in horticulture research by comparing different planting bed sizes and growing substrates illustrate an additional advantage from the use of co-copeat. "Plants grown in cocopeat produced the highest marketable fruit quantities (56.2%) per plant and yielded the greatest (445.6 g) marketable yield per plant. Plants grown in a cocopeat substrate produced higher fruit quantities (5.2%) and total yield (0.7%) than those with a rockwool substrate."[2] In this experiment it was also observed, that *L. esculentum* grown within cocopeat growing substrate produced the highest fruit weight.

In this study four F1-*L. esculentum* were used: Campari, Temptation, Annamay and Adoration, four cultivars with a similar fruit weight than the four F1-*L. esculentum* cultivated by "Zeiler"-greenhouse. "Two sets of experiments were conducted simultaneously under the same climate-controlled greenhouse. For the first experiment, planting beds were arranged parallel in a north-south direction and with a bed width of 20cm, 40cm, 60cm and 80 cm constructed by laying a wooden plank along both sides of the bed."[3] The bed height was 8 cm and the distance between the cultivation rows was 70 cm. The optimum distance between the plants with the highest (marketable) fruit yield is 60 cm.

In the second experiment, these cultivars as mentioned above were bedded in cocopeat, rockwool and masato.[4] All plants were drip irrigated, supplied with a standard nutrient solution and treated according to recommended cultural practices. The results of this study will be the basis for the following simulation model in terms of plant density and rasterization. The cocopeat slabs used in this study were 95 cm x 15 cm x 8 cm and placed over a Styrofoam slab (as the truss) with similar dimensions. At "Zeiler"-greenhouse the truss-slab was made of rockwool.

Cocopeat is a renewable organic natural material, with its low bulk density and obvious advantages for plant growth and development it can be considered a suitable substrate for the simulation model. Plant distances of 60 x 70 cm will be applied to configure the ground floors and plant arrangements.

---

1 http://ag.arizona.edu/ceac/sites/ag.arizona.edu.ceac/files/Comparing%20Media%201996%20 97.pdf, p.1, retrieved 15.10.2015
2 LIUTEL B.P. et al., „Yield and Fruit Quality of Tomato (*Lycopersicon esculentum* (Mill.) Cultivars Established at Different Planting Bed Size and Growing Substrates", Department of Horticulture, kangwon National University, Chuncheon 200-701, Korea, Hort.Environ. Biotechnol. 53(2):102-107.2012, ISSN (print): 2211-3452, ISSN (online): 2211-3460
3 ibid.
4 Unfortunately it wasn't possible to find any material about substrate „masato". To my knowledge it must be a specific soil, used in South-eastern Asia.

4.4. Setting up Vertical *L. esculentum* - Farm

4.4.1. Light availability in Vienna and Greenhouse Practises

The photosynthetically active radiation (PAR) represents from 44%[1] to 50%[2] of the visual light spectrum depending on different research results. Even though also PAR can vary by some percentage points, use of the factor 0.5 to obtain PAR from the total radiance in the eastern region of Austria has become established practice. To obtain the photon flux density and the daylight integral it is necessary to evaluate the exact amounts involved by using spectrometers on top of the greenhouse or the plant canopy. These values are dependent on geography, the latitude and longitude, the climate zone and also the influence of specific conditions the plant is growing under (air quality, microclimate etc.).

It is therefore necessary to work with specific climate data in order to achieve further precise estimates. For this purpose we continued our calculations with data from Vienna, Austria. For various reasons: *L. esculentum* cultivation in eastern Austria has a relatively short but successful tradition. Vienna has a high density in glasshouses which cultivate *L. esculentum*. As a result of the climatic conditions here with very substantial variations of temperature and humidity between summer and winter, glasshouses are largely in use to extend the cultivation time throughout the year. Most of the greenhouses stop *L. esculentum* production in October due to the low sunlight levels.

Vienna is located in northeastern Austria, on the foothills of the Alps in the Vienna basin. According to the Köppen-Geiger-Classification, Vienna lies within the Cfb-climate and the humid continental climate. Its summers are warm to hot with average temperatures between 24 - 31,7 degrees (dry bulb temperature). It has not been uncommon in recent years for temperatures to reach 40°C.

Winters are dry and cold with average temperatures around freezing point. In January and February very low temperatures are possible (down to - 18,30°C). Spring and autumn are mild. The average precipitation is relatively modest at around 600 - 620 mm annually. Snow is relatively uncommon compared to southern and western Aus-

---

1   GIACOMELLI, G. 1998. Components of Radiation Defined: Definition of Units, Measuring Radiation Transmission, Sensors. CCEA, Center for Controlled Environment Agriculture, rutgers university, Cook College.
2   http://www.landwirtschaftskammer.de/gartenbau/beratung/technik/artikel/lichtwerte-umrechnen.htm, retrieved 12.05.2015

tria. The elevation of the city ranges from 156,68 m.o.A (meters over the sea-level of the Adria) to 484 m (Kahlenberg).

Global irradiation in Vienna reaches around 1.120 kWh/m2/a whereas the ratio between direct and diffuse light is about 56.5% (direct irradiation) and 43.5% (diffuse irradiation). The annual distribution of global total radiation shows that the month with the highest total solar irradiation is July with 172.62 kWh/m2 or 621.42 MJ/m2 (dir=68.35%, diff=31.65%). The month with the lowest total solar irradiation is December with 18.93 kWh/m2 or 68.14 MJ/m2 (dir=72.58%, diff=27,42%).
The month with the most daylight hours is July with 248 hours, followed by August and May with 244 hours, respectively 238 hours. The month with the lowest daylight hours is December, followed by November and January with 52, 64 and 67 hours.

No statistics are currently available from ZAMG[1] or other institutions to provide the average DLI data for Vienna. To my knowledge there is only one single large-scale graphic available to evaluate DLI. The Institute of Floriculture from Michigan State University, Jim Faust, Clemson University, developed an "Outdoor Daily Light Integral (DLI) Map for the United States. Daily light integrals (mol/m2/d) are visible here from 5 to 60 DLI throughout all the climate zones from January to December. In Chapter 5.6 DLI is calculated on the basis of daylight availability taken from the annual solar radiation simulation from Vienna.

> DLI for Vienna for the simulation model got calculated as follows: W(solar radiation)*0.5[W$_{PAR}$]*4.57[µmol/m^{-2}/s^{-1}]*seconds[daylength]/1,000,000[2]

The photosynthetically active radiation (PAR) in Vienna ranges from 8.33 kWh/m2 in December to 75.95 kWh/m2 in July. In springtime and autumn goes from 36.47 kWh/m2 in March to 71.06 kWh/m2 in June and from 42.33 kWh/m2 in September to 12.10 kWh/m2 in November respectively. The following photosynthetic photon flux densities (PPFD) were evaluated (using the factor 4.57/W from Gene Giacomelli, Arizona University) in relation to the photosynthetically active radiation (PAR). The PPFD value is the quotient of the daylight integral (DLI) and the average length of the day in seconds. The values in µmol/m^{-2}/s^{-1} range from 720.82 in July to 188.23 in January with an average of 459.78 µmol/m^{-2}/s^{-1} throughout the year.

By overlaying the climate data and the average solar irradiation with specific responses of *L. esculentum* on sunlight we receive the time with enough sunlight for plant cultivation. The light compensation point for most of C3-plants ranges from 40 - 60 W/m2 sunlight or ca. 90 - 135 µmol/m^{-2}/s^{-1} respectively. The light saturation

---

[1] Zentralanstalt für Meteorologie und Geodynamik (Central Institution for meteorology and geodynamics), Austria
[2] Table 25 (Appendix): DLI, µmol/m-2/s-1 for Vienna

points of *L. esculentum* ranges from ca. 260 W/m2 to 350 W/m2 sunlight or 600 - 800 µmol/m^{-2}/s^{-1}.[1]

DIMINISHING FACTORS AND CURRENT RESEARCH TENDENCIES FOR GH

Several universities, firms and horticulturists are currently working on the "low energy greenhouse"[2][3], focusing on the reduction of heat loss and the optimization of photoperiods and light quality to reduce the energy requirement for artificial (supplemental) lighting. In the context of characteristics for greenhouse glazing, or better covering, materials, glass is still the material with the highest PAR-transmittance-capacity.[4] Current calculations minimize outside PAR through greenhouse covers up to 30%, taking the age of the glass, impurities and pollution, profiles and construction into account.

The objective of "ZINEG"[5] is to increase energy efficiency within the greenhouse-industry. Its system approach is to operate greenhouses "without fossil energy, without fossil CO2-emissions". One of the solutions lies in developing new covering materials. The fundamental requirements for a greenhouse-cladding-material are to obtain high light transmittance plus good insulation.

---

1 JONES J. Benton, 2007. Tomato Plant Culture : In the Field, Greenhouse, and Home Garden, Second Edition, Edition 2, CRC Press, p.58
2 „Low Energy Greenhouse and the Hot-Box Approach, kamer van Koophandel Haaglanden, Den Haag, 27.01.2009, Prof.Dr. hans-Jürgen Tantau, Institute of Biological Production Systems, Biosystems and Horticultural Engineering section, Leibniz Universität Hannover
3 https://www.wageningenur.nl/en/show/Three-fruitful-years-of-experience-in-low-energy-greenhouses, retreived 15.09.2014,
2006, Jon Kristinsson, „The Energy-producing Greenhouse, em.Prof. Faculty of Architecture, Delft University of Technology, Netherlands and
http://www.dhlicht.de/fileadmin/Downloads/pflanzenbroschuere.pdf, retreived 09.08.2014
4 https://hrt.msu.edu/Energy/Notebook/pdf/Sec1/AJ_Both_Greenhouse_Glazing.pdf, retreived 13.09.2014
5 http://www.zineg.de/, „Gesamtziel des Verbundvorhabens ist es, für die Pflanzenproduktion in Gewächshäusern den Verbrauch fossiler Energie für die Heizung und damit die (fossilen) CO2-Emissionen möglichst auf Null zu reduzieren. Zur Erreichung dieses Ziels ist ein systemorientierter Ansatz durch Kombination technischer und kulturtechnischer Maßnahmen erforderlich." , Laufzeit: 01.05.2009 - 30.04.2014, retreived 12.08.2014

4.4.2. Artificial Lighting - General Data

Assuming that artificial lighting is the key energy consumer in Vertical Farms, the choice for lamp types with a high luminous efficiency[1] is mandatory. Most lamps are produced to illuminate indoor or outdoor spaces for people. The measurement for color, temperature, light intensity and luminous flux are all calibrated on a wavelength of 555 nm. Complete datasets are available for these lamp types, including the emitted light spectrum and the light angle.

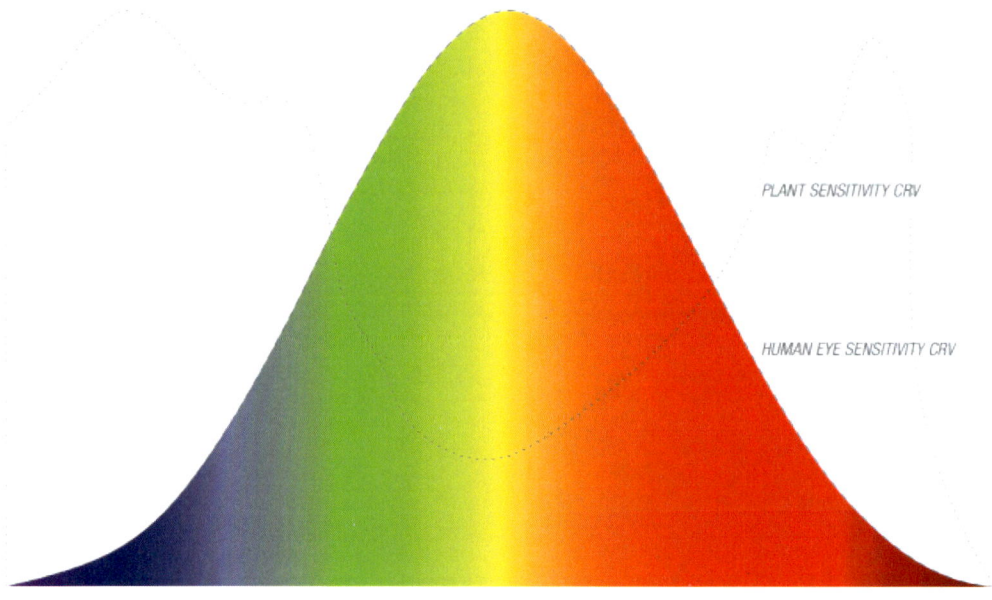

Fig.61.   Light spectrum - Human Eye Sensitivity

This is diminished by 10% for classic greenhouses due to the reflections from the building envelope.
Additional diminishment of daylight availability can be expected with stacking plant production in a Vertical Farm, conceived to the skyscraper typology.

Photosynthesis is at its most effective with blue and red light. The choice of lamp types in which the blue and red light spectral components are highest is necessary, not only for reasons of efficiency, but also to reduce operating costs.

---

[1] Luminous efficiency is the percentage of lumen per Watt compared to the luminous efficiency of sunlight. Per Watt the sun has a luminous flux of 683 lm.

By contrast with lamps conceived with the human eye luminous efficiency factor in mind, the choice here is for the photo efficiency μmol/J [Ws]. The following diagrams show the spectral distribution of the most common lamp types used in greenhouses. Additional lamp types and their spectral distribution ae listed in the Appendix.
The lamp types referred to in this work for in the Vertical Farm are LED types and this for the following reasons:

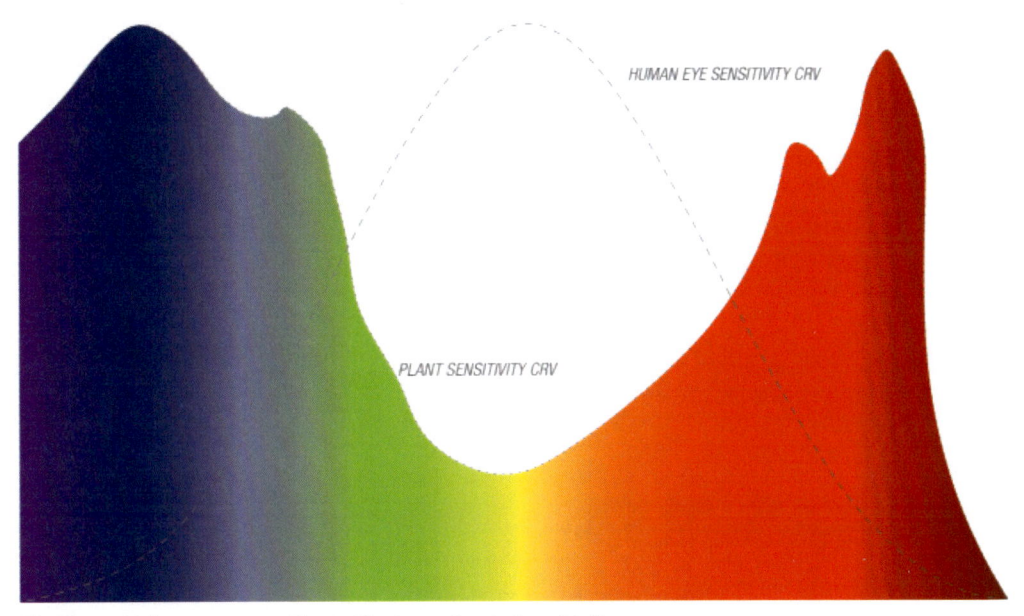

Fig.62.   Light spectrum - Plants Photosynthesis Sensitivity

- LEDs have the highest photon efficiency of all growth promoting lamps.
- LEDs are produced to emit the ideal spectrum for photosynthesis.
- LED technology has "rapidly advanced over the past decade (...) [. A] similar significant advances throughout the coming decades can be expected. (...) [LED] technology (...) progresses under what is known as Haitz's Law, which observes and predicts that the cost per (...) useful light emitted) of LEDs falls by a factor 10 every decade."[1][2][3]

---

1   GRAFF, G. 2011. Skyfarming. Master of Architecture, Waterloo, Ontario, Canada. p.94
2   http://www.nature.com/nphoton/journal/v1/n1/full/nphoton.2006.78.html, retrieved 12.01.2015
3   This is necessary to consider, because even though LEDs have the highest photon efficiency of all lamp types produced nowadays for plant growth, fixture costs are five to seven times higher.
     see: NELSON, N. 2009. Planning the productive city. Available: http://www.nelsonelson.com/DSA-Nelson-renewable-city-report.pdf.

- LEDs produce much less excess heat than any other lamp types used for plant growth[1] and therefor the lamp itself can be placed closer to the canopy without overheating leaf surfaces, which leads to water stress in the plant, and these lamps are thus ideal for plant production within the skyscraper building typology with lower floor heights.

Incomplete and difficult to compare descriptions and data for LED lamp types made it necessary to contact several manufacturers in order to obtain a precise data pattern, not only concerning photo-efficiency, but also about the light angles and the photon flux densities along the light radiation.

Two lamp types were initially considered for the simulation model setup, both produced by LumiGrow, Novato, California, USA. The support from the manufacturer side was required to define the optimum distance between the LED fixture and the plant canopy for supplying *L. esculentum* with the necessary PPFD on one hand, and to define a raster for the fixture to cover the complete cultivation area on each of the different levels on the other.

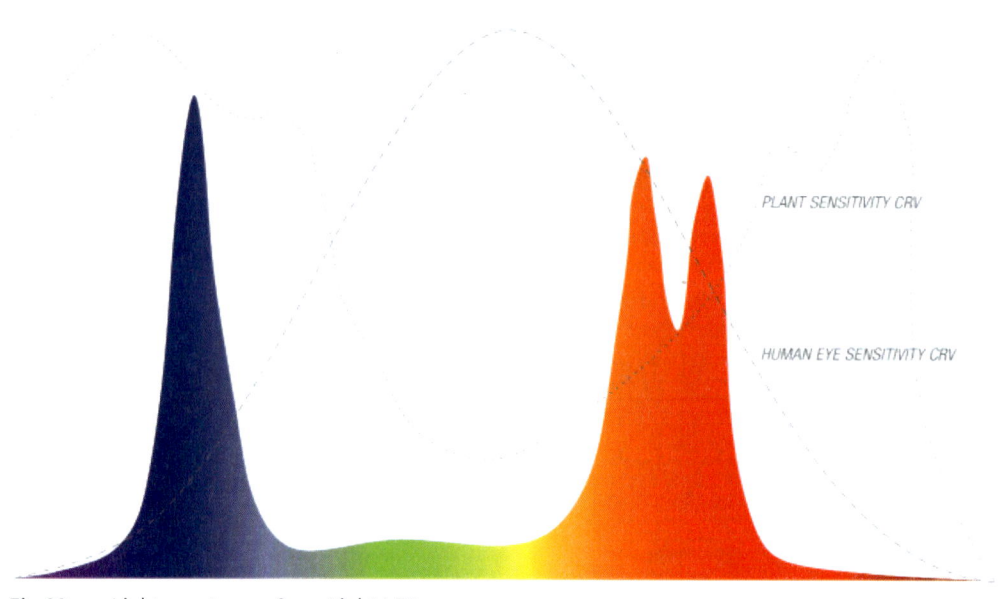

Fig.63.   Light spectrum - Grow Light LED

---

1   Zeiler for instance within his greenhouse uses HPS-lamps (High Pressure Sodium Lamps). Excess heat within the greenhouse is not considered as waste heat, but while enlarging the photoperiod during the day and enlarging crop rotation throughout the year, heat can be used and additional heat costs therefor can be diminished.

4.4.3. Artificial Lighting - LED LumiGrow 325 and LumiGrow 650

LumiGrow, Inc. is one a provider of horticultural lighting solutions, "enabling commercial growers and researchers to achieve operational efficiencies, reduce energy consumption and improve crop yields."[1]

Fig.64. LumiGrow Pro 325 and 650 LED light fixture

The LumiGrow Pro-series (325 and 650 [W, Ed.] are developed to "output more red and blue in the essential PAR range than the industry's most powerful conventional lighting systems, including any high-intensity discharge (HID) fixture. Peerless in the industry, the Pro 650 light delivers 2X the red and blue PAR of a 1000 Watt HID light while it consumes 40% less energy. The Pro 325 provides a red and blue PAR equivalent to that of a 1000 Watt HID light while it reduces energy consumption by a whopping 70%."[2]

The LumiGrow Pro-series lamps have a light angle of 120°. The PPFD for the plant is dependent on the distance of the lamp to the canopy of the *L. esculentum*. The PPFD output of LumiGrow 325 can thus be considered adequate for the simulation model. The table below shows the dependency of PPFD to height-distance between fixture and plant canopy. Furthermore the list highlights the amount of blue and red light from the overall PPFD within the PAR range and the amount of green and yellow light (between 500 nm and 600 nm). It is an extended table of values provided by the author, based on official data taken from the LumiGrow website. The following data were verified by the corporate research department of the LumiGrow. The DLI was added and calculated for a photoperiod of 16 hours, or 57,600 seconds with a constant of 300 $\mu mol/m^{-2}/s^{-1}$ of PPFD.

It is important to mention, that $\mu mol/m^{-2}/s^{-1}$ listed in Tab.10 must be defined as adjusted PAR-values. PAR weights the entire spectrum between 400 nm and 700 nm equally. The norm DIN-5031-10 defines the photobiological and thermal effect of optical radiation.[3] Based on this spectral sensitivity curve (DIN-5031-10) every wavelength gets absorbed differently from the leaves. This explains why the effective

---

1   www.lumigrow.com, retrieved 12.02.2015
2   http://www.lumigrow.com/products/pro-series-greenhouse-lights, retrieved 05.01.2015
3   DIN 5031-10:2013-12 (D) „Strahlungsphysik im optischen Bereich und Lichttechnik - Teil 10: Photobiologisch wirksame Strahlung, Größen, Kurzzeichen und Wirkungsspektren" http://www.beuth.de/de/norm/din-5031-10/195406448, retrieved 29.10.2015

output of 81 μmol/m^{-2}/s^{-1} [1] provided by Lumigrow Pro 325 is sufficient for optimum plant growth.

The specific conversion factor for Lumigrow Pro 325 from lux to μmol/m^{-2}/s^{-1} is 2*0.081. Simulation results provided by Lumigrow to evaluate the simulation results in Chapter 5 can be seen in the Appendix. Lighting schedule, developed in Chapter 5, therefor, is based on Adjusted PAR.

- The operating frequency is from 50 - 60 Hz
- Power consumption is 325 Watts
- Power Factor is 0.95
- Operating temperature from -20 to 60°C
- Dimensions: 254 mm x 279 mm x 140 mm)

		PRO 325					
height (m)	height (ft)	PPFD (μmol/m²/s)	blue	red	PAR b/r	DLI (16)	500-600 nm
0.61	2.00	510	86.70	341.70	428.40	24.68	81.60
0.76	2.50	326	55.42	218.42	273.84	15.77	52.16
0.80	**2.62**	**300**	**51.21**	**201.84**	**253.05**	**17.28**	**46.95**
0.84	2.75	276.50	47.01	185.26	232.26	13.38	44.24
0.91	3.00	227	38.59	152.09	190.68	10.98	36.32
1.07	3.50	167	28.39	111.89	140.28	8.08	26.72
1.22	4.00	128	21.76	85.76	107.52	6.19	20.48
1.37	4.50	101	17.17	67.67	84.84	4.89	16.16
1.52	5.00	82	13.94	54.94	68.88	3.97	13.12
1.68	5.50	67	11.39	44.89	56.28	3.24	10.72
1.83	6.00	57	9.69	38.19	47.88	2.76	9.12
1.98	6.50	48	8.16	32.16	40.32	2.32	7.68
2.13	7.00	42	7.14	28.14	35.28	2.03	6.72
2.29	7.50	36	6.12	24.12	30.24	1.74	5.76
2.44	8.00	21	3.57	14.07	17.64	1.02	3.36
2.59	8.50	28	4.76	18.76	23.52	1.35	4.48
2.74	9.00	25	4.25	16.75	21.00	1.21	4.00
2.90	9.50	23	3.91	15.41	19.32	1.11	3.68
3.05	10.00	20	3.40	13.40	16.80	0.97	3.20

Tab.11.   Lumigrow 325 PRO - PPFD, DLI values depending on fixture height

Tab. 11 shows the photon emmission of Lumigrow PRO 325. The table got generated based on official data retrieved by the company[2] and compared to results of the study on the „Economic Analysis of Greenhouse Lighting: Light Emitting Diodes vs. High Intensity Discharge Fixtures".[3] Micromoles, here, are interpolated to correspond to

---

1 See Appendix - Simulation provided by Lumigrow to evaluate simulation results in Chapter 5
2 http://www.lumigrow.com/products/pro-series-greenhouse-lights, retrieved 05.01.2015
3 NELSON, J. A. & BUGBEE, B. 2014. Economic Analysis of Greenhouse Lighting: Light Emitting Diodes vs. High Intensity Discharge Fixtures. PLOS ONE ; 9/6, 1-10. retrieved 15.01.2015

adjusted PAR-values (400 nm to 700 nm). These values are used for building up the lighting schedule used for tracing lighting demand of the Vertical Farm Simulation types.

4.4.4. *L. esculentum* - Organization of the vertical cultivation area

The results of chapter 4 now provide enough data to set up a generic arrangement of the cultivation area for *L. esculentum* suitable for *L. esculentum* cultivars which produce small to medium sized fruits and grow on and along trusses.[1][2]
The planting bed size is 60 cm x 70 cm, the growing substrate chosen for the theoretical model is "cocopeat".[3]

The maximum PPFD for fruit production will be 300 µmol/m^{-2}/s^{-1} (=65.64 W), starting with 150 µmol/m^{-2}/s^{-1} (=32,82 W) and ending with 100 µmol/m^{-2}/s^{-1} (=21.88 W) by the end of crop rotation.[4] Considering that daylight availability (average PPFD) for soil-based agriculture-*L. esculentum* in Vienna is only given from March to October, supplementary light will be needed for the photoperiod.

LED-light types will be used to cover the lacking PAR availability within the building.[5] The performance analysis has shown that LumiGrow PRO 325 is an LED lamp type which is useful for *L. esculentum*-Production within the stacked greenhouse, the Vertical Farm.[6] LED-fixture will be placed 80 cm upon the *L. esculentum* plant canopy.

All the *L. esculentum*-cultivars mentioned are indeterministic plant types and will grow in the z-axis until they reach -1.00 m from the under-edge of the room height (20 cm height of LED-lamp + 80 cm distance between the light and the plant canopy). The raster of the LED-fixtures is defined in such a manner that the overall surface of the plant canopies are completely covered with the photon flux given by the light angle of 120°.

Placement of the construction support, height of the growing substrate, placement of the CO2-pipes, drainage and additional support construction shown the next page are based on recommendations experienced during the excursion to „Tomaten Zeiler" and by the „Sächsische Landesanstalt für Landwirtschaft, Fachbereich

---

1. LUITEL, B. P., ADHIKARI, P. B., YOON, C.-S. & KANG, W.-H. 2011. Yield and Fruit Quality of Tomato (*Lycopersicon esculentum* Mill.) Cultivars Established at Different Planting Bed Size and Growing Substrates. Horticulture, Environment, and Biotechnology ; 53(2), 102-107.
Within this work the following cultivars were planted: „Campari", „Temptation", „Annamay" and „Adoration.
2. see Appendix: Tomaten Zeiler - Exkursionsbericht; „Zeiler" produces the following cultivars: „Avalantino", „Juanita", „Sunstream" and „Vesuvius San Marzano"
3. LUITEL, B. P., ADHIKARI, P. B., YOON, C.-S. & KANG, W.-H. 2011. Yield and Fruit Quality of Tomato (*Lycopersicon esculentum* Mill.) Cultivars Established at Different Planting Bed Size and Growing Substrates. Horticulture, Environment, and Biotechnology ; 53(2), 102-107.
4. see chapter 4
5. ibid.
6. ibid.

Gartenbau".[1] Based on findings within this chapter, recommendations of "Zeiler" and discussions at the Department of Crop Sciences at the University of Natural Resources and Life Sciences, Vienna three essential schedules for the simulation model can be implemented:

LIGHTING SCHEDULE:

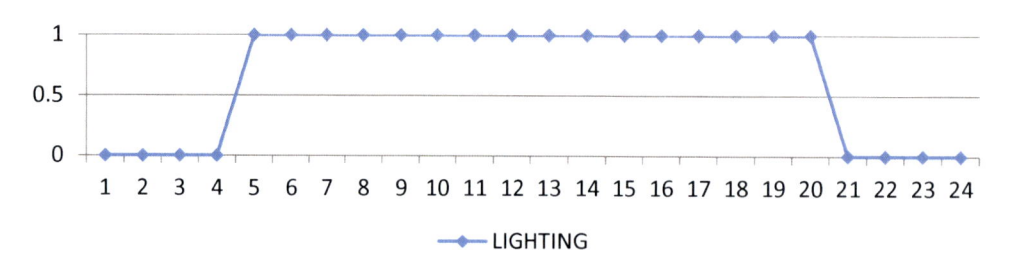

Fig.65.   Lighting schedule used for simulation model

16 h photoperiod, 8 h respiration phase

HEATING HIGH SCHEDULE / HEATING LOW SCHEDULE.

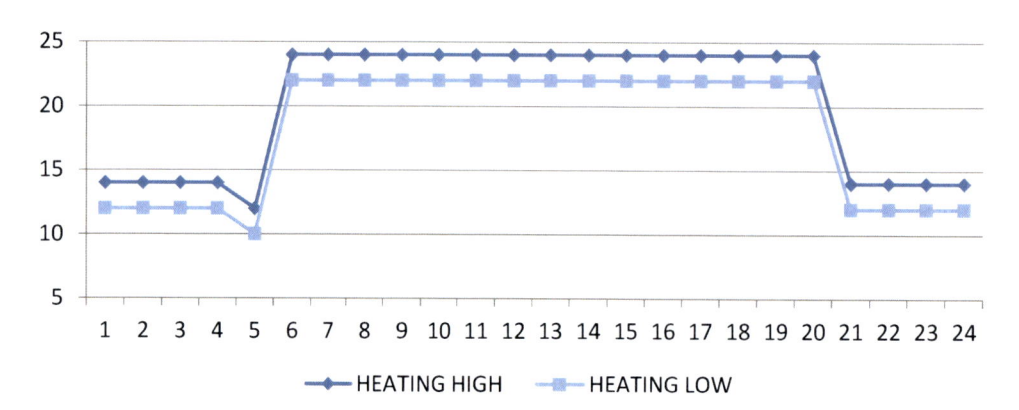

Fig.66.   Daily heating schedules used for simulation model

16 h photoperiod, 8 h respiration phase

Fig. 28 on page 208 represents the total solar radiation in Vienna for 8760 hours, the percentage of Photosynthetically Active Radiation, the needed radiation in PAR for *L. esculentum* with its saturation point - all these values overlapped by the individual crop rotation phases throughout a whole year in closed conditions.

---

1   SÄCHSISCHE LANDESANSTALT FÜR LANDWIRTSCHAFT, FACHBEREICH GARTENBAU 2004. Gewächshaustomaten. Hinweise zum umweltgerechten Anbau. Managementunterlage. [Dresden]: Sächsische Landesanstalt für Landwirtschaft, Fachbereich Gartenbau. p. 33 ff.

VERTICAL FARM - SUBSTITUTION OF NATURAL GROWTH FACTORS

150 µmol/m2/s
32,82 WPAR
8,62 DLI

300 µmol/m2/s
65,64 WPAR
17,28 DLI

ESTABLISHMENT seeding - (trans-) planting
VEGETATIVE GROWTH developement / photomorphogenesis
FLOWERINT TO FRUIT-SET blossom, pollination, first fruits
FRUIT GROWTH
RIPENING TO FIRST HARVEST fruit growth - lycopene production
FULL HARVEST TO END

week	2	4	6	9	12
day	14	28	42	63	84
hour	336	672	1,000	1,512	2,016

▨ TOTAL SOLAR RADIATION
▨ PHOTOSYNTHETICALLY ACTIVE RADIATION
▨ NEEDED RADIATION FOR OPTIMUM GROWTH
130 ── SATURATION POINT (130 W or 600 µmol/m2/s)

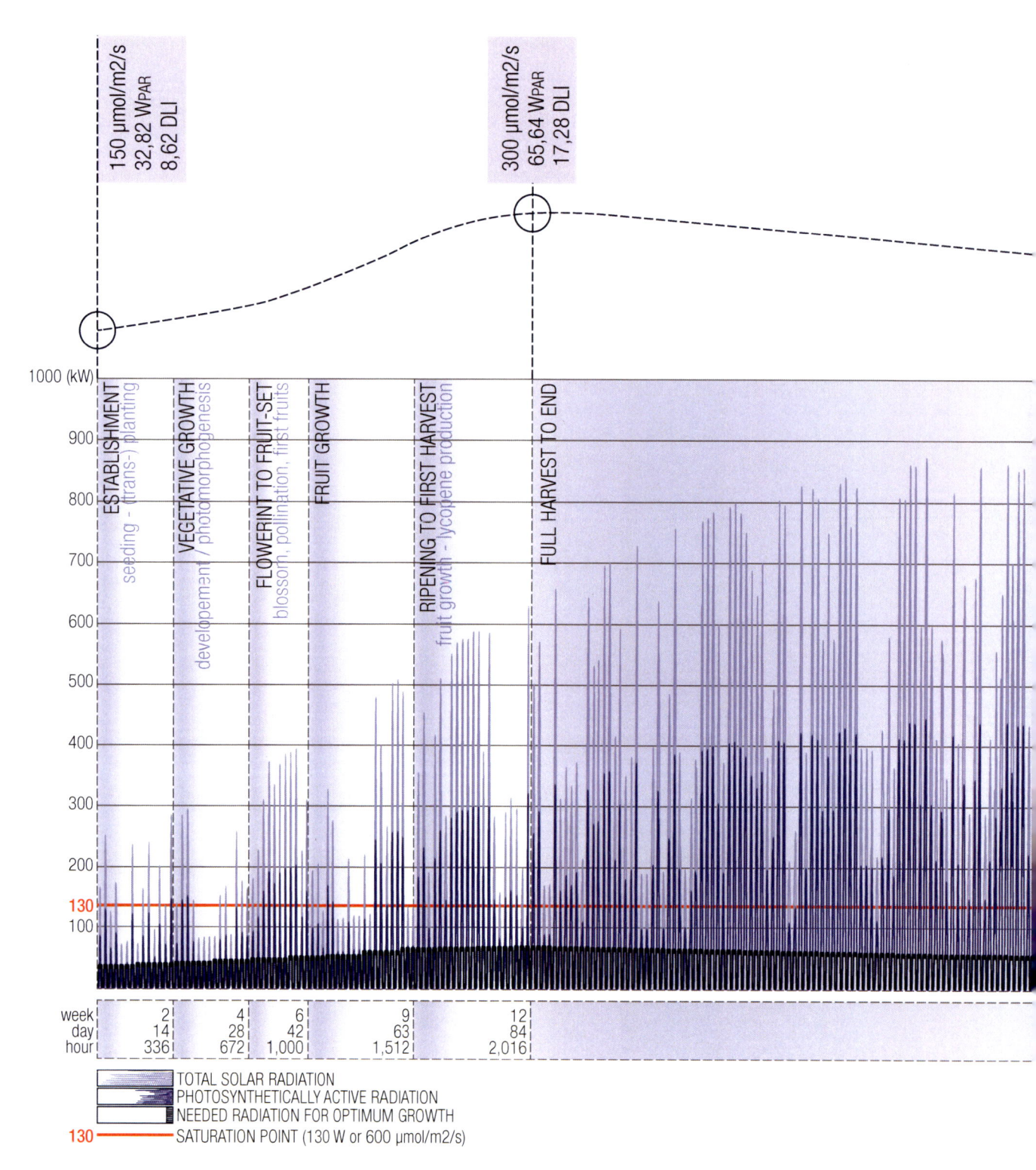

Fig.67.   Total solar radiation, ratio of PAR and needed radiation for *L. esculentum* in Vienna

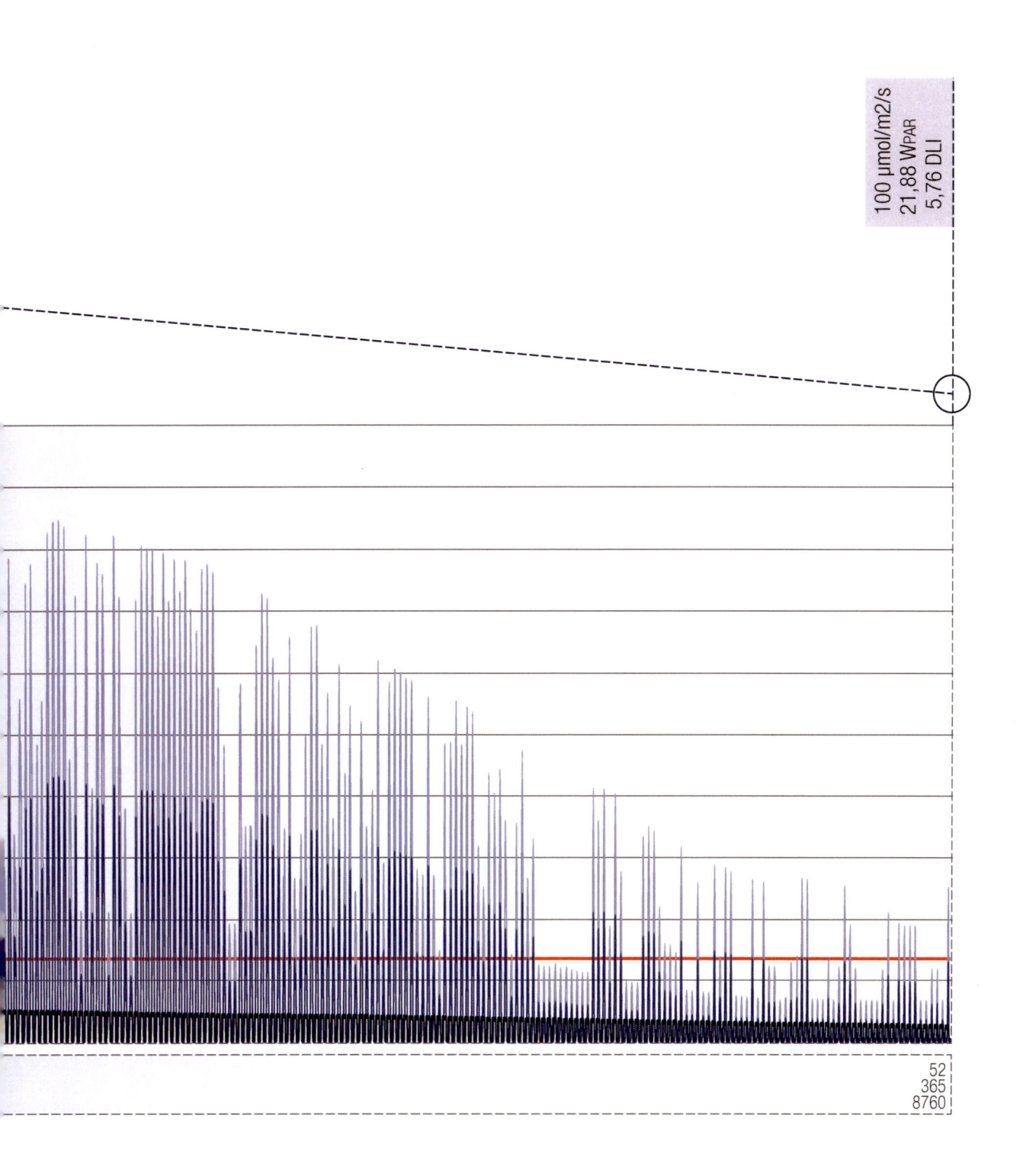

VERTICAL FARM - SUBSTITUTION OF NATURAL GROWTH FACTORS

100 μmol/m2/s
21,88 WPAR
5,76 DLI

52
365
8760

5. The Vertical Farm - Simulationmodel

5.1. Introduction

The Vertical Farm developed by Plantagon and Sweco[1] for Linköping, Sweden, is one of the most promising of all the different Vertical Farm proposals we have so far encountered, and this for many reasons: The production and cultivation methods employ the basic design principle resulting in the typology of the vertical greenhouse. The production conveyor belt, the building depth and orientation all indicate that they are the result of the necessary daylight optimization required here. Furthermore Plantagon with its Vertical Farm in Linköping combines two different programs (production of vegetables facility plus an office building) to increase the energy performance of the building by enabling synergy potentials between the two different uses.

The second Urban Agriculture Summit, organized by Plantagon in Linköping, Sweden, on January 30th in 2013, offered the author the opportunity to discuss recent findings in the context of horticulture, energy demand and essential issues for future research. Furthermore it was possible for the author to continue his doctoral thesis by using published data (both on building dimensions, cultivation and production method and energy consumption data) as reference.

The volume of the simulation model for this doctoral thesis thus leans substantially on the building volume of the Vertical Farm in Linköping, Sweden.[2] This volume will be the basis for daylight simulation and the heating demand, varying only for obtaining equal heights on the different levels. Vertical circulations and vertical cores for building services will be ignored; the focus of the simulation results is the creating of awareness for a theoretical possible daylight distribution through different building depths and the interdependency of this with the heating demand.

---

1   http://plantagon.com/about/business-concept-2/lead-project, retrieved 12.12.2014
2   The Volume of the parametric model refers only to the actual volume used for food production and processing, placed on the south side of the building and doesn't include the volume for the office building on the north, nor the volume between the two functions, used for vertical circulation.

5.2. Goals

The issue of primary interest for which the simulation models have been developed is evaluating the power demand for artificial lighting and the heating necessary for a year round crop rotation of *L. esculentum*.

For a more detailed investigation a volume which is oriented to a similar volume to Plantagon's Vertical Farm in Linköping, Sweden, is parametrically generated with three different A/V-ratios to evaluate the influence of the interdependency between daylight availability and/or power demand and energy demand for heating.

The different Vertical Farm typologies are analyzed on the basis of the annual availability of daylight on the different levels where the influence of different floor heights and the position (bottom, center and top of the building) are clear factors.

Furthermore the Vertical Farms are divided in zones to evaluate solar penetration and heating demand regarding their orientation (North, East, South, West and Centre zone).

Lighting and heating demand is influenced by two factors: The specific temperature and lighting demand throughout the changing phases of the crop rotation through the daily photoperiod and the available climatic conditions of the location. The results will thus be subdivided into monthly power and energy demands, following the sigmoid growth curve of *L. esculentum*.

The three different Vertical Farms with their specific A/V-ratios are additionally compared with the use of three different building envelopes: a single glazing facade, a double-layer ETFE facade and a special double glazing facade with a high visual transmittance.

5.3. Method

5.3.1. Simulation Software

The Vertical Farm simulation model is parametrically developed with Rhino/Grasshopper, one of the "most widely used platforms by designers today."[1] Beside different environmental plugins which already exist today for Rhino/Grasshopper the author decided to work with Ladybug for weather data analysis, daylight radiation simulation from EnergyPlus Weather Data and Honeybee for advanced indoor lighting simulation. Using Honeybee was of primary interest, because it offers, by contrast with other simulation software used in architecture, the possibility for advanced radiation studies within the building, or more precisely, getting power results instead of illuminance values from solar radiation.

LADYBUG

Ladybug is a „free and open source environmental plugin" for Rhino/Grasshopper. Grasshopper is a „graphical algorithm editor (...)."[2] „Ladybug benefits the parametric platform of Grasshopper to allow the disigner to explore the direct relationship between environmental data and the generation of the diesign through graphical data outputs that are highly integrated with the building geometry."[3]

HONEYBEE

Honeybee was used to run the daylight analysis based on the weather data within the building. Similar to Ladybug, Honeybee is a plugin for Rhino/Grasshopper, but for "more advanced studies."[4] The plugin is necessary for facilitating the simulation process for this study. It offers the possibility for interactively analyzing, e.g. radiation (W/m2) results at different control points on each level of the simulation model withhout being disconnected from different platforms. It "automates the process of intersecting the [building-, Ed.] masses, and finding adjacent surfaces (...).[5]

---

1   http://www.grasshopper3d.com/group/ladybug, retrieved 01.05.2015
2   ibid.
3   ibid.
4   ibid.
5   ibid.

Most of the currently available tools for energy- and daylight simulations only export geometries from a "design environment to simulation files and read the results back, however Ladybug provides a two way import/export connection [the author, Ed.] (...) can import back the simulation file and visualize it in Rhino/Grasshopper environment before executing the simulation."[1]

Both, Ladybug and Honeybee offer the possibility to adapt and customize the tools based on the needs during the simulation process. Furthermore Honeybee and Ladybug offer the additional capabilities of individually developing and designing the "graphical data outputs that are highly integrated with the building geometry."[2]

ENERGY PLUS

Components of Honeybee were used for thermal simulation of the different Vertical Farms, which connect grasshopper directly to the EnergyPlus energy simulation program[3]. Climate- and weather data for Vienna were retrieved from the US Department of Energy.[4]

---

1 http://www.grasshopper3d.com/group/ladybug, retrieved 01.05.2015
2 ibid.
3 http://apps1.eere.energy.gov/buildings/energyplus/, retrieved 21.03.2015
4 http://energy.gov/eere/office-energy-efficiency-renewable-energy, retrieved 06.08.2015

VERTICAL FARM - SIMULATION MODEL

5.3.2. Location

CLIMATE DATA, Vienna, Austria: 58°24'39"N 15°37'17"E

hours/year	8,760
dayhours	4,401
sunshine hours	(43.01%) 1,893
kWh/m2/a	1,119.32
diffuse sunlight (kWh/m2/a)	620.99
direct sunlight (kWh/m2/a)	498.32
GJ/m2/a	4.03
kCal/m2/a	962,442
l Oil Eq./m2/a	108.57
av. kWh/m2/a	0.25
av. MJ/m2/a	0.92
av. kCal/m2/a	218.69
av. l Oil Eq./m2/a	0.02

The Vertical Farm simulation models are located in Vienna, Austria (ASHRAE Climate Zone 5A, Köppen-Geiger-Klimate Zone Dfb) on 48.12 Latitude and 16.57 Longitude, Elevation 190 m. "AUT_vienna.Schwechat.110360.epw" is used for the simulation the EnergyPlus weather file.[1]

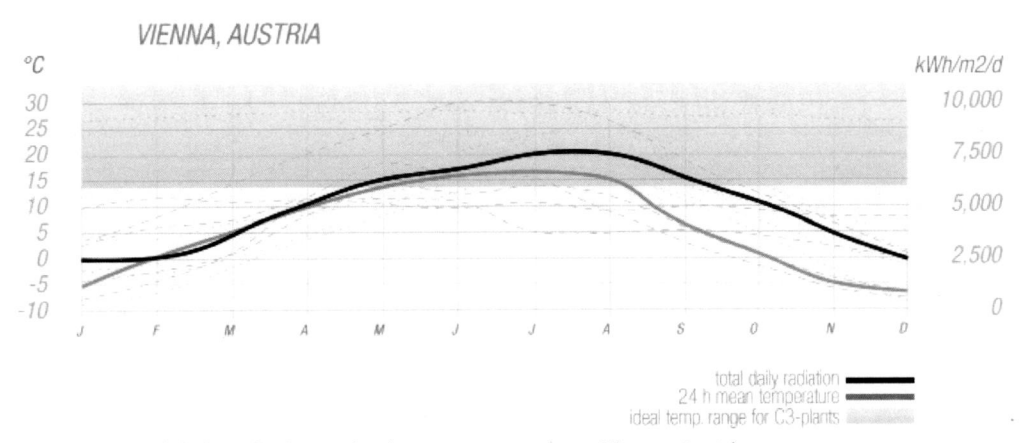

Fig.68.   Total daily radiation and 24 h mean temperature, Vienna, Austria

---

1   http://apps1.eere.energy.gov/buildings/energyplus/cfm/weather_data3.cfm/region=6_europe_wmo_region_6/country=AUT/cname=Austria, retrieved 12.12.2014

VERTICAL FARM - SIMULATION MODEL

ANNUAL PHOTOSYNTHETIC-PHOTONFLUX-DENSITY

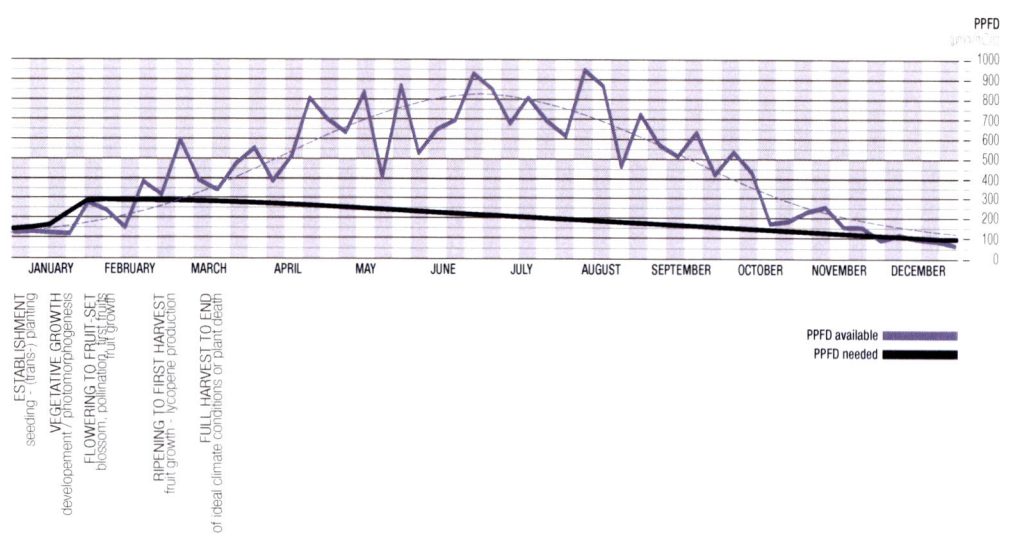

Fig.69.   PPFD availability in Vienna and PPFD needed along the crop rotation

The upper diagrams show the PPFD-distribution and the DLI in Vienna throughout the year and the needed amount of PPFD and needed DLI for *L. esculentum* through the whole crop rotation when the plant is established by the beginning of the year.

The diagram below shows the mean high and mean low temperatures and the possible range for growing *L. esculentum* in greenhouses (from 12 to 24°C), the two heating schedules used for the simulation model.[1]

ANNUAL TEMPERATURE VIENNA

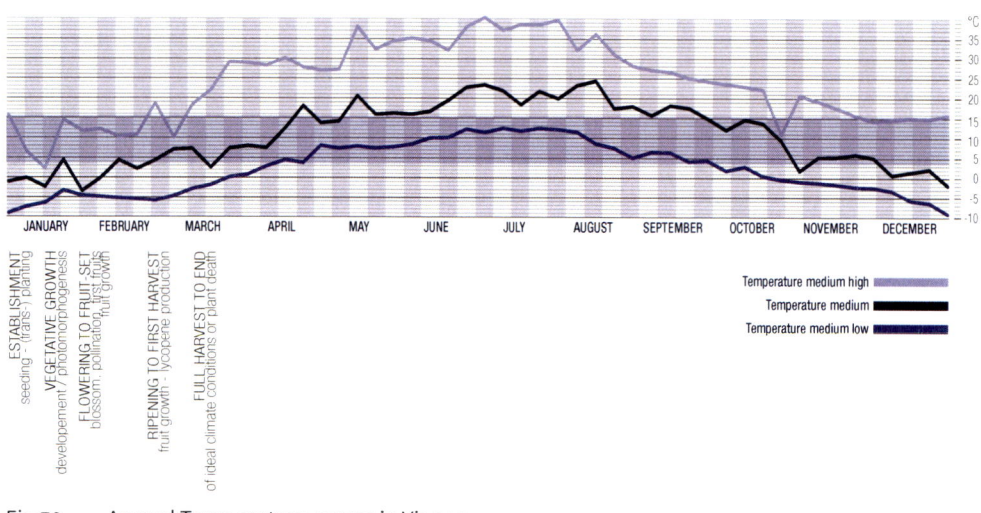

Fig.70.   Annual Temperature curves in Vienna

---

1   See Chapter 4.4.5 and Chapters before.

VERTICAL FARM - SIMULATION MODEL

PHOTOSYNTHETICALLY ACTIVE RADIATION - VIENNA

Fig.72. Watt (PAR) availability in Vienna

Looking at the available W/m2 of the photosynthetically active radiation we see that on the field in Vienna *L. esculentum* would be supplied with enough light for photosynthesis for three quarters of the year.

It is possible to extend the photoperiod to 16 h to optimize crop yield, taking into account that *L. esculentum* needs 8 h for respiration. See Chapter 4.4.1 for statistics data details. The diagram below shows that only June 21st reaches the maximum photoperiod length of 57,600 seconds.

The diagrammatic representation of the values for total global radiation, PAR and the radiation needed for optimum plant growth of *L. esculentum* can be found on the last page of Chapter 4.

AVAILABLE DAYLENGTH FOR PHOTOSYNTHESIS

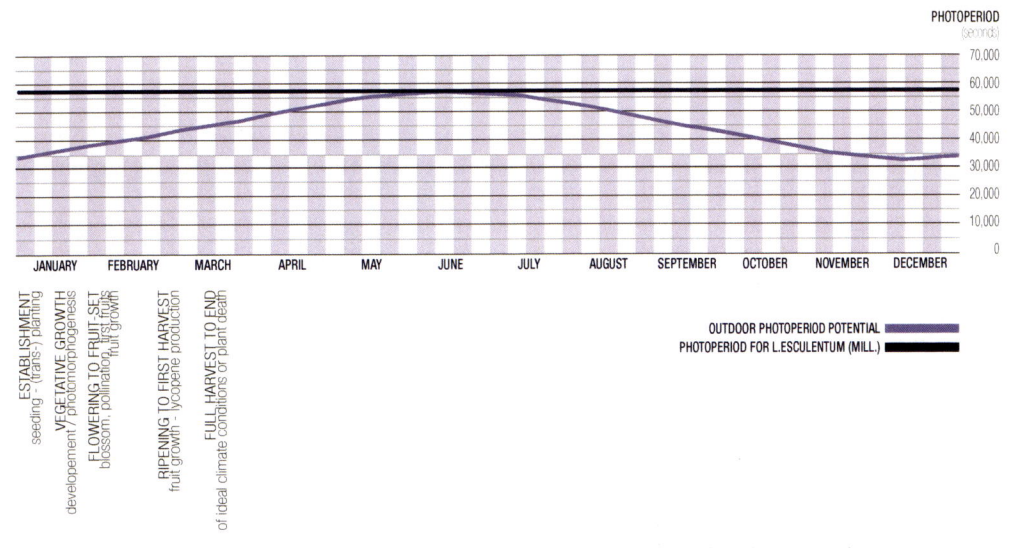

Fig.71. Daylength in Vienna and Photoperiod for Photosynthesis within the Vertical Farm

5.3.3. Parametric Simulation Building

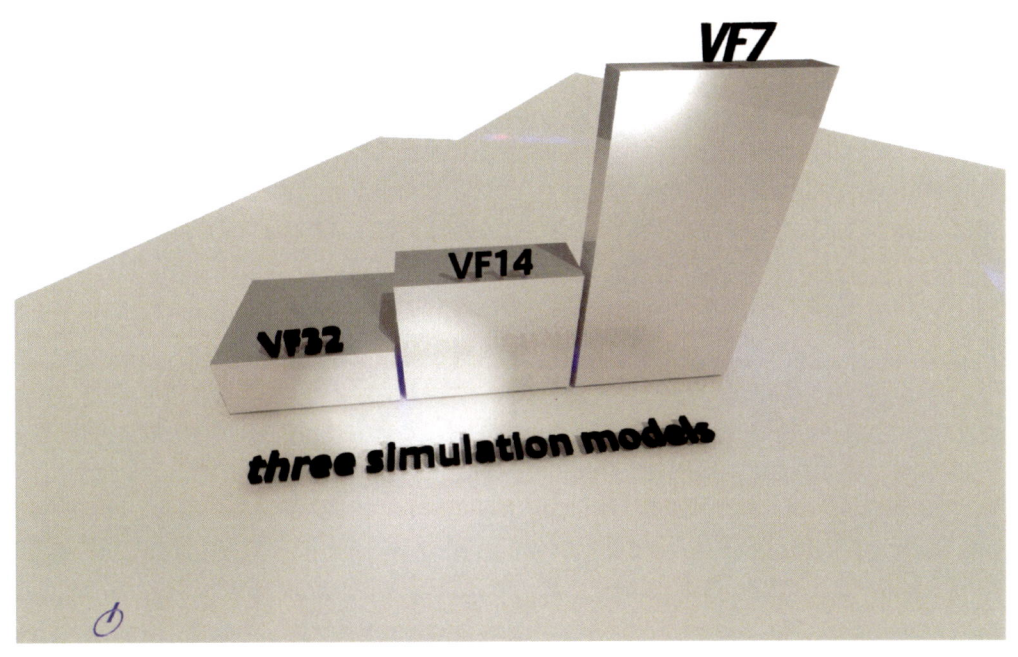

Fig.73.　　Three Simulation Models - VF32, VF14, VF7

Three different building types are developed and compared here. The volume, as already mentioned, is oriented to the volume of the Vertical Farm planned in Linköping, Sweden by Plantagon and Sweco. The reconstruction of the geometry developed on plans available on the official Plantagon website, leads to a volume of 15,003.00 m3. In order to make the results of parametric Vertical Farms with different building depth comparable the following decisions have been made:

The building-orientation is east-west, the x-axis constantly stays the same for every simulation model and is 36 m, a common length of greenhouses used in southern Vienna.
The y-axis defines the differences in the ground floor, namely 7.2 m, 14.4 m and 32 m. The dimensions were chosen to provide a constant reference to the most common greenhouse in central Europe, the Venlo-Greenhouse, whereas 3.6 m is a modular entity.

The ground floor in every simulation model is defined as being 5 m in height for two reasons: if food (partially) is also produced on the 0-level within the city are, it is also conceivable that part of the ground level will also be used as public- or semipublic space (market/trade areas etc.) In addition this decision offers the opportunity to compare daylight availability changes through different floor heights. Every typical floor and the top level present a floor height of 3.5 m.

VERTICAL FARM - SIMULATION MODEL

The floor heights on every simulation model must therefore stay the same so as to be able to see directly the difference of the simulation outputs in the context of the influence of different building depth. This led to the decision of referring to the exact volume of Linköping's Vertical Farm, which is still important, but on a secondary level. This leads to different building volumes of the simulation model from 13,824.00 m2 (VF x=32) to 15,811.20 m2 (VF x=7.2) to 15,292.80 m3 (VF x=14.4).

In the following these three Vertical Farms for simplification reasons get named VF32, VF14 and VF7.

The resulting building volume as building mass is divided into cultivation levels. The levels are not interrupted by any building entities such as walls or central core. The vertical circulation for building services, freight elevators, elevators for people and fire-escapes are not considered in this simulation model. The simulation is therefore reduced to an open-floor single volume building (comparable to the cultivation practice in greenhouses). The basements of Vertical Farms 7, 14 and 32 have storey heights of 5 m, while every other level has a storey height of 3.5 m.

Dimensions of VF32:   x = 36 m;   y = 32 m,   z = 3 levels, incl. groundfloor
Dimensions of VF14:   x = 36 m;   y = 14.4 m,   z = 8 levels, incl. groundfloor
Dimensions of VF7:   x = 36 m;   y = 7.2 m,   z = 17 levels, incl. groundfloor.

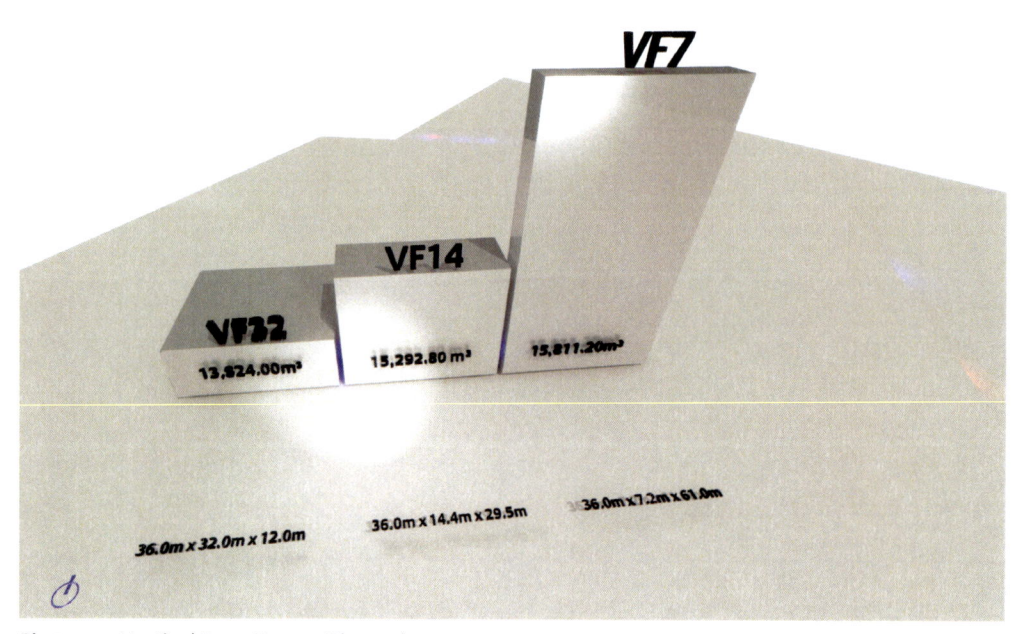

Fig.74.   Vertical Farm Types - Dimensions

LEVELS

Three levels are of basic interest for the daylight simulation to obtain the sunlight penetration throughout the year in W/m2 which are fundamental to create the lighting schedule for the thermal heating simulation, these are:

> the ground floor with a floor height of 5 m
> the first floor with a floor height of 3.5 m
> the last floor with a floor height of 3.5 m.

The difference between the ground floor and the first floor gives us results which show the effect of the different floor heights. The lighting schedule of the first floor can be used for all the typical levels, and the results for the top floor can be seen and read as the daylight demand for greenhouses.

THREE BUILDING LEVELS SIGNIFICANT FOR LIGHTING DEMAND

Every simulation transition now captures three identical areas with the same control points, but with different spatial properties for every Vertical Farm typology:

- The ground floor has a ceiling with a light transmissivity of 10%, the exterior walls are 90% covered with glass.
- The standard floor (=all intermediate floors).
- The top level of the Vertical Farm can be read as the result for a typical greenhouse. The exterior walls have the same properties as those of the penultimate level, the ceiling is defined as the roof in Honeybee, however, and has the same material properties and the same ratio between opaque and transparent material as those of the exterior walls.

SIMULATION GRID AND LUMIGROW 325PRO LED LIGHTING FIXTURE

The planting bed size or the distance from every *L. esculentum* (stems) is 0.6 m x 0.7 m. The density of the planted bed size is explained in more detail in Chapter 4.3.6. "Planting the tomato cultivars in single row at 60 cm bed width is a better approach to optimize the production space in greenhouse (...) where marketable fruit was highest."[1]

In the control points for daylight simulation the grid size is 1. Every control point is spaced at a distance 1 m from the next one. The grid is 1.1 m distant from the base surface, the 0.00 position of every level simulated. This is, where most of the fruits are ripening in practice (especially from blossom to fruit where 300 µmol/m^{-2}/s^{-1} or 65.64 W$_{PAR}$ are needed).

The placement of the LED fixture is placed 80 cm above the canopy and it is therefore theoretically moving in the z-axis during the first third of the crop rotation.
LumiGrow 325 has a light angle of 120° at which the light fixture points were foreseen and marked with additional control points such as the controls points for the plant canopy position. This guarantees the coverage of the necessary PPFD amount for the plants on the entire cultivation area in each Vertical Farm on every level throughout the whole crop rotation. For more information on lighting fixtures see p. 203.

Simulation grid and light fixtures are generated parametrically and adapted to all three Vertical Farm types.

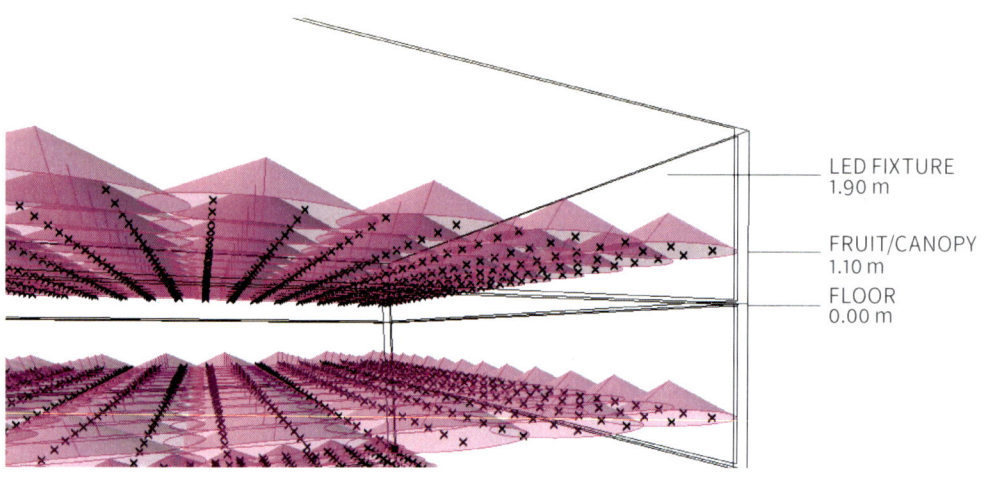

Fig.75.   VF32 - Cultivation levels with LED-fixtures, screenshot perspective

---

1   LUITEL, B. P., ADHIKARI, P. B., YOON, C.-S. & KANG, W.-H. 2011. Yield and Fruit Quality of Tomato (*Lycopersicon esculentum* Mill.) Cultivars Established at Different Planting Bed Size and Growing Substrates. Horticulture, Environment, and Biotechnology ; 53 (2), 102-107.

LED LUMIGROW 325 PRO - LIGHTING FIXTURE

VF32
Floor area: 1,152 m2
Levels: 3
Cultivation area: 3,456.00 m2
LED-fixtures/floor: 224
LED Total: 672

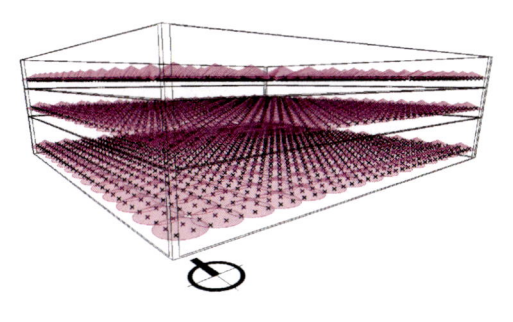

Fig.76.   VF32 - Cultivation levels with LED-fixtures

VF14
Floor area: 518.4 m2
Levels: 8
Cultivation area: 4,147.20 m2
LED-fixtures/floor: 96
LED Total: 768

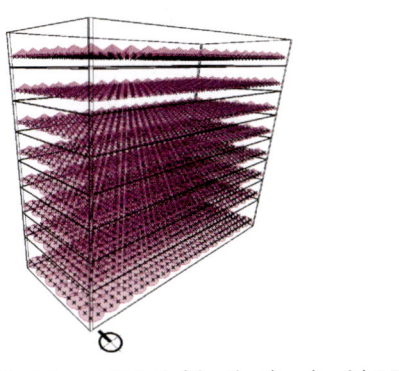

Fig.77.   VF14 - Cultivation levels with LED-fixtures

VF7
Floor area: 259.2 m2
Levels: 17
Cultivation area: 4,406.40 m2
LED-fixtures/floor: 48
LED Total: 768

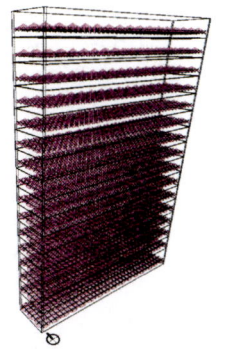

Fig.78.   VF7 - Cultivation levels with LED-fixtures

ZONINGS AND CULTIVATION AREA

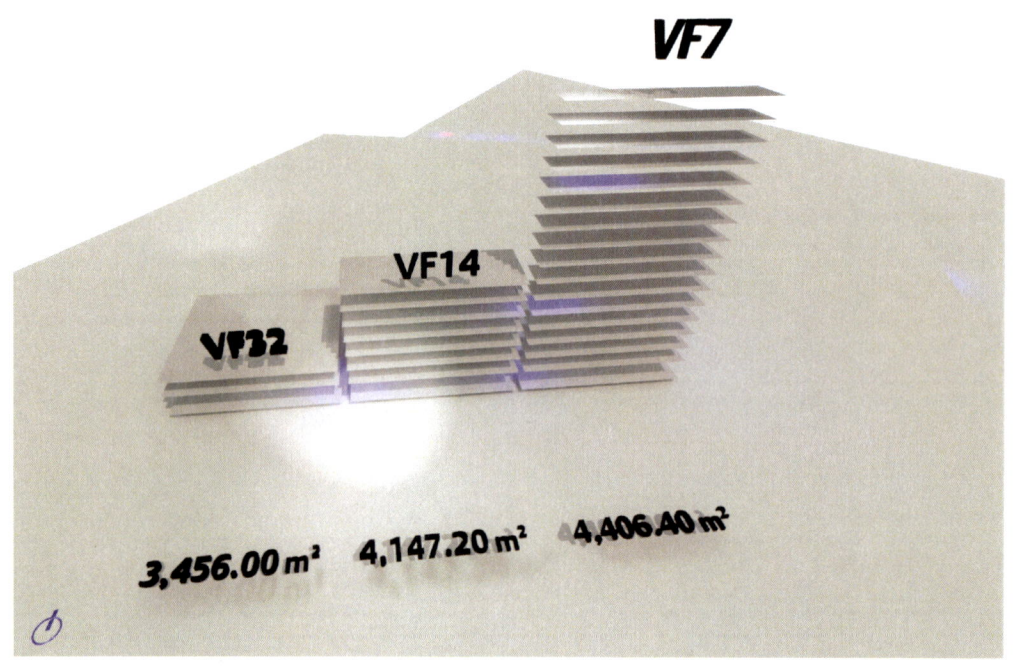

Fig.79.   Vertical Farm Resulting Cultivation Area

The different depths in the levels of each Vertical Farm type can be split into different zones, where necessary. This is necessary basically for Vertical Farm 32 and Vertical Farm 14 to obtain more specific values. The zoning is dependent on the decision to offset the outline of the Vertical Farms on the inside at a distance of 5m.

The difference in cultivation ares between VF7, VF14 and VF32 results from the ratio of groundfloor's volume. VF32 has a cultivation area of 3,456 m2, VF14 4,147.2 and VF7 with its 4,406.40 m2 has the highest number of the cultivation area.

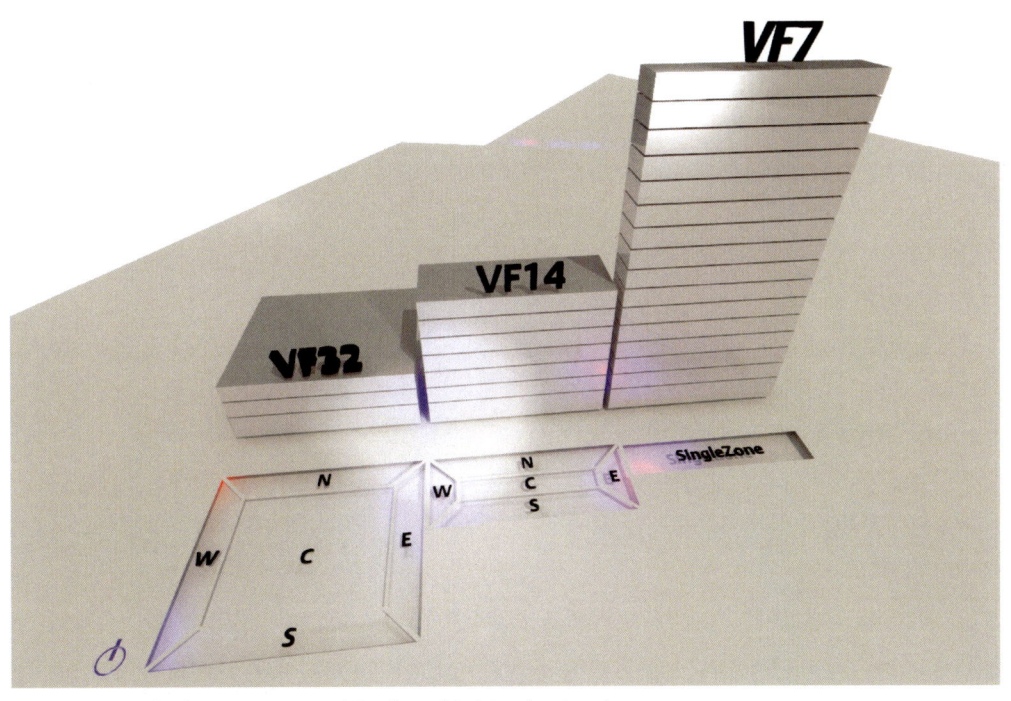

Fig.80. Vertical Farm Types and Zoning of Cultivation levels

DAYLIGHT SIMULATION ZONING - VF32

- At the center zone the outline is offset by 5 m from the building envelope,
- the north zone connects the northern edge points from the center zone to the building envelope,
- the east zone connects the eastern edge points from the center zone to the building envelope,
- the west zone connects the western edge points from the center zone to the building envelope,
- the south zone connects the southern edge points from the center zone to the building envelope.   Ground floor and first floor;
- Top floor zone (single zone)
- = 11 annual daylight simulation results (kWh/m²) for each building envelope material.

DAYLIGHT SIMULATION ZONING - VF14

- At the center zone, the outline is the offset by 5 m from the building envelope,
- the north zone connects the northern edge points from the center zone to the building envelope,
- the east zone connects the eastern edge points from the center zone to the building envelope,
- the west zone connects the western edge points from the center zone to the building envelope,
- the south zone connects the southern edge points from the center the zone to the building envelope.
- Ground floor and First Floor;
- The analysis results for the first floor are used for the intermediate floors between ground floor
- Top floor zone
- = 11 annual daylight simulation results (kWh/m^2) for each building envelope material.
- The analysis results for the first floor are used for the typical floors between ground floor and top floor

DAYLIGHT SIMULATION ZONING - VF7

- Ground floor, first floor, top floor
- = 3 annual daylight simulation results (kWh/m^2) for each building envelope material
- The analysis results for the first floor are used for intermediate floors between Ground floor

VERTICAL FARM - SIMULATION MODEL

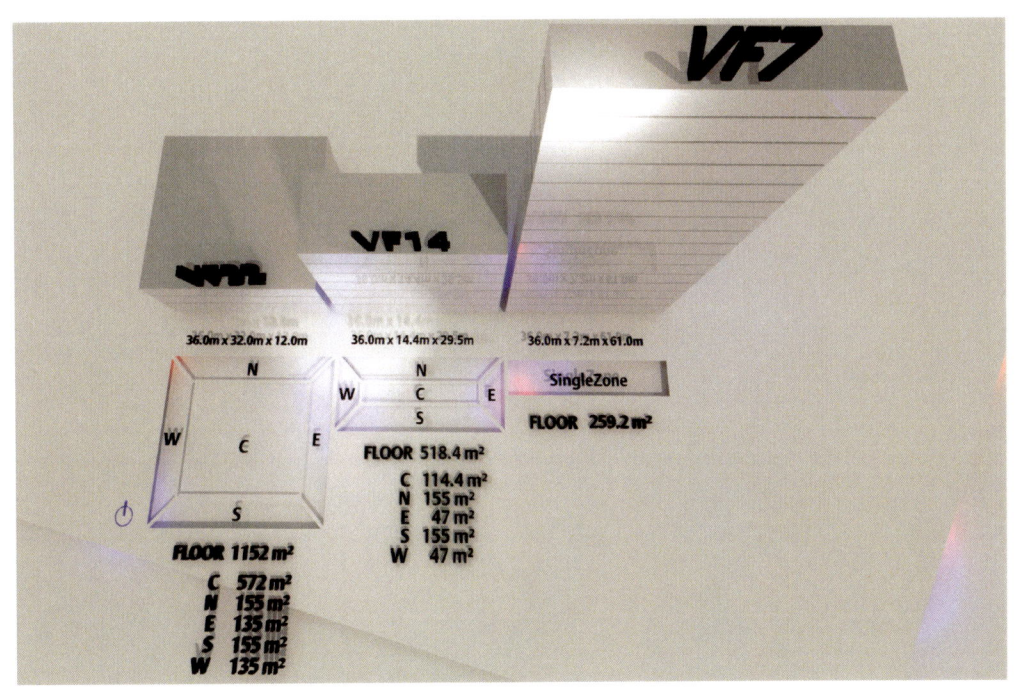

36.0m x 32.0m x 12.0m   36.0m x 14.4m x 29.5m   36.0m x 7.2m x 61.0m

```
      N                    N                  SingleZone
   W  C  E              W  C  E              FLOOR  259.2 m²
      S                    S
FLOOR 1152 m²         FLOOR 518.4 m²
    C  572 m²             C  114.4 m²
    N  155 m²             N  155 m²
    E  135 m²             E   47 m²
    S  155 m²             S  155 m²
    W  135 m²             W   47 m²
```

Fig.81.   Vertical Farm Types and Zoning Dimensions

233

5.4. Lighting Parameters

5.4.1. Lighting demand and Lighting schedule

As explained in Chapter 4.4.5 photoperiodism for *L.esculentum* growth is 16 h or 57,600 seconds/day. Lighting throughout the year starts at 4 am and ends at 8 pm. The lighting demand in kWh/m2 will be the indoor difference between the available daylight through the different facades and the optimum light supply throughout the crop rotation.

RADIANCE MATERIALS:

Each Vertical Farm is simulated for daylight availability with three different facade materials.
- Single glazing, standard E+ single glazing material
- Double glazing[1],[2]
- Double-ETFE

The resulting values for the building envelope radiance materials for daylight simulation are as follows:

SG: R-Transmittance: 0.85
      G-Transmittance: 0.85
      B-Transmittance: 0.85
      Refractive Index[3]: 1.52

DG: R-Transmittance: 0.91
      G-Transmittance: 0.91
      B-Transmittance: 0.91
      Refractive Index: 1.52

---

[1] https://www.groglass.com/img/media/Artglass_for_Architecture_Technical.pdf, retrieved 13.01.2015
[2] This double glazing material was selected because of its high transmission value and its low U-value, despite the author is conscious about the fact that these values physically are difficult to achieve.
[3] [The, Ed.] refractive index, also called index of refraction, measure of the bending of a ray of light when passing from one medium into another. (...). http://www.britannica.com/science/refractive-index, retrieved 15.07.2015

VERTICAL FARM - SIMULATION MODEL

ETFE:
- R-Transmittance: 0.85
- G-Transmittance: 0.85
- B-Transmittance: 0.85
- Refractive Index: 1.42

OTHER MATERIALS:

Facade construction:
- R-Reflectance: 0.90
- G-Reflectance: 0.90
- B-Reflectance: 0.90
- Roughness: 0.10
- Specularity: 0.10

Floors:
- R-Reflectance: 0.70
- G-Reflectance: 0.70
- B-Reflectance: 0.70
- Specular Reflection: 0.10
- Diffuse Transmission: 0.10
- Specular Transmission: 0.10
- Roughness: 0.10

Ceilings:
- R-Reflectance: 0.70
- G-Reflectance: 0.70
- B-Reflectance: 0.70
- Specular Reflection: 0.10
- Diffuse Transmission: 0.10
- Specular Transmission: 0.10
- Roughness: 0.10

Ground floor:
- R-Reflectance: 0.90
- G-Reflectance: 0.90
- B-Reflectance: 0.90
- Roughness: 0.10
- Specularity: 0.10

Simulation type:

The simulation type for the daylight simulation formula is the annual daylight simulation. Different setups were tested to refine the results files. Thereafter the following radiance-parameters were set:
Quality: 1
The number of ambient bounces was set to 2.
Ambient division, ambient supersamples, ambient resolution and ambient accuracy were kept untouched, or respectively worked with standard preferences.
Daysim parameters were set to 1 to obtain radion results (W/m2).

DAYLIGHT SIMULATION FILES:

The daylight simulation results in W/m2 for every zone of the different Vertical Farm typologies were combined in a single excel-file. The different results of every set-up control point (varying from 47 to 1,152 whereas the control points correspond to the area of the specific zones) were summarized and divided by the number of control points to obtain the average available daylight radiation number for every defined Vertical Farm-zone for every hour of the year (8,760).

These result files were then multiplied by 0.5 to obtain the PAR-value W_{PAR}/m2 from 400 - 700 nm) out of the total solar radiation values (W/m2 from 30 - 3,000nm).
Based on the PPFD requirement ($\mu mol/m^{-2}/s^{-1}$) (Chapter 4.4.4) the DLI need was determined as weekly average from which value the W_{PAR}/m2-value required per day was distributed through the photoperiod of 16 hours starting at 4 am and ending at 8 pm. The time schedule from 8 pm to 4 am (respiration period) was set to 0.

The difference between the value needed in W_{PAR}/m2 and the available value in W_{PAR}/m2 was divided by the lighting power demand for Lumigrow 325 Pro and additionally divided by 100 to obtain values between 0.00 and 1.00 to give the lighting control for every single hour of a year.

The lighting control for the top floor levels was kept on 0 if the available DLI exceeded the lighting amount needed within a specific day during the crop rotation. This exception only was adapted on the top floor because daylight availability in all the other zones was not distributed evenly on the surface of the control points, e.g. sunlight availability close to the facade sometimes exceeds the saturation point while plants at 5 m distance to the facade (the building depth of VF7 is 7.2m) were not provided with all the daylight they needed.

Two different csv[1]-schedules for lighting control were thus created to be used for thermal heating simulation in EnergyPlus:

csv-schedules for the top floor, setting lighting control to 0 if DLI was higher than the light quantity needed on a specific day during the crop rotation

csv-schedules for the remaining zones, not taking a possible DLI-excess into consideration.

---

[1] CSV, comma separated values, files are commonly used to transport large amounts of tabular data between either companies or applications that are not directly connected. The files are easily editable using common spreadsheet applications like Microsoft Excel. http://www.csvreader.com/csv_format.php, retrieved 20.07.2015

5.5. Heating Parameters

5.5.1. EnergyPlus Material Parameters

Starting from the building mass and the zone Honeybee finds the adjacencies of the geometry and divides it into the following building components:

- exterior walls
- abstract interior zoning walls
- ground floorfloors (of the storeys)
- ceiling (of the storeys)
- windows
- roof.

All the exterior walls and the roof obtain the same glazing parameters in every direction (north, east, south and west), with this at the maximum theoretical value of 99% (provided by Honeybee) whereas 1 % is opaque construction material. The glass material has a transmissivity of between 85% (single glazing) and 91% (a special coated double glazing window, see p. 237 and Appendix).

The solar heat gain coefficient (SHGC) indicates the quantity of energy from the sun transmitted through the window as heat. As the SHGC increases, the solar gain through the window increases. A value of 0.6 or higher is desired for greenhouses. Visible transmittance (VT) indicates the percentage of the visible portion of the electromagnetic solar spectrum transmitted through the glass. The visual transmittance of 0.9 is the value of a 3 mm single layer glass. A VT-value of 0.8 was chosen for the simulation model. The result is then within the range of a double-layer glass and an alternative polycarbonate glass.

For reasons of simplification construction materials (exterior walls and floors) all have a reflectance factor of 0.85 and a speculatory factor of 0.09.

Three values here are of basic interest:

- U-value (heat transfer coefficient) represents the quantity of energy transferring through an area of 1m2 per second, when air temperature on both sides of the material differs by 1K.[1]
- SHGC (Solar heat gain coefficient) or g-value is the fraction of solar radiation admitted through a window (...) either transmitted directly and/or absorbed, and subsequently released as heat inside (...). The lower the SHGC, the less solar heat it transmits and the greater the shading ability. A product with a high SHGC rating is more effective at collecting solar heat during the winter. A product with a low SHGC rating is more effective at reducing cooling loads during the summer by blocking heat gain from the sun. (...).[2]
- VT (Visual Transmittance): It is the fraction of the visible spectrum of sunlight, from 380 to 720 nm (close to PAR values which is 400 - 700 nm), "weighted by the sensitivity of the human eye that is transmitted through the glazing of a window (...). A product with a higher VT transmits more visible light. VT is expressed as a number between 0 and 1. The VT needed for a window (...) should be determined by the applicable (...) daylight requirements."[3]

---

[1] Bauphysik (VO), Prof. Brian Cody, p.52 ff, Institute for Buildings and Energy, Graz University of Technology, Austria, 1st edition, 2009
[2] ibid. retrieved 05.05.2014
[3] http://energy.gov/energysaver/articles/energy-performance-ratings-windows-doors-and-skylights, retrieved 05.05.2014

SINGLE GLAZING:

Irrespective of the fact that single glazing cannot be used for multi-storey buildings, the author nevertheless examined the results for this type of building envelope to obtain values comparable to those for traditional greenhouses. The material properties of single glazing used for daylight simulation are:

U-Value: 5,88 W/m2 K
SHGC: 0,8
VT: 0,85

ETFE:

This building envelope consists in double-layered ETFE-membranes with a high visual transmittance for natural daylight and a low g-value. Triple-layered ETFE-membranes would optimize the U-value but additionally reduce the VT-value.

U-Value: 2,9 W/m2 K
SGHC: 0,65
VT: 0,85

DOUBLE GLAZING:

U-Value: 1,7 W/m2 K
SGHC: 0,7
VT: 0,91

For this type of facade material a glazing type was used to get a building envelope primarily with a high VT-value and an optimized U-value. A thin film layer is sputtered on a glass surface to reduce visible reflection and to increase transition at the same time. "A side effect of this AR stack is increased reflectance in the near IR part of the spectrum. That being said, our coating does not affect glass emissivity value, therefore U-value is the same as that of untreated glass."[1] The following values are for double glazed units (DGU's) for Artglass ARTM DGU's with NIR-Block/Low-E[2] (argon gas filled)[3]:

---

[1] Producer infromation provided by Groglass Customer Service, retrieved 14.07.2015. Further technical information on http://www.groglass.com/img/media/Artglass_for_Architecture_Technical.pdf, retrieved 09.02.2015, „What regards the materials used, our coating first of all is intended for performance in visible light spectrum. The most commonly used materials for this purpose are silicon dioxide and titanium dioxide. (...)."
[2] https://www.groglass.com/img/media/diamond_nir-block_uv_insert.pdf, retrieved 14.07.2015
[3] see Appendix for product data sheet

OTHER MATERIALS:

1. FACADE MATERIALS (FROM OUTSIDE TO INSIDE):

Facade Material: Steelconstruction
- Roughness — Smooth
- Thickness(m) — 0,15
- Conductivity (W/m-K) — 44,95
- Density (kg/m3) — 7.650
- Specific Heat (J/kg-K) — 410
- Thermal Absorptance — 0,9
- Solar Absorptance — 0,6
- Visible Absorptance — 0,6

Facade Material: Insulation
- Roughness — Medium rough
- Thickness(m) — 0,7
- Conductivity (W/m-K) — 0,03
- Density (kg/m3) — 265
- Specific Heat (J/kg-K) — 836
- Thermal Absorptance — 0,9
- Solar Absorptance — 0,7
- Visible Absorptance — 0,7

Facade Material: Gypsum (double layered)
- Roughness — Smooth
- Thickness(m) — 0,0125
- Conductivity (W/m-K) — 0,16
- Density (kg/m3) — 785,9
- Specific Heat (J/kg-K) — 836
- Thermal Absorptance — 0,9
- Solar Absorptance — 0,4
- Visible Absorptance — 0,4

Facade Material: Metal Siding
- Roughness — Smooth
- Thickness(m) — 0,0015
- Conductivity (W/m-K) — 44,95
- Density (kg/m3) — 7.688,86
- Specific Heat (J/kg-K) — 410
- Thermal Absorptance — 0,9
- Solar Absorptance — 0,6
- Visible Absorptance — 0,6

Resulting Facade Construction:
U-Value: 0,4
R-Value: 2,49

2: FLOOR AND CEILING MATERIALS:

Floor and Ceiling Materials: Metal Siding
- Roughness — Smooth
- Thickness(m) — 0,0015
- Conductivity (W/m-K) — 44,95
- Density (kg/m3) — 7.688,86
- Specific Heat (J/kg-K) — 410
- Thermal Absorptance — 0,9
- Solar Absorptance — 0,6
- Visible Absorptance — 0,6

Floor and Ceiling Materials: Lightweight Concrete
- Roughness — Medium Rough
- Thickness(m) — 0,01
- Conductivity (W/m-K) — 0,53
- Density (kg/m3) — 1280
- Specific Heat (J/kg-K) — 840
- Thermal Absorptance — 0,9
- Solar Absorptance — 0,5
- Visible Absorptance — 0,5

Floor and Ceiling Materials: Insulation
- Roughness — Medium Rough
- Thickness(m) — 0,05
- Conductivity (W/m-K) — 0,03
- Density (kg/m3) — 43
- Specific Heat (J/kg-K) — 1210
- Thermal Absorptance — 0,9
- Solar Absorptance — 0,7
- Visible Absorptance — 0,7

Floor and Ceiling Materials: Steelconstruction
- Roughness — Smooth
- Thickness(m) — 0,15
- Conductivity (W/m-K) — 44,95
- Density (kg/m3) — 7.650
- Specific Heat (J/kg-K) — 410
- Thermal Absorptance — 0,9

Solar Absorptance 0,6
Visible Absorptance 0,6

Facade Material: Gypsum (double layered)
- Roughness — Smooth
- Thickness(m) — 0,0125
- Conductivity (W/m-K) — 0,16
- Density (kg/m3) — 785,9
- Specific Heat (J/kg-K) — 836
- Thermal Absorptance — 0,9
- Solar Absorptance — 0,4
- Visible Absorptance — 0,4

Floor and Ceiling Materials: Metal Siding
- Roughness — Smooth
- Thickness(m) — 0,0015
- Conductivity (W/m-K) — 44,95
- Density (kg/m3) — 7.688,86
- Specific Heat (J/kg-K) — 410
- Thermal Absorptance — 0,9
- Solar Absorptance — 0,6
- Visible Absorptance — 0,6

Resulting Floor and Ceiling Construction (=Adiabatic):
U-Value: 0,5
R-Value: 2,01

GROUNDFLOOR MATERIALS (FROM OUTSIDE TO INSIDE):
Groundfloor Material: Concrete
- Roughness — Medium Rough
- Thickness(m) — 0,1
- Conductivity (W/m-K) — 0,53
- Density (kg/m3) — 1.280
- Specific Heat (J/kg-K) — 840
- Thermal Absorptance — 0,9
- Solar Absorptance — 0,5
- Visible Absorptance — 0,5

Groundfloor Material: Insulation
- Roughness — Medium Rough
- Thickness(m) — 0,1

Conductivity (W/m-K) 0,03
Density (kg/m3) 43
Specific Heat (J/kg-K) 1210
Thermal Absorptance 0,9
Solar Absorptance 0,7
Visible Absorptance 0,7

Groundfloor Material: Heavyweight Concrete
Roughness Medium Rough
Thickness(m) 0,2
Conductivity (W/m-K) 1,95
Density (kg/m3) 2.240
Specific Heat (J/kg-K) 900
Thermal Absorptance 0,9
Solar Absorptance 0,5
Visible Absorptance 0,5

Groundfloor Material: Insulation
Roughness Medium Rough
Thickness(m) 0,05
Conductivity (W/m-K) 0,03
Density (kg/m3) 43
Specific Heat (J/kg-K) 1210
Thermal Absorptance 0,9
Solar Absorptance 0,7
Visible Absorptance 0,7

Groundfloor Material: Lightweight Concrete
Roughness Medium Rough
Thickness(m) 0,1
Conductivity (W/m-K) 0,53
Density (kg/m3) 1.280
Specific Heat (J/kg-K) 840
Thermal Absorptance 0,9
Solar Absorptance 0,5
Visible Absorptance 0,5

Groundfloor Material: Metal Siding
Roughness Smooth
Thickness(m) 0,0015
Conductivity (W/m-K) 44,95
Density (kg/m3) 7.688,86
Specific Heat (J/kg-K) 410

VERTICAL FARM - SIMULATION MODEL

Thermal Absorptance		0,9
Solar Absorptance		0,6
Visible Absorptance		0,6

Resulting Groundfloor Construction:
U-Value: 0,18
R-Value: 5,48

Floors:	R-Reflectance:	0.70
	G-Reflectance:	0.70
	B-Reflectance:	0.70
	Specular Reflection:	0.10
	Diffuse Transmission	0.10
	Specular Transmission	0.10
	Roughness:	0.10
Ceilings:	R-Reflectance:	0.70
	G-Reflectance:	0.70
	B-Reflectance:	0.70
	Specular Reflection:	0.10
	Diffuse Transmission	0.10
	Specular Transmission	0.10
	Roughness:	0.10
Ground floor:	R-Reflectance:	0.90
	G-Reflectance:	0.90
	B-Reflectance:	0.90
	Roughness:	0.10
	Specularity:	0.10
Interior walls[1].	U-Value:	5.88
	SHGC:	0.99
	VT:	0.99

---

[1] Abstract interior zoning walls to achieve specific heating demands for each zone

5.5.2. EnergyPlus Simulation Parameters

Occupancy schedules (=growing period =photoperiod):
    16 hours starting at 4 am and ending at 8 pm
Equipment schedules:
    16 hours starting at 4 am and ending at 8 pm
Heating schedules:
    16 hours starting at 4 am and ending at 8 pm

Two different heating schedules are compared:

HeatingHigh (HH-schedule)
HeatingLow (HL-schedule), explained in more detail in Chapter 4.3.5 and 4.4.4.

Occupancy schedules:
    16 hours starting at 4 am and ending at 8 pm
Cooling schedules:
    no cooling considered

Loads:
    5W equipment loads/m2
    325 W lighting load (Lumigrow 325 PRO)
    65.64 W lighting load/m2[1]
    0,0144 people/m2
    Infiltration rate is set to 0
    Ventilation per area and person is set to 0

Weather file:
    epw-file of Vienna[2]

HB-Context:
    Standalone Vertical Farm with no urban context

HB-Zones:
    Vertical Farm VF32:   11
    Vertical Farm VF14:   36
    Vertical Farm VF7:   16

---

1   See Chapter 4.4.4 Tab.1 Lumigrow 325 PRO - PPFD-DLI values depending on fixture height; 65.64 W = 300 µmol/m-2/s-1
2   http://apps1.eere.energy.gov/buildings/energyplus/cfm/weather_data3.cfm/region=6_europe_wmo_region_6/country=AUT/cname=Austria, retrieved 12.12.2014

5.5.3. Limitations and Restrictions

Total solar radiation in W/m2, the inquiries into missing light in W_{PAR} for constructing the lighting schedule and heating demand simulation are set up by not considering the following parameters, which surely distort actual values that are to be expected within a fully equipped stacked greenhouse. A primary limitation of this simulation model is the fact of plant growth and its influence in daylight distribution. A parametric plant growth model to simulate auto-shading through crop rotation was developed but did not find a place in the work of this dissertation.

L. esculentum has a high water requirement, up to 1.39 m3/m2/a[1]. The transpiration of the plants and its impact on air humidity and as a consequence the resultant heating demand is not represented in the final results.
During crop rotation the biomass of the plants increases and also has an influence on the heating demand (heat absorption). This effect, which most probably influences the heating demand, was not considered.

On the other hand potentials also exist for increasing the energy efficiency of a Vertical Farm, which will not be integrated in the final statistics by the end of this chapter. They are seen as factors which probably balance out the above mentioned factors of influence, which would increase the energy demand. The following decreasing factors for energy demand are:

- The shutter sequence of LED-lights allows a reduction of the electricity demand by 10%. This means "that the LED are frequently turned on and off with a defined frequency. (…) [S]huttering of LED do not affect the development and growing of plants (…)."[2]
- Energy screens applied in greenhouses can reduce heating demand by up to 95%, depending on the heat requirement for the specific plant types.[3]
- The simulation analysis period extends over an entire year. The fact that yield is diminishing by the end of the crop rotation and, like all greenhouses, the Vertical Farm must be cleaned and disinfected to prepare for the next crop rotation, a cleaning period of one month could be assumed. This time is equivalent to the procedure of "Zeiler"-greenhouse practices. This will additionally reduce hea-

---

[1] Calculating WUE of 66 kg m3 as a value of highest efficiency obtained in high tech greenhouses in the Netherlands. 2013, Food and Agriculture Organization of the United Nations, Plant Production and Protection Division, W.Baudoin et al. „Good Agricultural Practices for greenhouse vegetable crops - Principles for Mediterranean climate areas", p.130

[2] BANERNJEE, C. 2012. Market Analysis for Terrestrial Application of Advanced Bio-Regenerative Modules: Prospects for Vertical Farming. Master of Science Masterarbeit, Rheinische Friederichs-Wilhems-Universität, Hohe Landwirtschaftliche Fakultät. p.78

[3] http://www.zineg.net/html/hannover.html, retrieved 12.09.2015

ting demand up to 25% on a farm average if the "cleaning month" is reserved for December.

Future research is necessary, especially by adapting additional simulation parameters like plant growth throughout the crop rotation and its effects on lighting demand, changes in CO2-rates through the photoperiod and plant respiration, changes of air humidity related to photosynthesis (both evaporation and transpiration).

5.6. Lighting Demand - Simulation Results

5.6.1. Annual Results

Based on the assumptions of the needed PPFD for *L. esculentum* in Chapter 4.3.3 and the daylight availability taken from climate data of Vienna in Chapter 5.3.2 we can create a sigmoid growing curve throughout the year and all the phases of the crop rotation defined in Chapter 4.3.4. The table on the next page parses the available total solar radiation in Vienna throughout the year[1] and the day length in seconds to obtain the WPAR/m2, DLI and PPFD needed.

On examining the results of the annual solar radiation we see that from the available 561.19 kWhPAR/m2 only 268.35 kWhPAR/m2 is needed for optimum *L. esculentum* growth.

On a first look one might be inclined to say that the total availability would appear to be better than assumed. For optimum growing conditions (Chapter 4.4). DLI must be distributed over 16 hours. Artificial lighting must, therefore, substitute the missing daylight within the photoperiod of 57600 seconds/d. Comparing typical days in representative spaces of VF32 explains the effective artificial lighting demand to cover the plants with their specific DLI needs, seen on p. 246 sand 247.

---

[1] for average weekly data see Appendix „Additional Simulation Results"

5.6.2. Lighting Schedule

To guarantee optimum plant growth and crop yield *L. esculentum* is supplied with light throughout for 16 hours/day to make full use of the maximum photoperiod and to reserve 8 h/day for respiration. The artificial lighting therefore covers the entire period from 4 am to 8 pm and dimmed by the amount of the hourly penetrating daylight simulation resulting for each zone.

The following diagrams show typical days in the representative zones of VF32. The data show the reaction of the artificial lighting to daylight availability, the lighting schedule and the light supply needed to assure the DLI required for optimum plant growth.

Fig. 83 below visualizes a typical winter's day. It shows the daylight supply and artificial light requirement on the top level of VF32. Artificial lighting starts at 4 am (WPAR/m2 [i] LED) and supplies the area with a constant radiation in WPAR/m2 to reach the needed DLI (WPAR/m2 [i] for DLI) for the specific crop rotation period. At sunrise (W/m2 [o]) the power supply for artificial lighting is reduced by the percentage of available daylight. This interdependency between daylight availability and power demand for the LED growth lamps can be compared with the following diagrams, showing a typical equinox-day in Vienna and a summer day, all in the same zone.

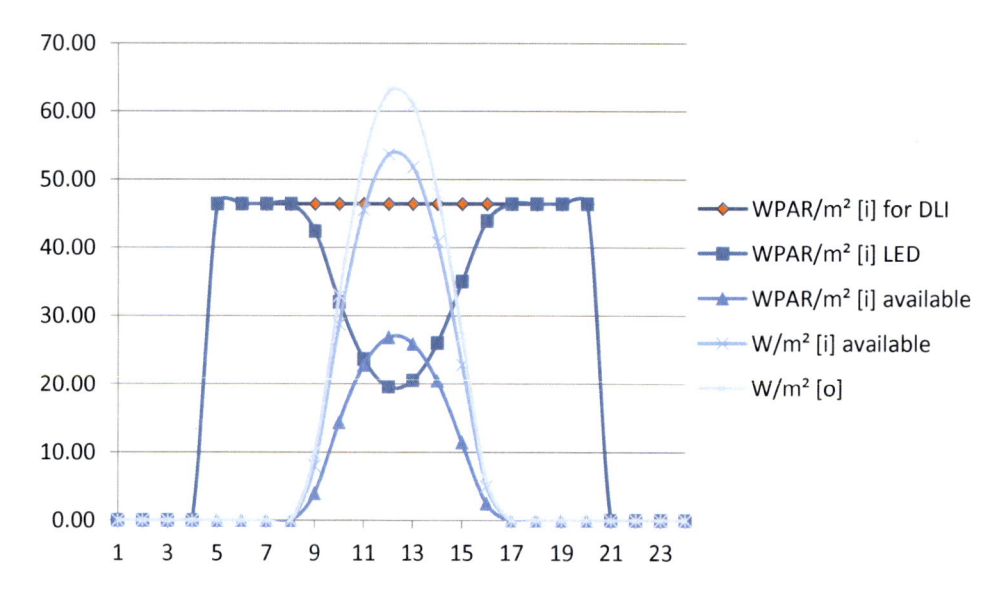

Fig.82.   Typical winter's day, Radiation availability, DLI and LED coverage on the top level of VF32

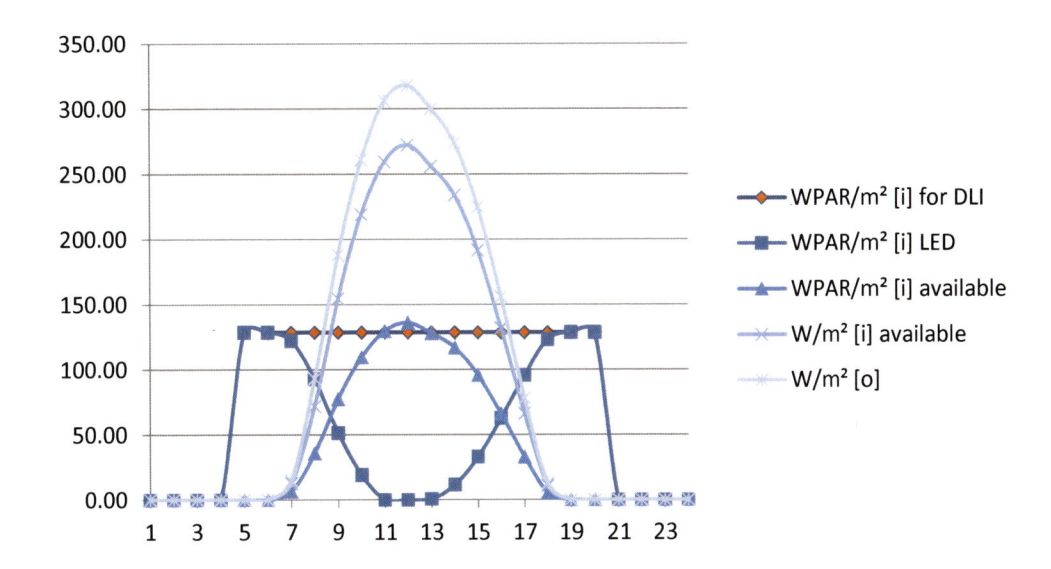

Fig.83. Typical equinox day, Radiation availability, DLI and LED coverage on the top level of VF32

On a typical spring day LED lighting supply again starts at the same hour and is dimmed after sunrise, in this diagram between 6 and 7 am, when sunlight is already reaching the zone floor area. Around noon the available PAR surpasses the radiation level requirement and the LED lighting is reduced to zero.

In summer (diagram below), when the sum of PAR reaching the zone surface exceeds the DLI needed for optimum plant growth, lighting schedule is also reduced to 0. In this case, in VF32 on the top level and power demand for the entire day also is 0.

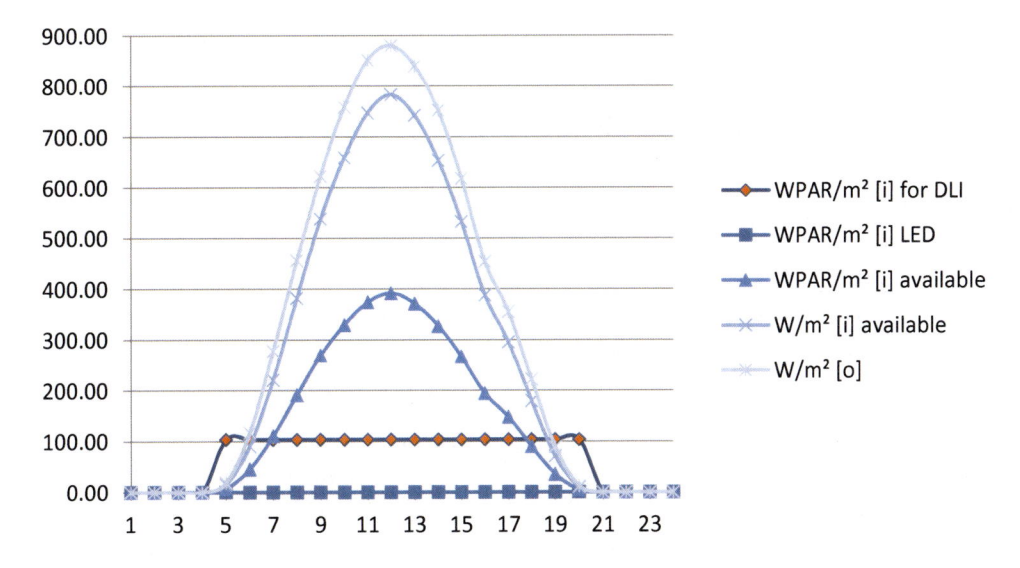

Fig.84. Typical summer's day, Radiation availability, DLI and LED coverage on the top level of VF32

Corresponding to Fig. 83, 84 and 85 on this page, Tab. 12, 13 and 14 visualize the procedure and method for creating csv-schedules to create the lighting schedule for the simulation model for specific days based on EnergyPlus epw-weather files.

The first column of the figures represented on these two pages shows the specific annual hours of the selected day, HOY (Hour Of the Year). The next two columns contain the specific hour or the time step. Wh/m2 are the retrieved solar radiation results of a specific zone at this hour. Detailed annual radiation results of representative levels and zones can be found in the Appendix (Additional Simulation Results).
Wh/m2 is then converted to WPAR/m2. This corresponds to the average available daylight penetration of a specific zone. Hourly data of WPAR/m2 needed are retrieved from DLI needed for the specific period of the crop rotation (p.208-209 and also see Appendix p.401 average DLI/week).

Lighting demand for December 21st: 6.33 [DLI needed in mol/m2/d]
* 1,000,000 [mol to μmol] / 4.57 [μmol to WPAR] / 57,600 [photoperiod in seconds] =
24.074 WhPAR/m2 needed

The difference between WPAR/m2 needed and WPAR/m2 (daylight availability) is the hourly lighting demand supplied by LED. Power demand for Lumigrow LED PRO 325 is 325 W. The csv-schedule now controls and dims LED power supply based on the difference between the demand for photosynthesis for optimum plant growth (WPAR/m2 needed) and the solar radiation results for each specific zone.

HOY	Start	End	Wh/m²	WhPAR/m²	lighting	WhPAR/m² needed	csv	DLI (mol/m2/d)	
8496	00:00	01:00	0.00	0.00	0.00	0.00	0.00	6.33	needed
8495	01:00	02:00	0.00	0.00	0.00	0.00	0.00	0.45	achieved
8494	02:00	03:00	0.00	0.00	0.00	0.00	0.00	5.88	LED cover
8493	03:00	04:00	0.00	0.00	0.00	0.00	0.00		
8492	04:00	05:00	0.00	0.00	24.07	24.07	0.07		
8491	05:00	06:00	0.00	0.00	24.07	24.07	0.07		
8490	06:00	07:00	0.00	0.00	24.07	24.07	0.07		
8489	07:00	08:00	0.00	0.00	24.07	24.07	0.07		
8488	08:00	09:00	8.00	4.00	20.07	24.07	0.06		
8487	09:00	10:00	28.76	14.38	9.69	24.07	0.03		
8486	10:00	11:00	45.58	22.79	1.28	24.07	0.00		
8485	11:00	12:00	53.70	26.85	0.00	24.07	0.00		
8484	12:00	13:00	51.83	25.91	0.00	24.07	0.00		
8483	13:00	14:00	40.88	20.44	3.63	24.07	0.01		
8482	14:00	15:00	22.84	11.42	12.65	24.07	0.04		
8481	15:00	16:00	5.00	2.50	21.57	24.07	0.07		
8480	16:00	17:00	0.00	0.00	24.07	24.07	0.07		
8479	17:00	18:00	0.00	0.00	24.07	24.07	0.07		
8478	18:00	19:00	0.00	0.00	24.07	24.07	0.07		
8477	19:00	20:00	0.00	0.00	24.07	24.07	0.07		
8476	20:00	21:00	0.00	0.00	0.00	0.00	0.00		
8475	21:00	22:00	0.00	0.00	0.00	0.00	0.00		
8474	22:00	23:00	0.00	0.00	0.00	0.00	0.00		
8473	23:00	00:00	0.00	0.00	0.00	0.00	0.00		

Tab.12. CSV-schedule for lighting, Top Level, VF32, for the 21st december

HOY	Start	End	Wh/m²	WhPAR/m²	lighting	WhPAR/m² needed	csv	DLI (mol/m2/d)	
1896	00:00	01:00	0.00	0.00	0.00	0.00	0.00	17.28	needed
1897	01:00	02:00	0.00	0.00	0.00	0.00	0.00	8.60	achieved
1898	02:00	03:00	0.00	0.00	0.00	0.00	0.00	8.68	LED cover
1899	03:00	04:00	0.00	0.00	0.00	0.00	0.00		
1900	04:00	05:00	0.00	0.00	65.65	65.65	0.20		
1901	05:00	06:00	0.00	0.00	65.65	65.65	0.20		
1902	06:00	07:00	10.99	5.49	60.15	65.65	0.19		
1903	07:00	08:00	57.73	28.87	36.78	65.65	0.11		
1904	08:00	09:00	94.84	47.42	18.23	65.65	0.06		
1905	09:00	10:00	105.25	52.63	13.02	65.65	0.04		
1906	10:00	11:00	324.10	162.05	0.00	65.65	0.00		
1907	11:00	12:00	493.27	246.63	0.00	65.65	0.00		
1908	12:00	13:00	557.58	278.79	0.00	65.65	0.00		
1909	13:00	14:00	447.63	223.81	0.00	65.65	0.00		
1910	14:00	15:00	264.96	132.48	0.00	65.65	0.00		
1911	15:00	16:00	74.62	37.31	28.33	65.65	0.09		
1912	16:00	17:00	38.73	19.36	46.28	65.65	0.14		
1913	17:00	18:00	6.96	3.48	62.17	65.65	0.19		
1914	18:00	19:00	0.00	0.00	65.65	65.65	0.20		
1915	19:00	20:00	0.00	0.00	65.65	65.65	0.20		
1916	20:00	21:00	0.00	0.00	0.00	0.00	0.00		
1917	21:00	22:00	0.00	0.00	0.00	0.00	0.00		
1918	22:00	23:00	0.00	0.00	0.00	0.00	0.00		
1919	23:00	00:00	0.00	0.00	0.00	0.00	0.00		

Tab.13. CSV-schedule for lighting, Top Level, VF32, for the 21st march

L.esculentum requires most light during the period of ripening to first harvest. A DLI of 17.28 mol/m2 has to be provided, which corresponds to 65.54 WPAR/m2 within a photoperiod of 57,600 seconds a day. What we see in Tab. 13 and 14, although DLI required is much higher compared to the end of the crop rotation (Tab.12), DLI achieved through the facade and the rooftop of the top level of VF32 comes close to the light quantity needed for the specific period of the crop rotation. In June, where the needed quantity of the light is already diminishing and daylight availability is the highest, only around 25% of the light has to be supplied artificially with LED compared to March 21st, or less than 50% compared to December 21st. On typical summer days with 0-10% coverage, csv is turned to 0 when DLI available exceeds DLI needed.

HOY	Start	End	Wh/m²	WhPAR/m²	lighting	WhPAR/m² needed	csv	DLI (mol/m2/d)	
4104	00:00	01:00	0.00	0.00	0.00	0.00	0.00	13.53	needed
4105	01:00	02:00	0.00	0.00	0.00	0.00	0.00	11.27	achieved
4106	02:00	03:00	0.00	0.00	0.00	0.00	0.00	2.26	LED cover
4107	03:00	04:00	0.00	0.00	0.00	0.00	0.00		
4108	04:00	05:00	7.84	3.92	47.50	51.42	0.15		
4109	05:00	06:00	45.90	22.95	28.47	51.42	0.09		
4110	06:00	07:00	109.22	54.61	0.00	51.42	0.00		
4111	07:00	08:00	153.68	76.84	0.00	51.42	0.00		
4112	08:00	09:00	170.89	85.45	0.00	51.42	0.00		
4113	09:00	10:00	159.96	79.98	0.00	51.42	0.00		
4114	10:00	11:00	236.30	118.15	0.00	51.42	0.00		
4115	11:00	12:00	305.98	152.99	0.00	51.42	0.00		
4116	12:00	13:00	355.30	177.65	0.00	51.42	0.00		
4117	13:00	14:00	379.53	189.76	0.00	51.42	0.00		
4118	14:00	15:00	367.94	183.97	0.00	51.42	0.00		
4119	15:00	16:00	322.56	161.28	0.00	51.42	0.00		
4120	16:00	17:00	274.44	137.22	0.00	51.42	0.00		
4121	17:00	18:00	175.92	87.96	0.00	51.42	0.00		
4122	18:00	19:00	70.98	35.49	15.93	51.42	0.05		
4123	19:00	20:00	10.99	5.49	45.93	51.42	0.14		
4124	20:00	21:00	0.00	0.00	0.00	0.00	0.00		
4125	21:00	22:00	0.00	0.00	0.00	0.00	0.00		
4126	22:00	23:00	0.00	0.00	0.00	0.00	0.00		
4127	23:00	00:00	0.00	0.00	0.00	0.00	0.00		

Tab.14. CSV-schedule for lighting, Top Level, VF32, for the 21st june

5.6.3. VF32 - Lighting Demand - Daysim Results

VF 36 - SG - ANNUAL DAYLIGHT AVAILABILITY (kWh/m2/a)

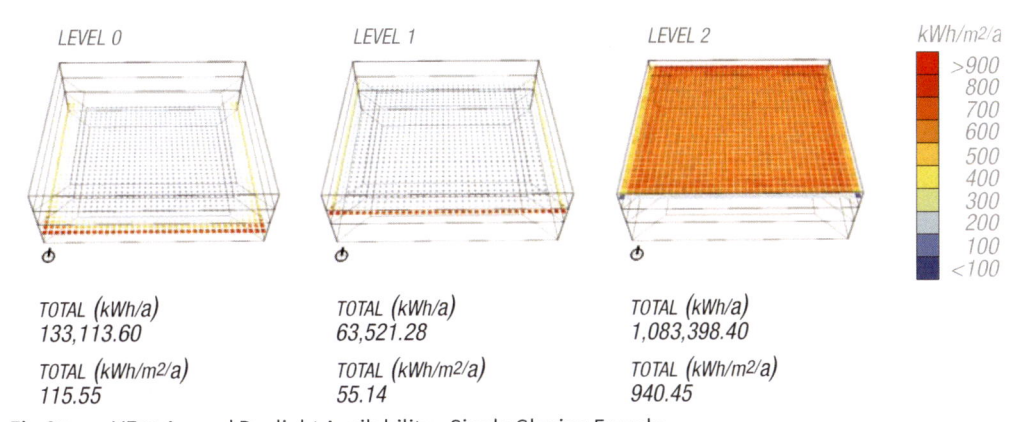

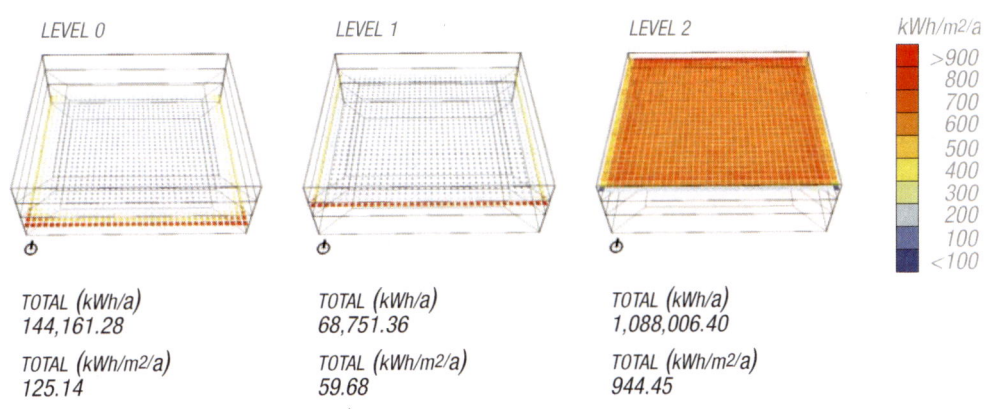

LEVEL 0

TOTAL (kWh/a)
133,113.60

TOTAL (kWh/m2/a)
115.55

LEVEL 1

TOTAL (kWh/a)
63,521.28

TOTAL (kWh/m2/a)
55.14

LEVEL 2

TOTAL (kWh/a)
1,083,398.40

TOTAL (kWh/m2/a)
940.45

Fig.85. VF32 Annual Daylight Availability - Single Glazing Facade

VF 36 - ETFE - ANNUAL DAYLIGHT AVAILABILITY (kWh/m2/a)

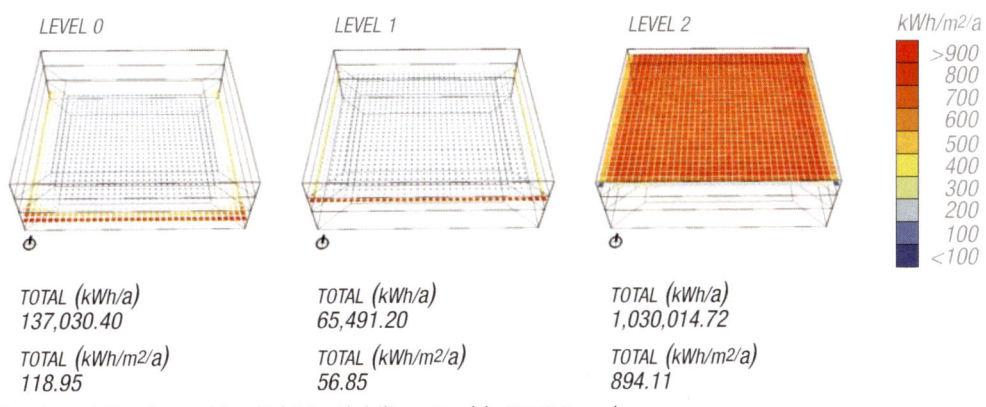

LEVEL 0

TOTAL (kWh/a)
137,030.40

TOTAL (kWh/m2/a)
118.95

LEVEL 1

TOTAL (kWh/a)
65,491.20

TOTAL (kWh/m2/a)
56.85

LEVEL 2

TOTAL (kWh/a)
1,030,014.72

TOTAL (kWh/m2/a)
894.11

Fig.86. VF32 Annual Daylight Availability - Double ETFE Facade

VF 36 - DG - ANNUAL DAYLIGHT AVAILABILITY (kWh/m2/a)

LEVEL 0

TOTAL (kWh/a)
144,161.28

TOTAL (kWh/m2/a)
125.14

LEVEL 1

TOTAL (kWh/a)
68,751.36

TOTAL (kWh/m2/a)
59.68

LEVEL 2

TOTAL (kWh/a)
1,088,006.40

TOTAL (kWh/m2/a)
944.45

Fig.87. VF32 Annual Daylight Availability - Double Glazing Facade

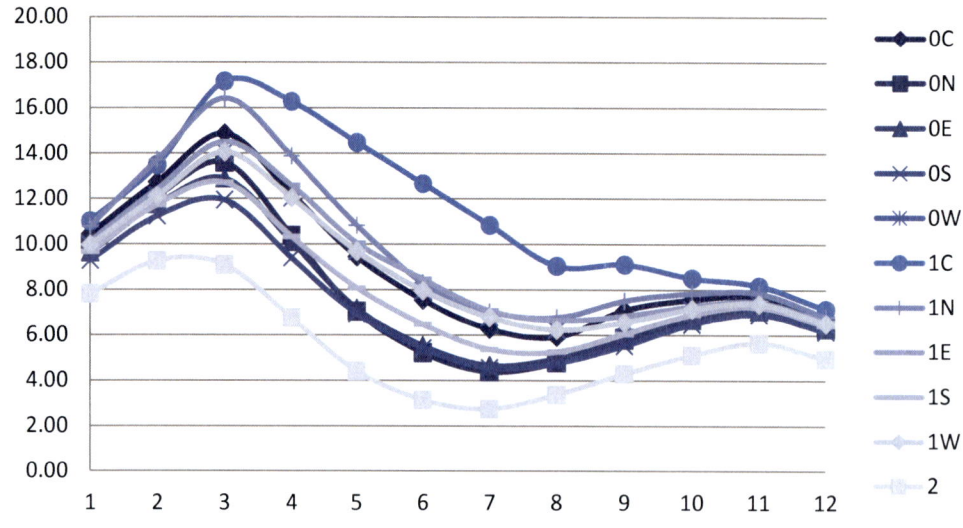

Fig.88. VF32 - Lighting Demand (kWh/m2/month), Single Glazing Facade, all zones

The power demand of the simulation results renders the sigmoid growing curve readable. Only after the summer power demand starts to rise again through until November when it declines once again.

The biggest differences of energy need are between the central zone 1C and Level 2, with values from 17.15 kWh in March during ripening to first harvest, while on Level 2 only 9.09 kWh are needed.

In this period of crop rotation on the ground floor in the central zone 0C approx. 3 kWh less is needed (14.86 kWh). This difference is explained through the floor height which is 1.5 m higher than on level 1.

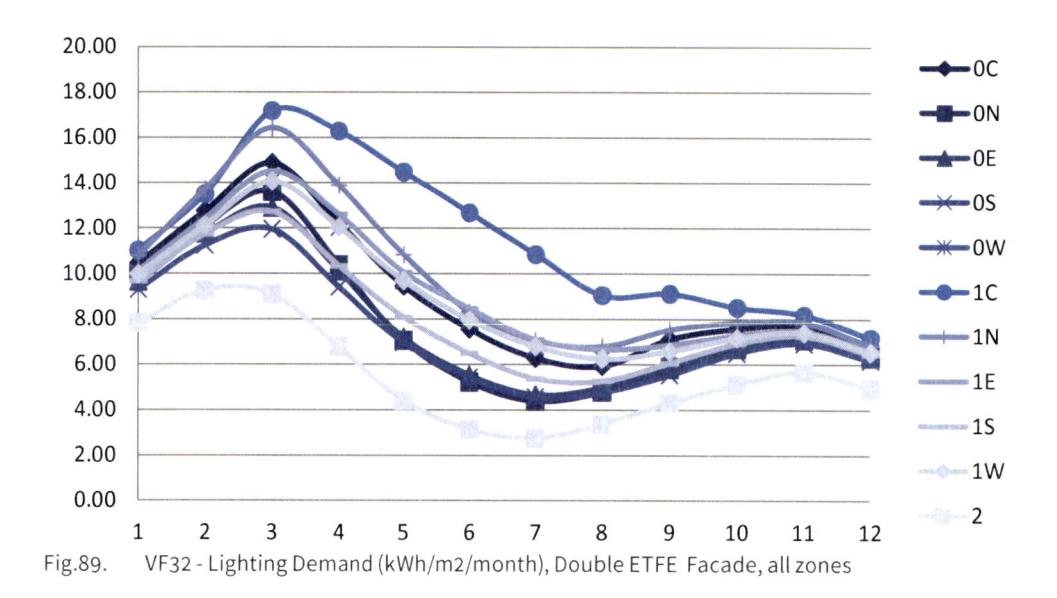

Fig.89. VF32 - Lighting Demand (kWh/m2/month), Double ETFE Facade, all zones

All the other zones range during March from 12.72 kWh (zone 1S) to 16.41 kWh (zone 1N).
The ETFE facade shows similar values with the same VT-value. They differ slightly, because of a higher refractive index (1.52 vs. 1.42).

Comparing the same zones covered by a single glazing facade zone 1C needs a power supply of 17.15 kWh in March, level 2 needs 7.83 kWh to provide *L. esculentum* with PAR light. Although we have a different power demand for lighting in every zone, comparing the single glazing facade with double-layer ETFE-facade, the difference is no higher than 1 to 4% due to the higher refractive index.

There is a margin of difference between the double glazing facade and former building envelopes, but this is not of a remarkable extent.
On the fully covered level two a difference can be defined which is 2.2% lower compared to Double-ETFE-facade. There are also similar differences to the center zones.

Looking at the annual results, as broken down at the end of this Chapter, we see an annual reduction by 3.9 % comparing the double-glazing facade to the standard single glazing facade, a reduction of 5.4 kWh/m2/a.
Anticipating the simulation results evaluation on the next pages we can assume that the building typology has a stronger influence on daylight availability than the building envelope. It is self-evident that when choosing a facade for a greenhouse or a Vertical Farm the highest VT-values will be selected and also that the differences between the values these have will only vary by a small percentage.

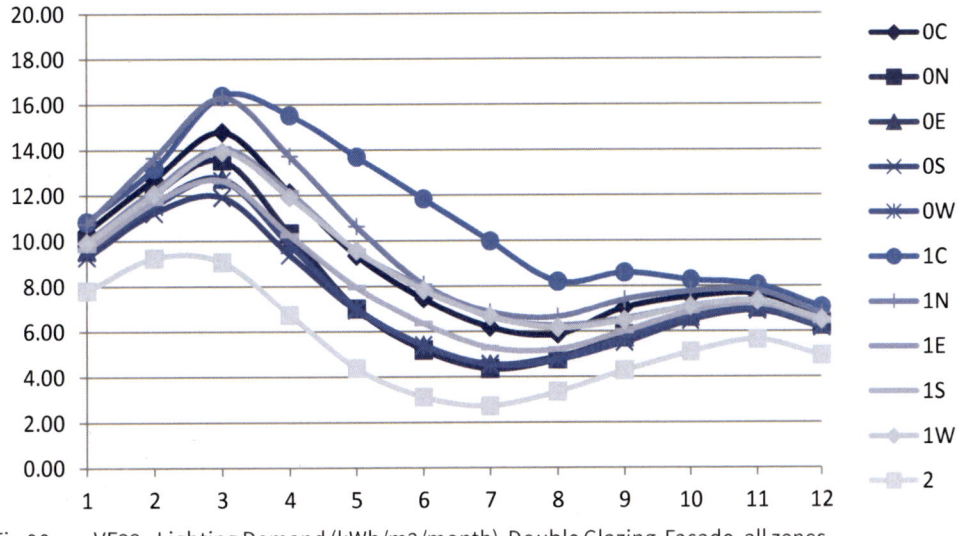

Fig.90. VF32 - Lighting Demand (kWh/m2/month), Double Glazing Facade, all zones

5.6.4. VF14 - Lighting Demand - Daysim Results

VF 14 - SG - ANNUAL DAYLIGHT AVAILABILITY (kWh/m2/a)

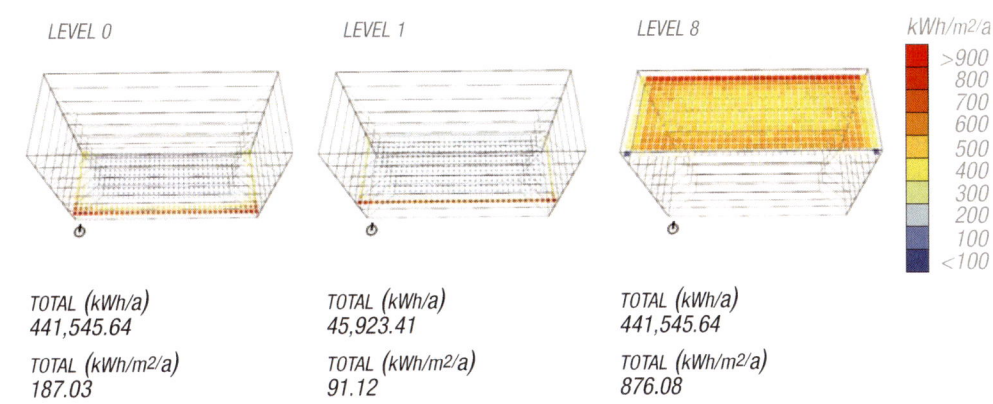

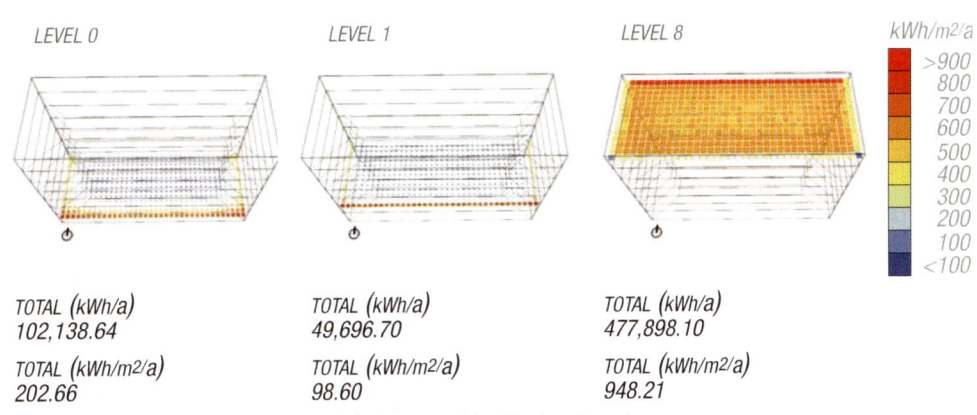

LEVEL 0	LEVEL 1	LEVEL 8
TOTAL (kWh/a) 441,545.64	TOTAL (kWh/a) 45,923.41	TOTAL (kWh/a) 441,545.64
TOTAL (kWh/m²/a) 187.03	TOTAL (kWh/m²/a) 91.12	TOTAL (kWh/m²/a) 876.08

Fig.91.   VF14 Annual Daylight Availability - Single Glazing Facade

VF 14 - ETFE - ANNUAL DAYLIGHT AVAILABILITY (kWh/m2/a)

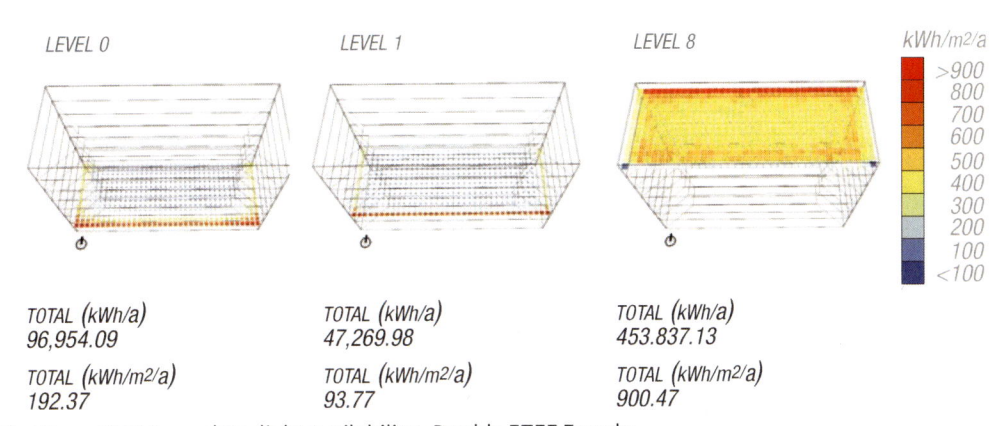

LEVEL 0	LEVEL 1	LEVEL 8
TOTAL (kWh/a) 96,954.09	TOTAL (kWh/a) 47,269.98	TOTAL (kWh/a) 453.837.13
TOTAL (kWh/m²/a) 192.37	TOTAL (kWh/m²/a) 93.77	TOTAL (kWh/m²/a) 900.47

Fig.92.   VF14 Annual Daylight Availability - Double ETFE Facade

VF 14 - DG - ANNUAL DAYLIGHT AVAILABILITY (kWh/m2/a)

LEVEL 0	LEVEL 1	LEVEL 8
TOTAL (kWh/a) 102,138.64	TOTAL (kWh/a) 49,696.70	TOTAL (kWh/a) 477,898.10
TOTAL (kWh/m²/a) 202.66	TOTAL (kWh/m²/a) 98.60	TOTAL (kWh/m²/a) 948.21

Fig.93.   VF14 Annual Daylight Availability - Double Glazing Facade

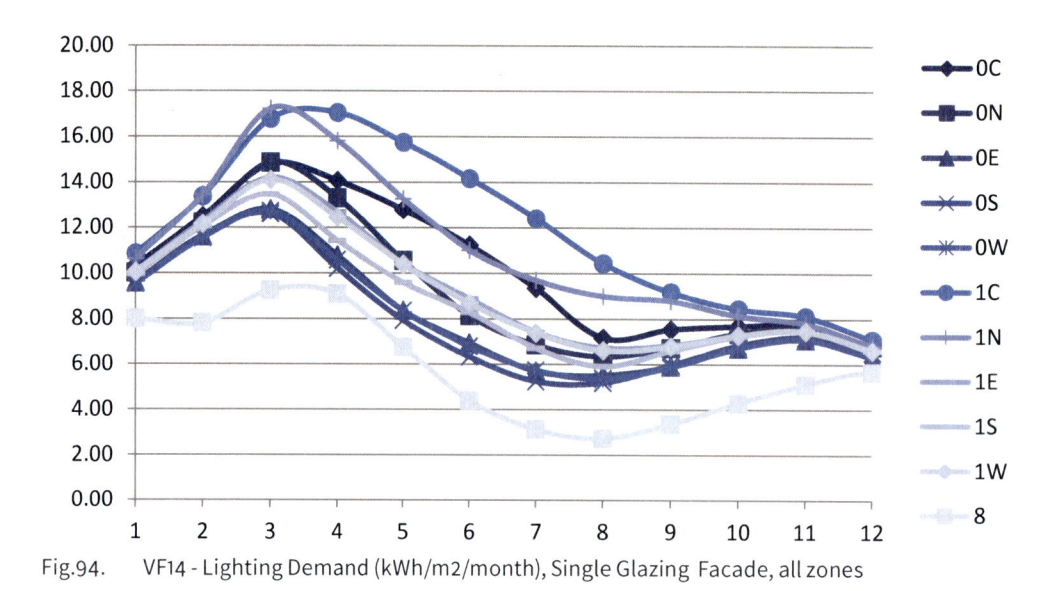

Fig.94. VF14 - Lighting Demand (kWh/m2/month), Single Glazing Facade, all zones

Comparing the results of VF14 to the former VF32 on the top level we see similar results in terms of the differences between each of the building envelopes. All facades with a high VT-value lead to the clear insight that the difference in lighting demand to be achieved is only negligible.

The diagrams on this page show the differences of each zone between VF32 and VF14. The center zone on level 1, has the highest lighting demand in both cases (level height 3.00 m).
A similar picture is also provided in VF32 where we see the impact of the different level height. In March, where power demand peaks because of highest light requirement during the fruit ripening of *L. esculentum*, and the values differ by more than 4 kWh/m2.

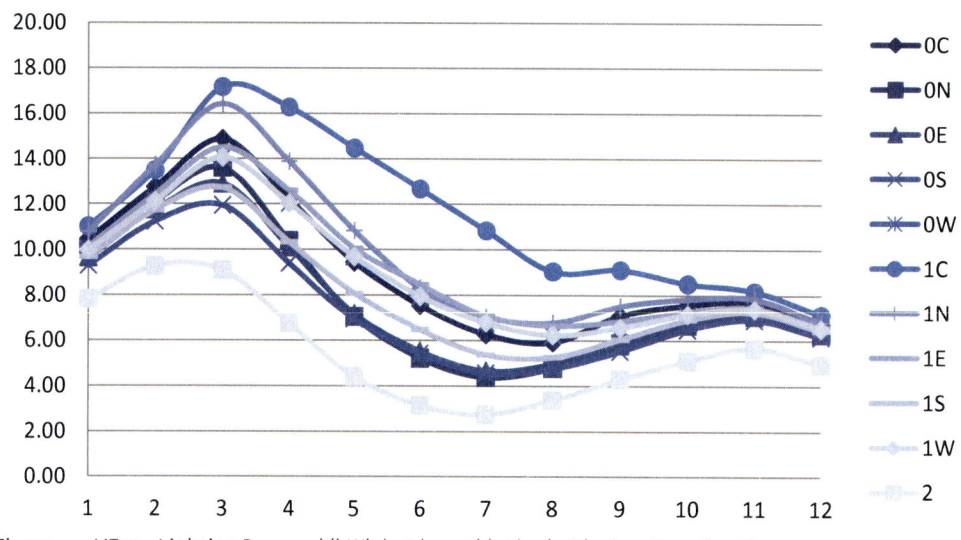

Fig.95. VF32 - Lighting Demand (kWh/m2/month), Single Glazing Facade, all zones

These diagrams make the influence of daylight availability comprehensible and make the sigmoid growing curve clearly readable, showing the effect of the different floor heights and the consequences of the sun exposure on the different zones. By the end of the chapter a deeper investigation will be elaborated by comparing effective values in power demand for the different zones with all three building envelopes.

We now already see a positive effect in daylight availability by minimizing the building depth on one the one hand, by the end of this chapter and taking the sun-exposed top level of all the farms into consideration we see that the area exposed to sunlight from the top is weighting the positive effect to a certain degree.

Before we do so we will first take a look at V7, the typology with the lowest value for the building depth (7.2 m) which did not require additional zoning.

5.6.5. VF7 - Lighting Demand - Daysim Results

VF 7 - SG - ANNUAL DAYLIGHT AVAILABILITY (kWh/m²/a)

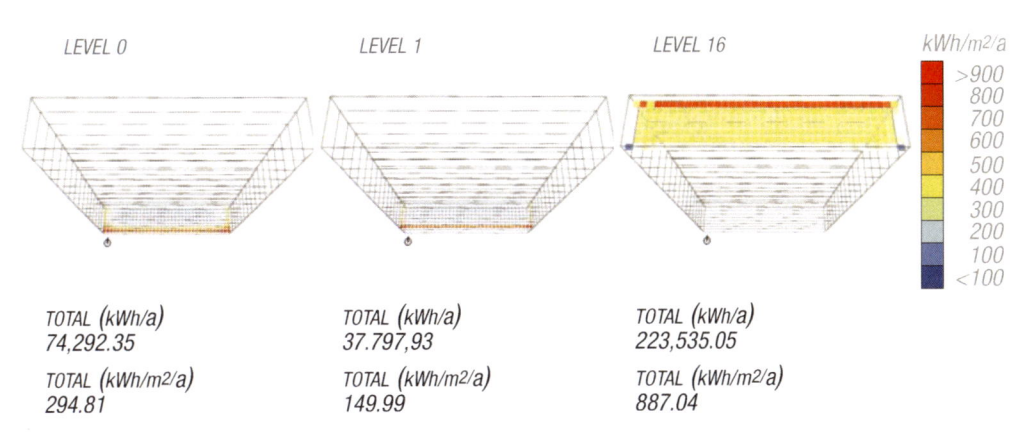

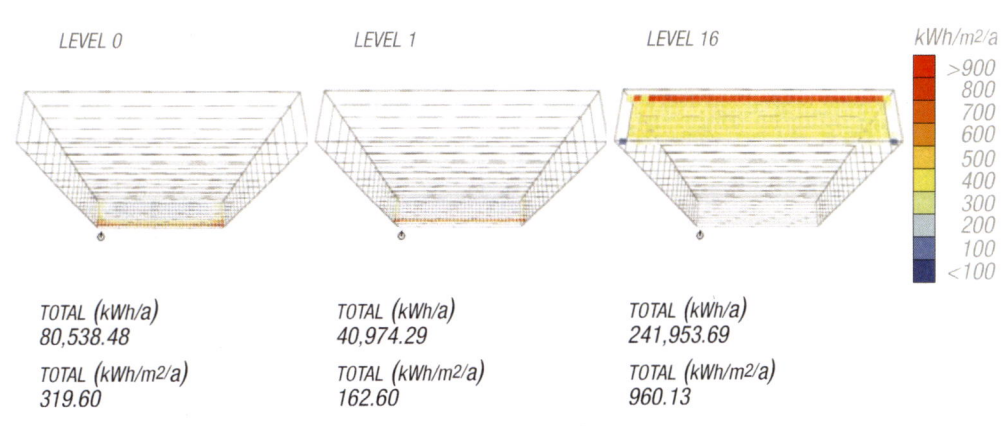

LEVEL 0	LEVEL 1	LEVEL 16
TOTAL (kWh/a) 74,292.35	TOTAL (kWh/a) 37.797,93	TOTAL (kWh/a) 223,535.05
TOTAL (kWh/m²/a) 294.81	TOTAL (kWh/m²/a) 149.99	TOTAL (kWh/m²/a) 887.04

Fig.96.   VF7 Annual Daylight Availability - Single Glazing Facade

VF 7 - ETFE - ANNUAL DAYLIGHT AVAILABILITY (kWh/m²/a)

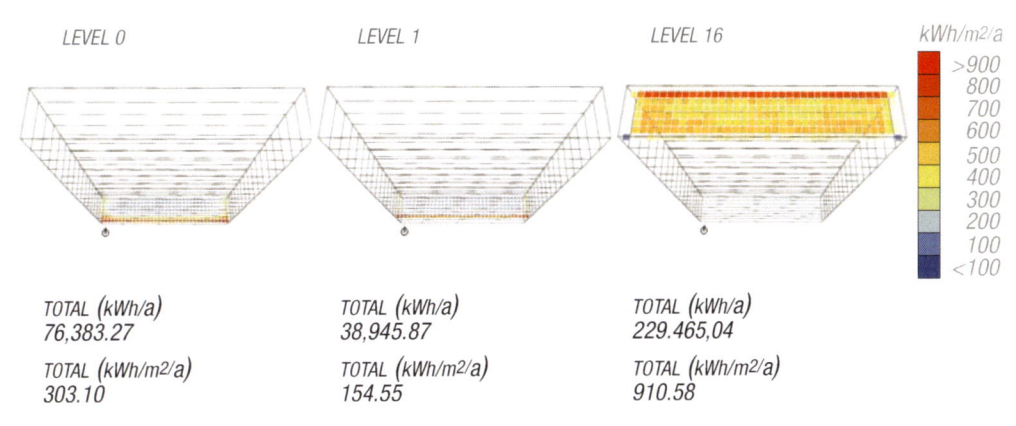

LEVEL 0	LEVEL 1	LEVEL 16
TOTAL (kWh/a) 76,383.27	TOTAL (kWh/a) 38,945.87	TOTAL (kWh/a) 229.465,04
TOTAL (kWh/m²/a) 303.10	TOTAL (kWh/m²/a) 154.55	TOTAL (kWh/m²/a) 910.58

Fig.97.   VF7 Annual Daylight Availability - Double ETFE Facade

VF 7 - DG - ANNUAL DAYLIGHT AVAILABILITY (kWh/m²/a)

LEVEL 0	LEVEL 1	LEVEL 16
TOTAL (kWh/a) 80,538.48	TOTAL (kWh/a) 40,974.29	TOTAL (kWh/a) 241,953.69
TOTAL (kWh/m²/a) 319.60	TOTAL (kWh/m²/a) 162.60	TOTAL (kWh/m²/a) 960.13

Fig.98.   VF7 Annual Daylight Availability - Double Glazing Facade

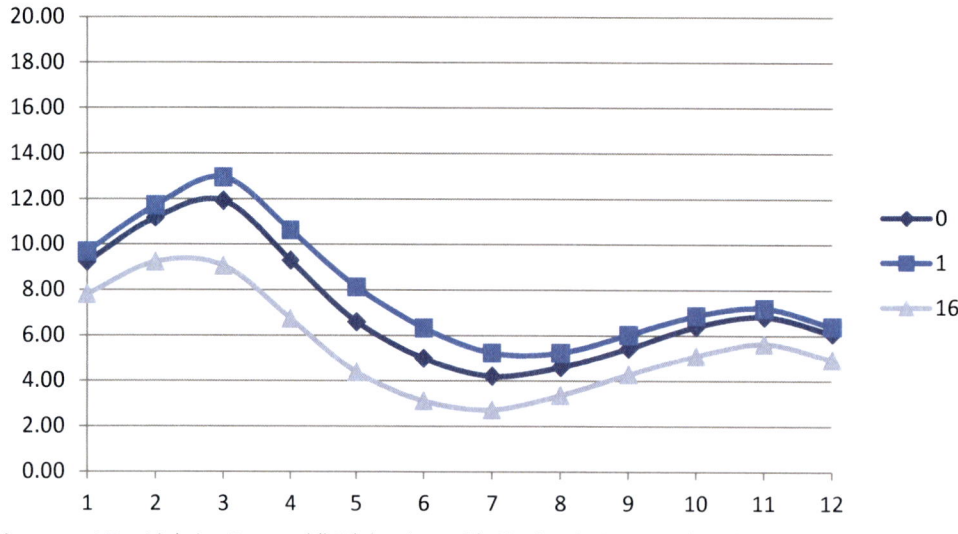

Fig.100.   VF7 - Lighting Demand (kWh/m2/month), Single Glazing Facade, all zones

The daylight results of this typology show the positive effects through the reduced building depth of only 7.2 m, corresponding to two standard -Venlo greenhouses. Light distribution is more evenly distributed throughout the different levels. We clearly see the effect of the different heights of level 0 (5.00 m) and level 1 (3.50 m) and a clear difference to level 16, comparable to the lighting requirement of a traditional greenhouse.

Predictably the lighting demand strongly increases in the first two months of the crop rotation until it peaks in March while *L. esculentum* moves from the phase of establishing blossoms to fruit growth and fruit ripening on to the first harvest. The increasing daylight availability from spring to summer more than halves the power demand for LED-lighting fixtures until it reaches its lowest level in July before it rises

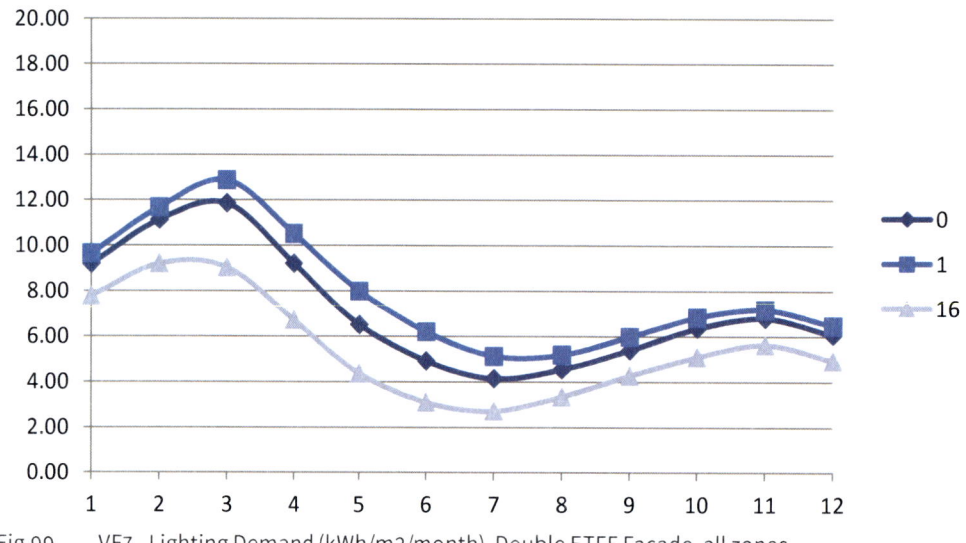

Fig.99.   VF7 - Lighting Demand (kWh/m2/month), Double ETFE Facade, all zones

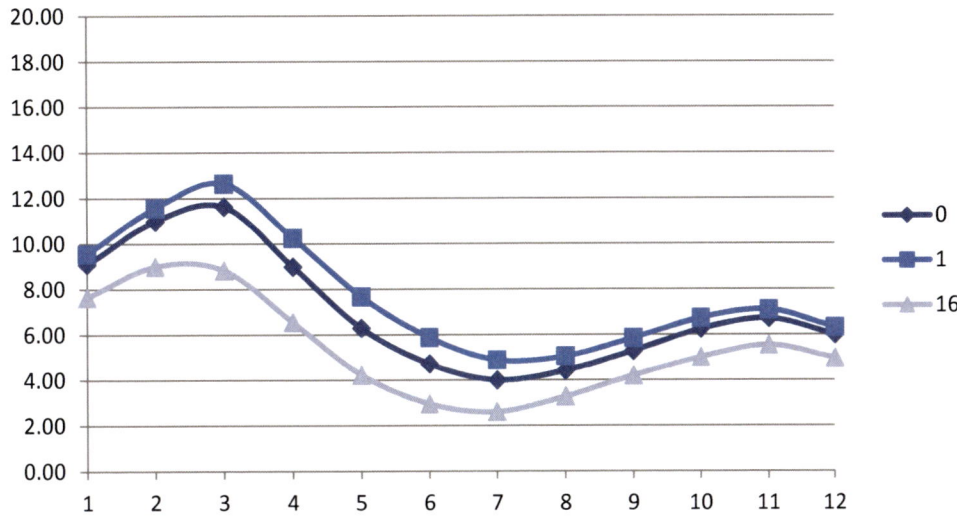

Fig.101. VF7 - Lighting Demand (kWh/m2/month), Double Glazing Facade, all zones

again until November. Throughout the last month, in the last phase of harvest and the end of the crop rotation, when lighting demand decreases once again.

So far, for VF7 the total annual daylight availability on the representative levels is from 887.04 kWh/m2/a on level 16 with Single Glazing material (lowest value) to 910.57 kWh/m2/a on the same level for Double-ETFE material and lastly 960.13 kWh/m2/a for double-glazing material. Comparing this value with the 268.35 kWhPAR/m2/a as mentioned, which is needed for optimum growth for *L. esculentum* and considering PAR is 50% of the total solar radiation we can assume that the total light availability of photosynthetically active radiation is more than is needed, and we can thus rule out a power demand at this level.

LIGHTING AND HEATING DEMAND (TPES) FOR L.ESCULTENTUM AND TWO THEORETICAL CROPS (kWh/m2/month)

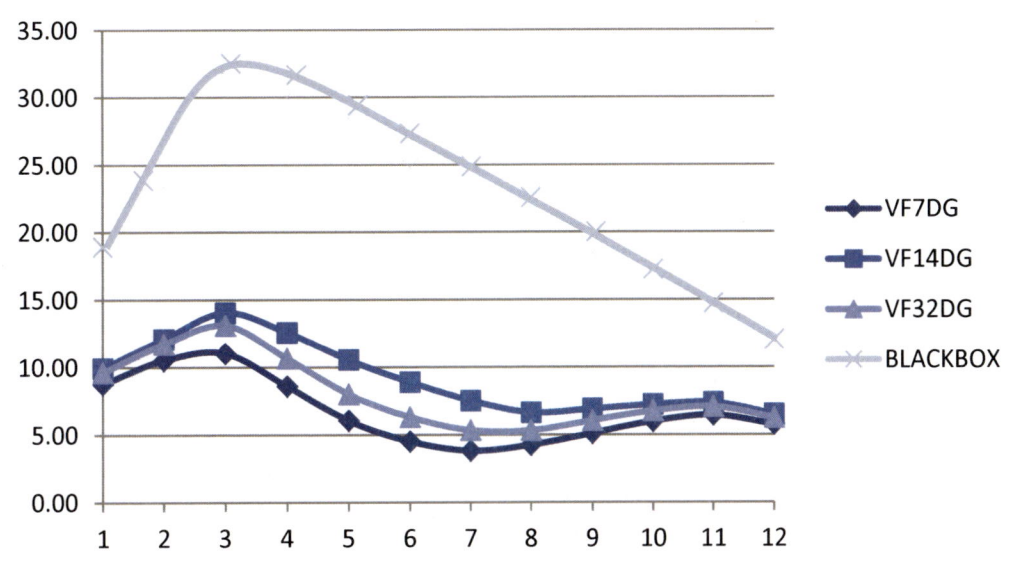

Fig.102. Comparison of monthly mean lighting demand for VF7, 14, 32 to a theoretical black box

5.7. Heating Demand - Simulation Results

5.7.1. VF32 - Heating demand

VF 32 - SG - HH Heating Demand (kWh/m2/a)

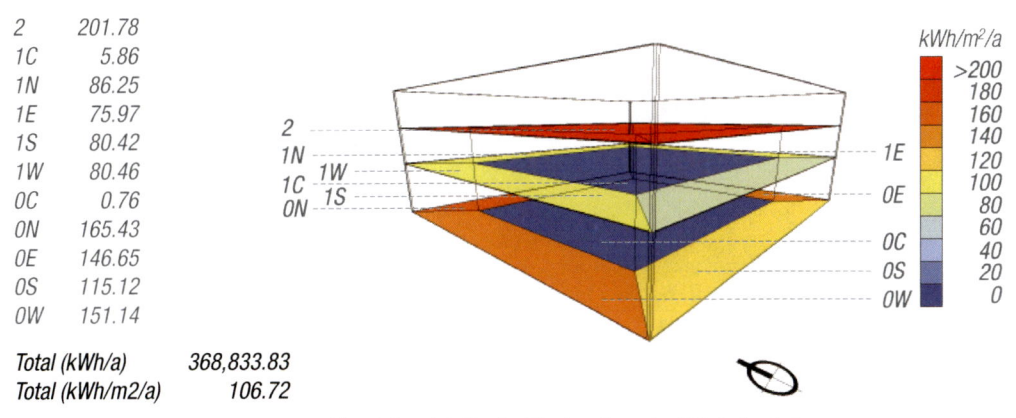

Zone	Value
2	201.78
1C	5.86
1N	86.25
1E	75.97
1S	80.42
1W	80.46
0C	0.76
0N	165.43
0E	146.65
0S	115.12
0W	151.14

Total (kWh/a)   368,833.83
Total (kWh/m2/a)   106.72

Fig.103.   VF32 Heating Demand (kWh/m2/a) Single Glazing Facade, HH-Schedule

SINGLE GLAZING FACADE

VF32 with a single glazing facade with a U-value of 5.88 W/m2/K and a SHGC of 0.8 and referring to the HH schedule the total heating demand throughout the whole crop rotation is 368,833.83 kWh/a or 106.72 kWh/m2/a on average throughout all the zones. The same facade for VF32 using the HL schedule reduces the heating demand to 285,286.09 kWh or 82.26 kWh/m2/a.

VF 32 - SG - HL Heating Demand (kWh/m2/a)

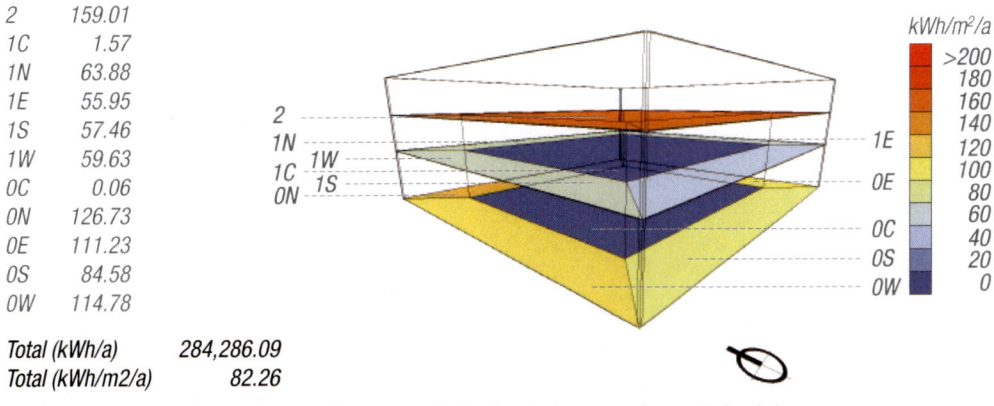

Zone	Value
2	159.01
1C	1.57
1N	63.88
1E	55.95
1S	57.46
1W	59.63
0C	0.06
0N	126.73
0E	111.23
0S	84.58
0W	114.78

Total (kWh/a)   284,286.09
Total (kWh/m2/a)   82.26

Fig.104.   VF32 Heating Demand (kWh/m2/a) Single Glazing Facade, HL-Schedule

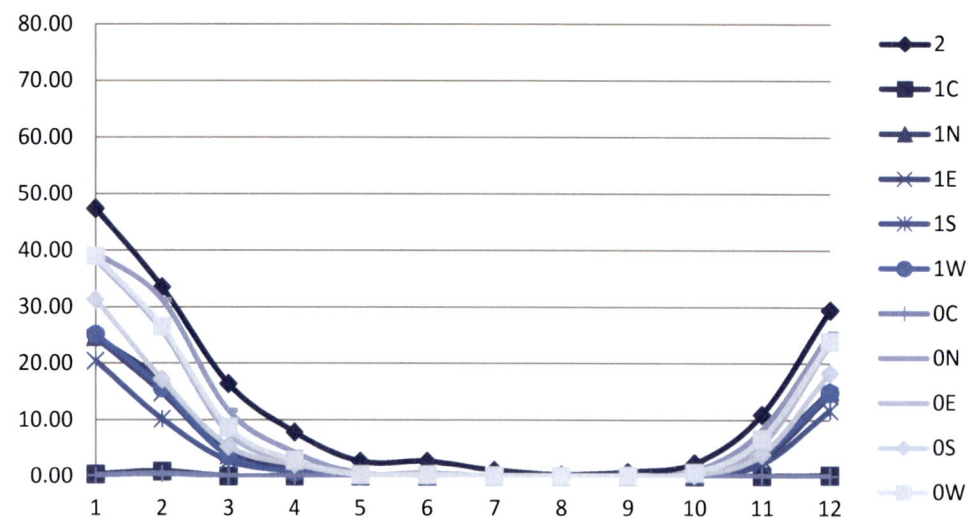

Fig.105. VF32 - Heating Demand (kWh/m2/month), HH-schedule, Single Glazing Facade, all zones

The highest heating demand in respect to the different the levels is at the top with over 200 kWh/m2/a followed by the exposed zones of the ground floor with its level height of 5.0 m compared to the zones on the first level with a height of 3.5 m, whereas the central zone on level 0 and level 1 have a heating demand close to 0.
Comparing the exposed zones, the highest heating demand is found as expected in the north, followed by the zones exposed to east and west and lastly the south zone.

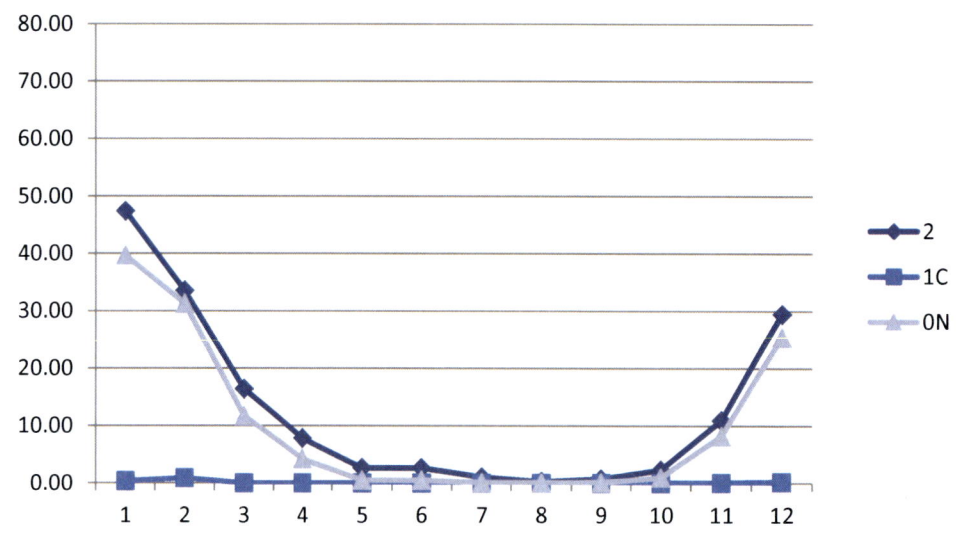

Fig.106. VF32 - Heating Demand (kWh/m2/month), HH-schedule, Single Glazing Facade, selected zones

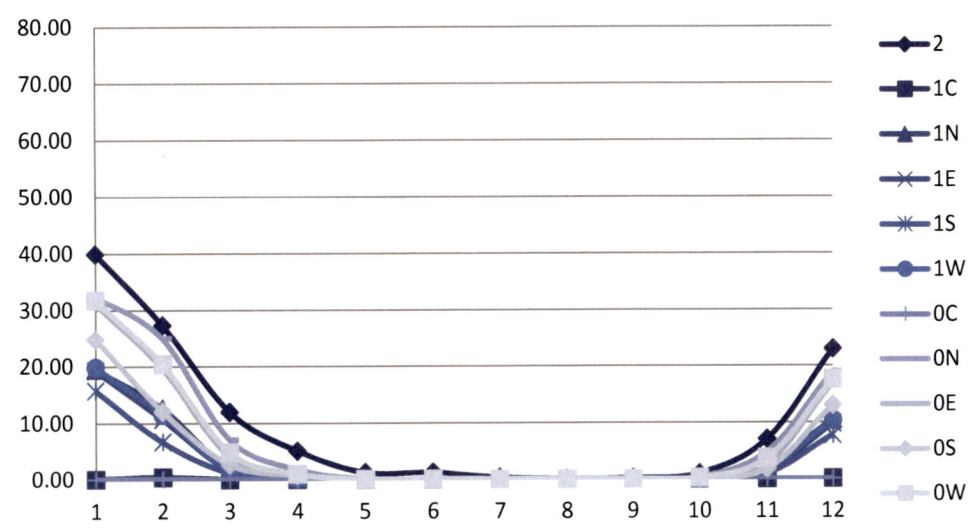

Fig.107. VF32 - Heating Demand (kWh/m2/month), HL-schedule, Single Glazing Facade, all zones

Greater differences in heating demand between the orientations are found on level 0, the ground floor with its height of 5 m.

By replacing the HH schedule with the HL schedule, which represents a reduction of 10%, the overall heating demand can be reduced by approx. 22.65%. The greatest difference we can observe is on the top level, where the annual heating demand is reduced from 20178 kWh/m2/a to 159.01 kWh/m2/a.

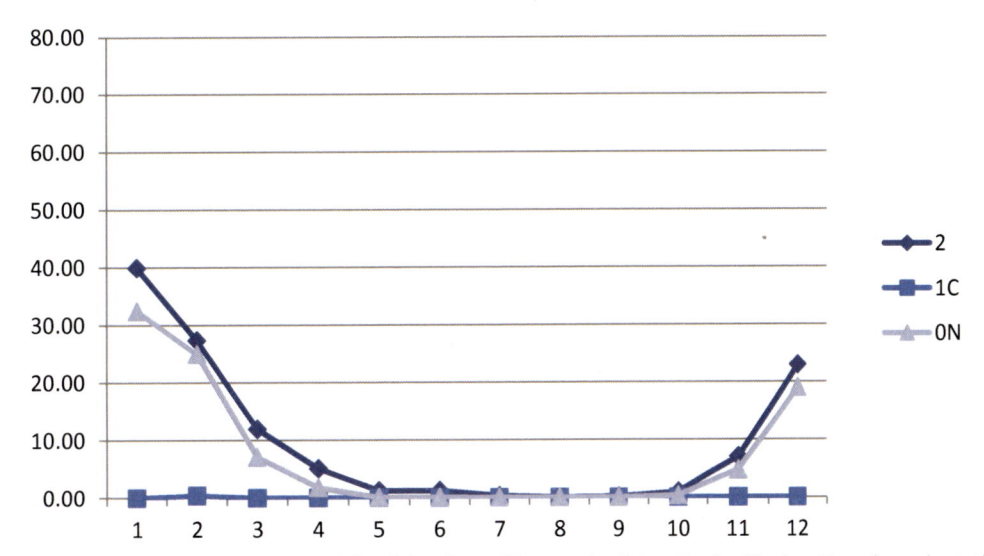

Fig.108. VF32 - Heating Demand (kWh/m2/month), HL-schedule, Single Glazing Facade, selected zones

DOUBLE LAYER ETFE FACADE

VF 32 - ETFE - HH Heating Demand (kWh/m2/a)

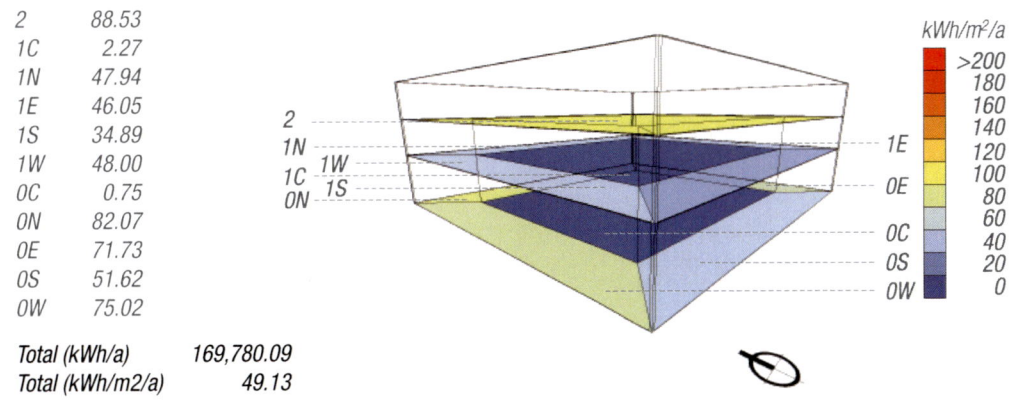

2	88.53
1C	2.27
1N	47.94
1E	46.05
1S	34.89
1W	48.00
0C	0.75
0N	82.07
0E	71.73
0S	51.62
0W	75.02

Total (kWh/a) 169,780.09
Total (kWh/m2/a) 49.13

Fig.110. VF32 Heating Demand (kWh/m2/a) Double ETFE Facade, HH-Schedule

VF 32 - ETFE - HL Heating Demand (kWh/m2/a)

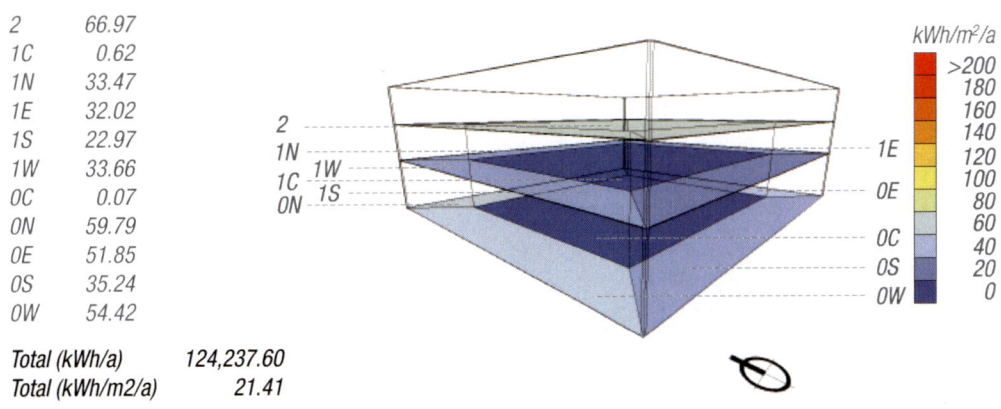

2	66.97
1C	0.62
1N	33.47
1E	32.02
1S	22.97
1W	33.66
0C	0.07
0N	59.79
0E	51.85
0S	35.24
0W	54.42

Total (kWh/a) 124,237.60
Total (kWh/m2/a) 21.41

Fig.109. VF32 Heating Demand (kWh/m2/a) Double ETFE Facade, HL-Schedule

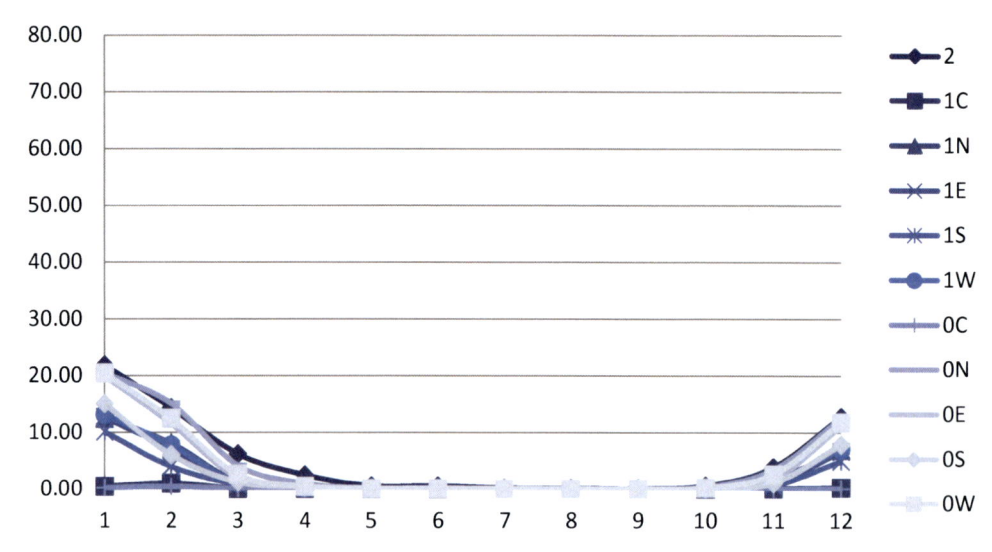

Fig.111. VF32 - Heating Demand (kWh/m2/month), HH-schedule, Double ETFE Facade, all zones

VF32 with a double ETFE facade (U-Value of 2.9 W/m2/K and a SHGC of 0.65) by using the HH schedule the total heating demand throughout the whole crop rotation is 368,833.83 kWh/a or on average throughout all the zones of 106.72 kWh/m2/a. The same facade for VF32 using the HL schedule reduces the heating demand to 124,237.60 kWh/a. By changing the heating schedule from HH to HL an overall reduction for heating demand of 33.68% per year is achievable.

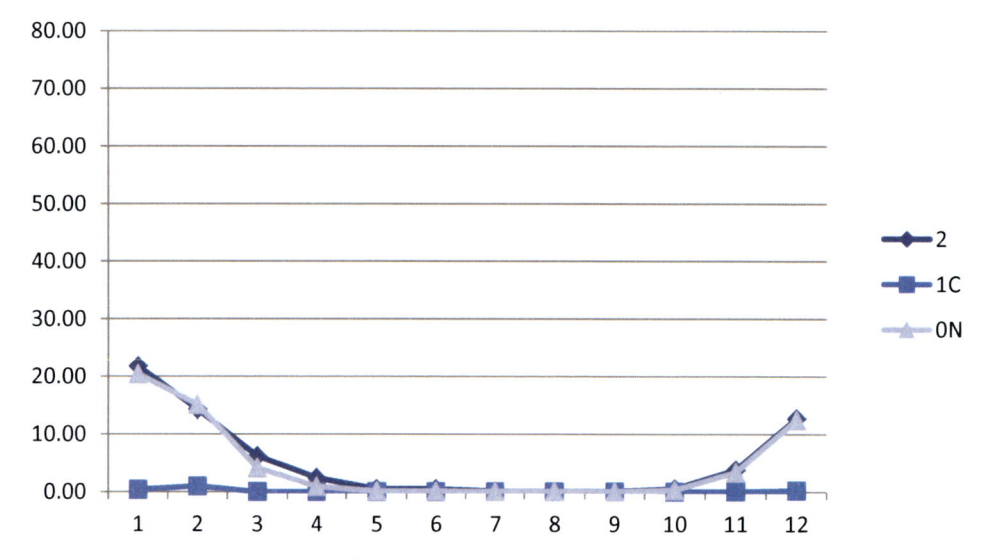

Fig.112. VF32 - Heating Demand (kWh/m2/month), HH-schedule, Double ETFE Facade, selected zones

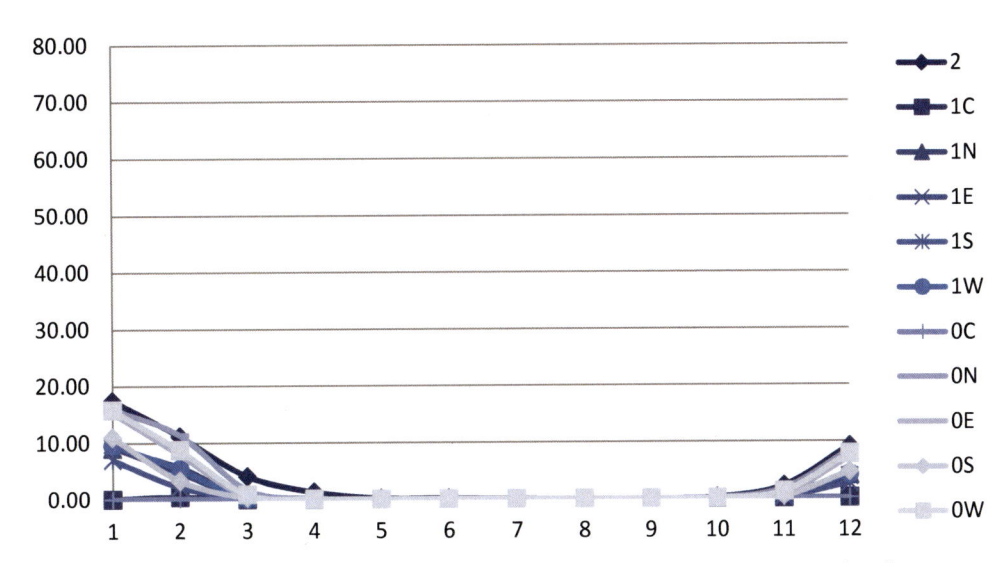

Fig.113. VF32 - Heating Demand (kWh/m2/month), HL-schedule, Double ETFE Facade, all zones

Compared to VF32 with a single glazing-envelope heating demand can be more than halved, from 106.72 kWh/m2/a to 49.13 kWh/m2/a (HH-schedule) or from 82.26 kWh/m2/a to 35.95 kWh/m2/a (HL-schedule) respectively.

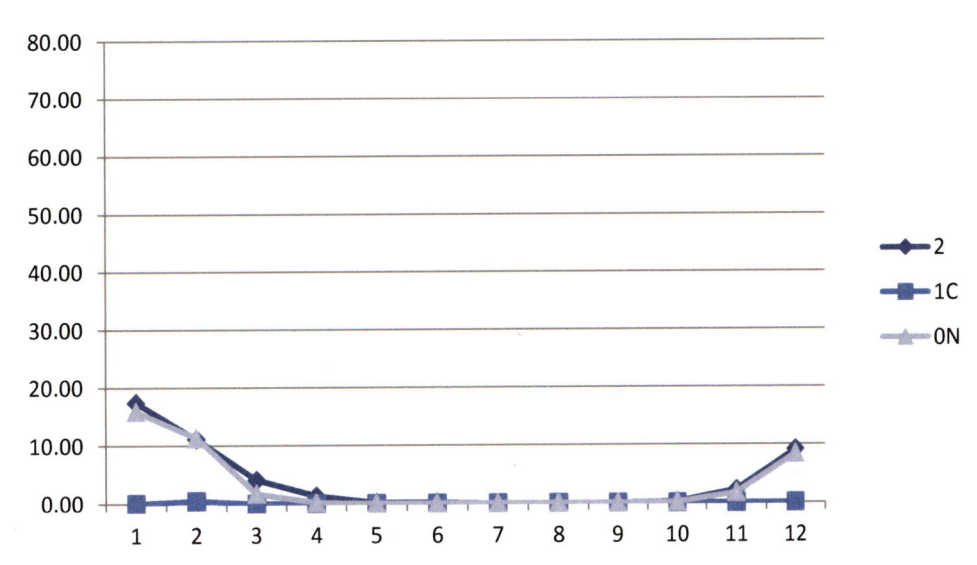

Fig.114. VF32 - Heating Demand (kWh/m2/month), HL-schedule, Double ETFE Facade, selected zones

Compared to VF32 with a single glazing-envelope heating demand can be more than halved, from 106.72 kWh/m2/a to 49.13 kWh/m2/a (HH-schedule) or from 82.26 kWh/m2/a to 35.95 kWh/m2/a (HL-schedule) respectively.

Heating demand for all the zones with the HH schedule is already below 10 kWh/m2/a by the end of February. Zone 2 with the highest heating demand, with a double ETFE facade has similar values to zone 1N with a single glazing-facade. Stronger differences for heating demand are notable for the exposed zones on level 1 as well as on the ground level if we compare the results between the envelopes SG and ETFE, especially by comparing south-oriented zones. VF32 ETFE using the HH schedule still has lower values than VF32 SG with the HL schedule. Comparing the total heating demand for the same building volume only 59.72% is needed.

Comparing the results for both schedules the total heating demand with the HL schedule is 73% of the annual heating demand that results when using the HH schedule. The monthly results clearly show that heating demand is touching the bottom level already by the end of April, and starts to rise by the end of September. VF32 SG heating demand comes close to zero only during the hottest weeks during July and August, after which it rapidly starts to rise again.

DOUBLE GLAZING FACADE

VF 32 - DG - HH Heating Demand (kWh/m2/a)

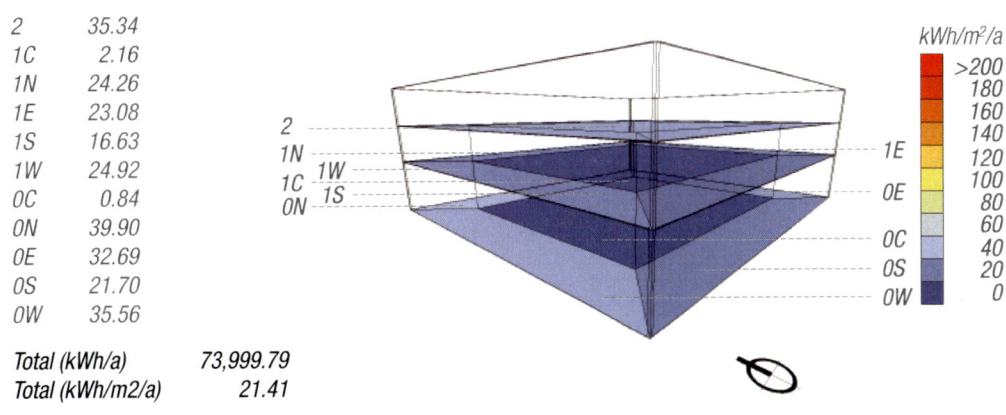

Zone	kWh/m²/a
2	35.34
1C	2.16
1N	24.26
1E	23.08
1S	16.63
1W	24.92
0C	0.84
0N	39.90
0E	32.69
0S	21.70
0W	35.56
Total (kWh/a)	73,999.79
Total (kWh/m2/a)	21.41

Fig.115.   VF32 Heating Demand (kWh/m2/a) Double Glazing Facade, HH-Schedule

VF32 with a double glazing facade (U-Value of 1.7 W/m2/K and a SHGC of 0.7) by using the HH schedule the total heating demand throughout the whole crop rotation is reaching 73,999.79 kWh/a or 21.41 kWh/m2/a.

The same facade for VF32 using HL-schedule reduces the heating demand to 48,654.25 kWh/a. Calculated per square meter this is a reduction of 30.9% reaching 14.08 kWh/m2/a. Compared to VF32 with single glazing the differences for heating demand between the zones, are notably levelled out. Interestingly the zone with the highest heating demand no longer is zone 2 but 0N, and this both by using HH schedule and HL schedule.

VF 32 - DG - HL Heating Demand (kWh/m2/a)

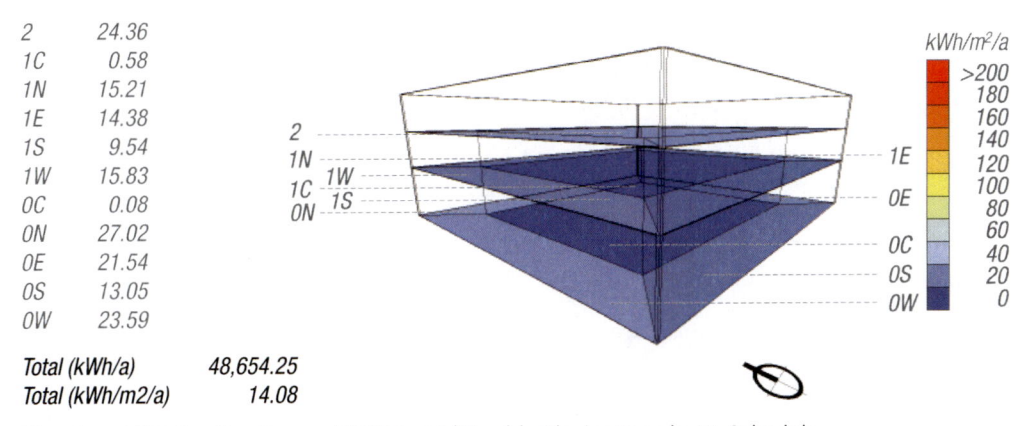

Zone	kWh/m²/a
2	24.36
1C	0.58
1N	15.21
1E	14.38
1S	9.54
1W	15.83
0C	0.08
0N	27.02
0E	21.54
0S	13.05
0W	23.59
Total (kWh/a)	48,654.25
Total (kWh/m2/a)	14.08

Fig.116.   VF32 Heating Demand (kWh/m2/a) Double Glazing Facade, HL-Schedule

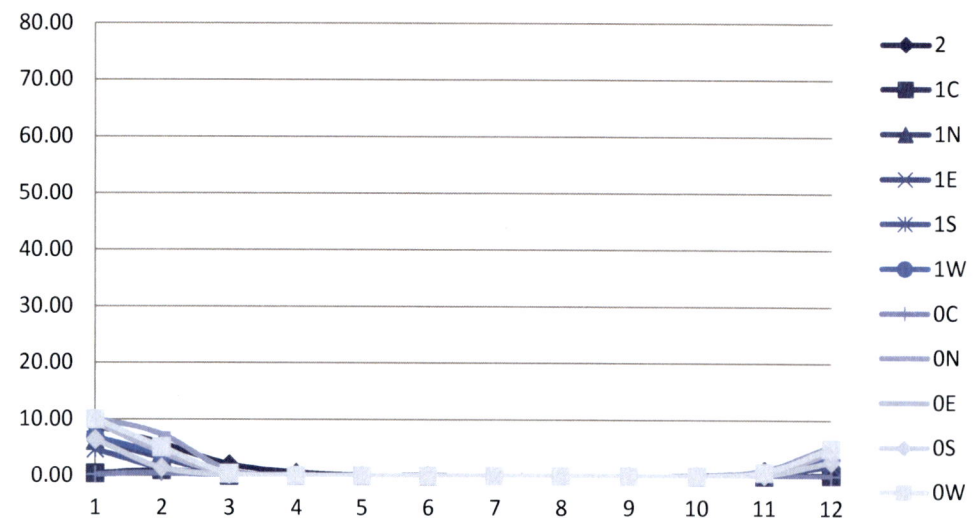

Fig.117. VF32 - Heating Demand (kWh/m2/month), HH-schedule, Double Glazing Facade, all zones

Regarding the heating demand throughout the crop rotation and reading the diagrams on these page we see that differences in heating demand between the zones are strongly levelled. Only by using HH schedule zone 2 and 0W are slightly over 10 kWh/m2/month, whereas all the other zones are below this value. Already by the end of February heating demand for every zone is below 5 kWh/m2/month. To anticipate the comparison of the overall energy consumption of the different VF-types, the heating demand values for VF32 with its double glazing facade are the lowest, although

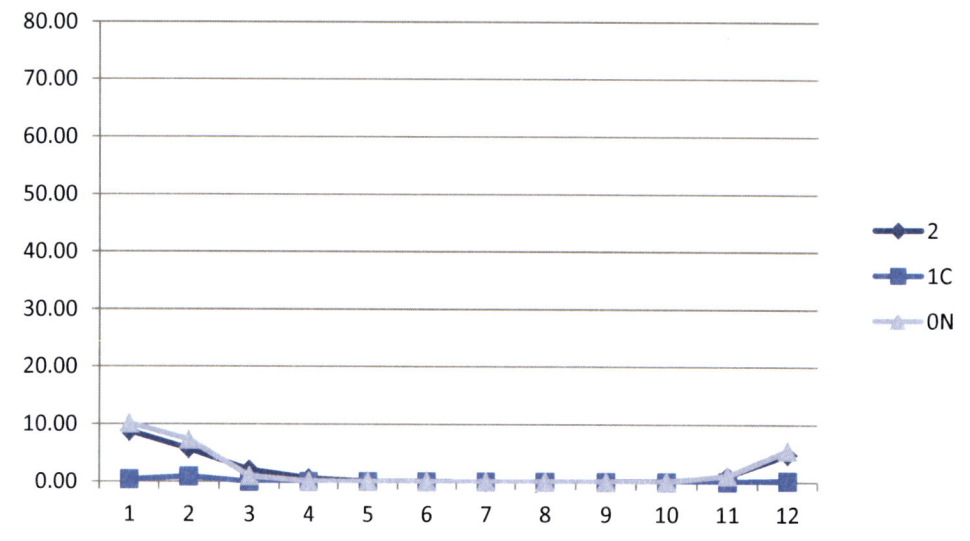

Fig.118. VF32 - Heating Demand (kWh/m2/month), HH-schedule, Double Glazing Facade, selected zones

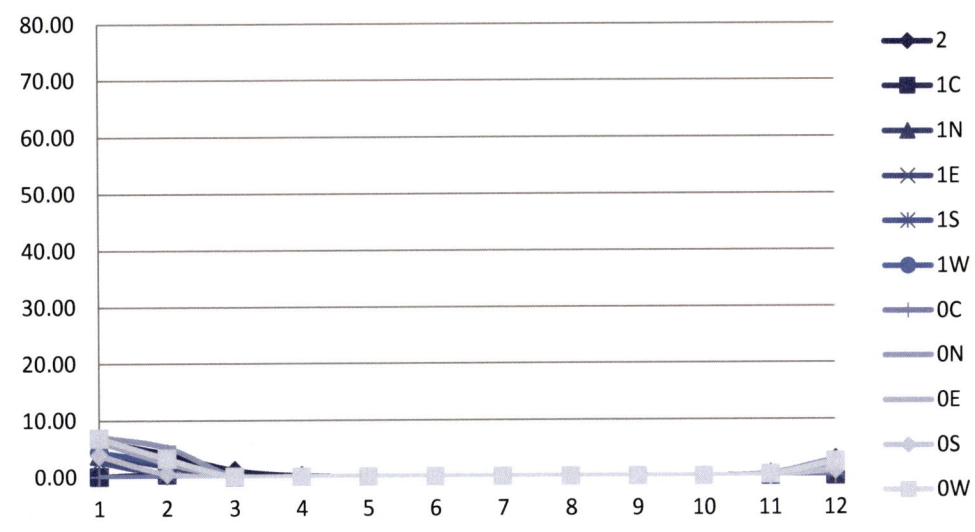

Fig.119. VF32 - Heating Demand (kWh/m2/month), HL-schedule, Double Glazing Facade, all zones

its A/V-ratio of 0.28 is slightly higher than the ratio of VF14 (0.26). This is due to the volume differences in the simulated volume. As explained in the pages describing the simulation method, in order to make the values more readily comparable, the priority in having the same level heights was more important than having exactly the same building volume. VF32 with a volume of 13,824.00 m3 is some 1,468.80 m3 smaller than VF14 and 1,987.20 m3 smaller than VF7.

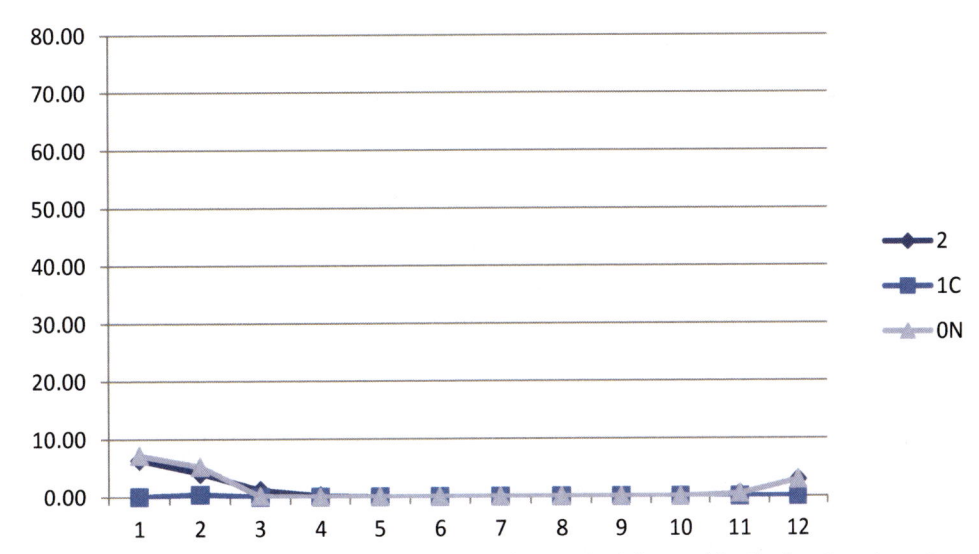

Fig.120. VF32 - Heating Demand (kWh/m2/month), HL-schedule, Double Glazing Facade, selected zones

5.7.2. VF14 - Heating demand

SINGLE GLAZING FACADE

VF 14 - SG - HH Heating Demand (kWh/m2/a)

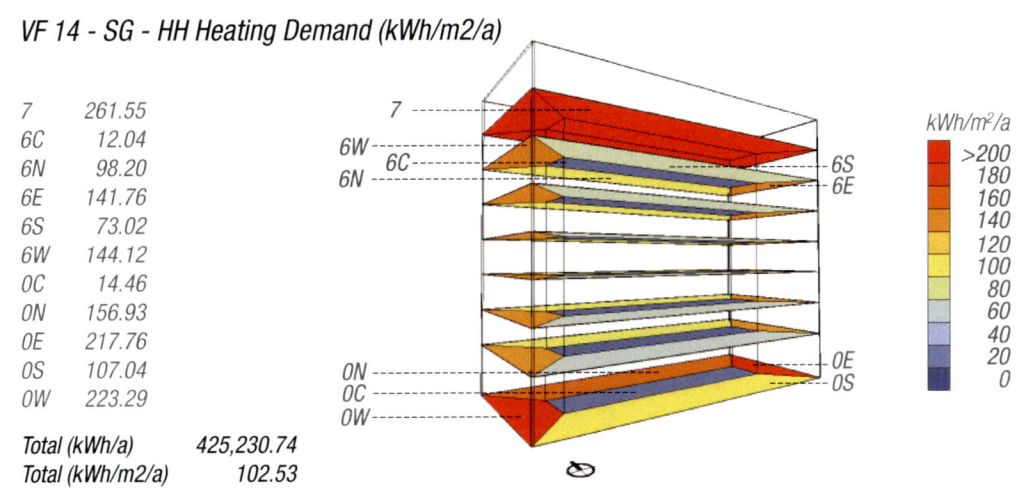

Zone	kWh/m²/a
7	261.55
6C	12.04
6N	98.20
6E	141.76
6S	73.02
6W	144.12
0C	14.46
0N	156.93
0E	217.76
0S	107.04
0W	223.29
Total (kWh/a)	425,230.74
Total (kWh/m2/a)	102.53

Fig.121.   VF14 Heating Demand (kWh/m2/a) Single Glazing Facade, HH-Schedule

The results of the HH schedule and HL schedule for VF14 vary by more than 100,000 kWh/a. By contrast with VF32 on the top level (zone 7) both values are over 200 kWh/m2/a. We find similar values in the zones exposed to the north and the south compared to VF32, because the dimensions are the same, while the values for heating demand in zones exposed to the east and the west are almost twice as high as the values for VF32, both for the HH and HL schedules.

VF 14 - SG - HL Heating Demand (kWh/m2/a)

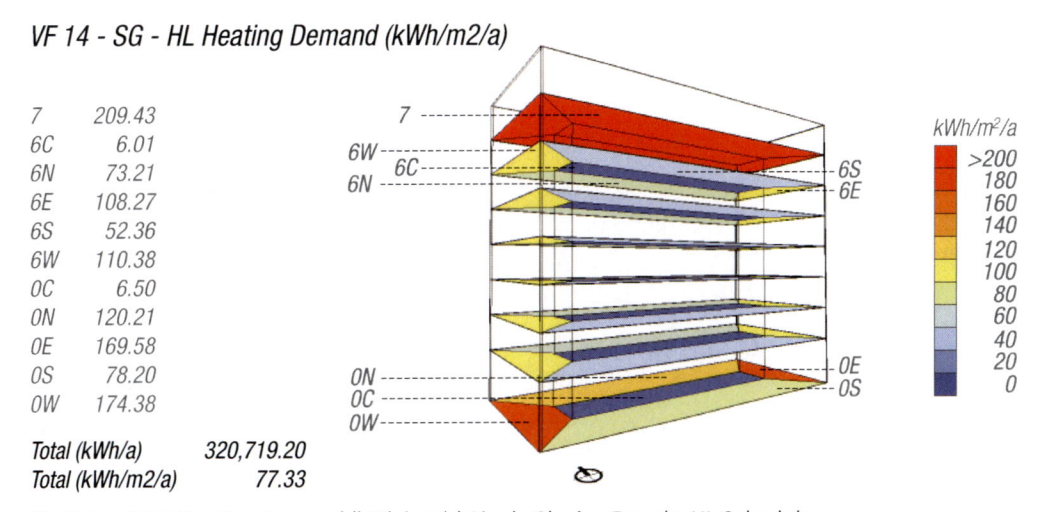

Zone	kWh/m²/a
7	209.43
6C	6.01
6N	73.21
6E	108.27
6S	52.36
6W	110.38
0C	6.50
0N	120.21
0E	169.58
0S	78.20
0W	174.38
Total (kWh/a)	320,719.20
Total (kWh/m2/a)	77.33

Fig.122.   VF14 Heating Demand (kWh/m2/a) Single Glazing Facade, HL-Schedule

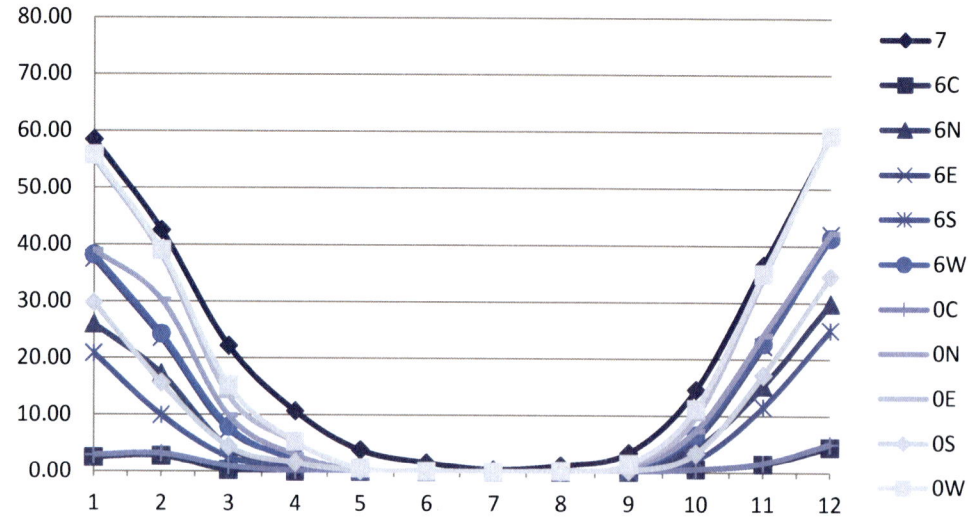

Fig.123. VF14 -Heating Demand (kWh/m2/month), HH-schedule, Single Glazing Facade, all zones

On a monthly basis we see the high demand per square meter for heating for west and east oriented zones. In absolute numbers, however, the values are around 50%, compared e.g. to the zones in the south.

We see a strong decline and rise for heating demand by the beginning and the end of the crop rotation similar to the situation with VF32. The exposed zones, especially zone 7 and the east and west zones, strongly contribute to the high heating demand

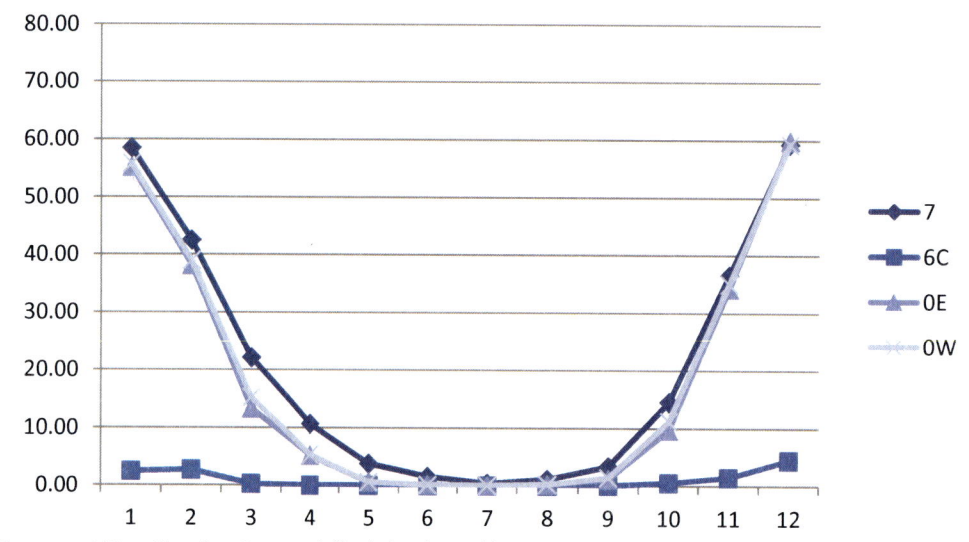

Fig.124. VF14 -Heating Demand (kWh/m2/month), HH-schedule, Single Glazing Facade, selected zones

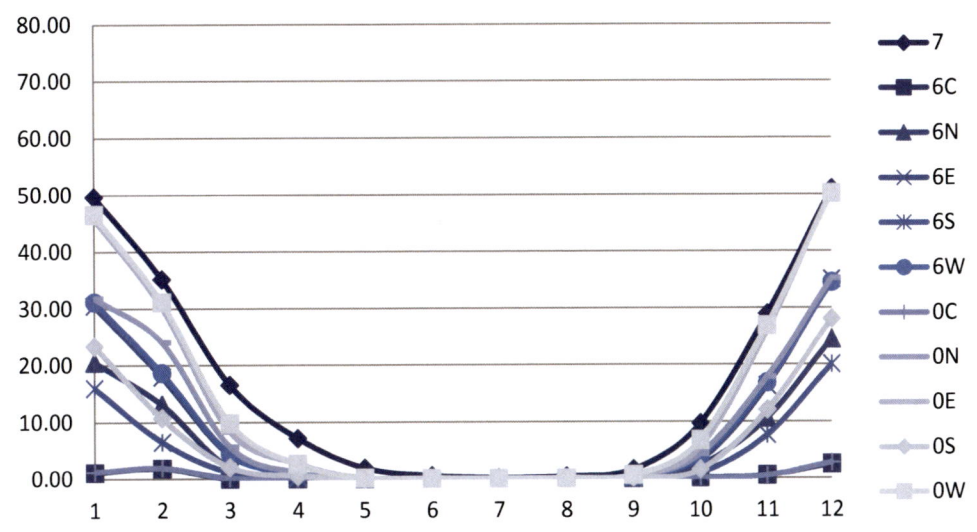

Fig.125. VF14 -Heating Demand (kWh/m2/month), HL-schedule, Single Glazing Facade, all zones

per square meter. Only during the hottest weeks from July to August does heating demand in all zones reach its lowest level. The reduction achieved by replacing the HH schedule with the HL schedule contributes to 24.58% of the annual heating demand. Heating demand for VF14 with the SG building envelope is 15% higher compared to VF32, because of its reduced west and east orientation and the top level, which accounts only for 11.81% of the overall building volume.

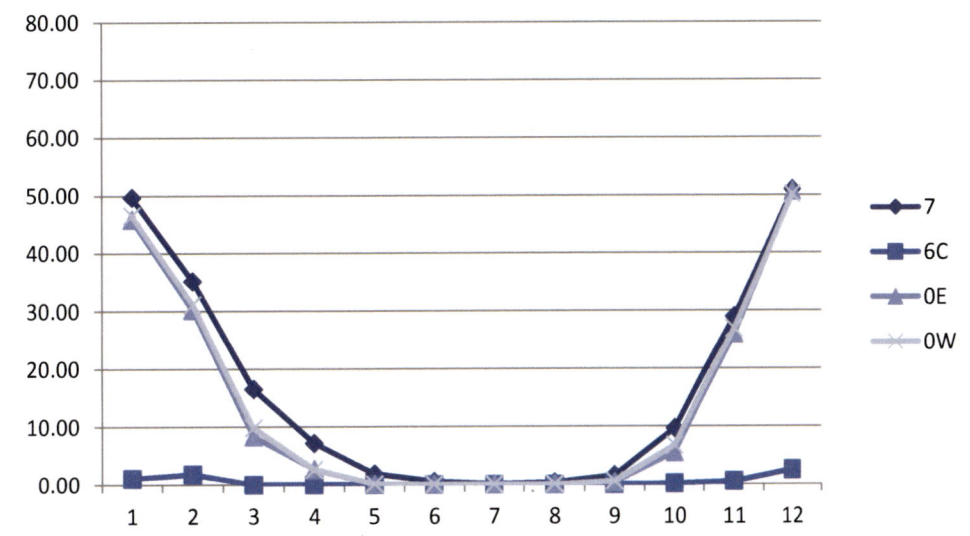

Fig.126. VF14 -Heating Demand (kWh/m2/month), HL-schedule, Single Glazing Facade, selected zones

DOUBLE ETFE FACADE

VF 14 - ETFE - HH Heating Demand (kWh/m2/a)

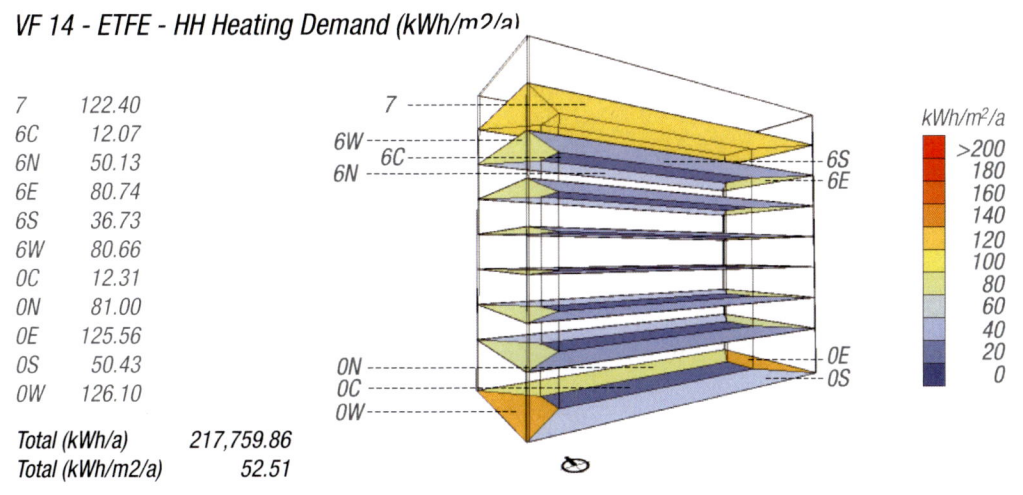

7	122.40
6C	12.07
6N	50.13
6E	80.74
6S	36.73
6W	80.66
0C	12.31
0N	81.00
0E	125.56
0S	50.43
0W	126.10

Total (kWh/a)   217,759.86
Total (kWh/m2/a)   52.51

Fig.127.   VF14 Heating Demand (kWh/m2/a) Double ETFE Facade, HH-Schedule

The heating demand for VF14 can be halved, or more precisely reduced by 48.8% using HH schedule or 51.5% using the HL schedule by changing the building envelope from SG to Double-ETFE. As with the results for VF14 SG, the central zones show the lowest values, and interestingly these are very similar values for both double ETFE and SG (12.07 kWh/m2/a on zone 6C with double ETFE facade compared to 12.04 kWh/m2/a with SG facade). Heating demand on the top level now shrinks to 122.40 kWh/m2/a using HH schedule and went below the 100 mark with the HL schedule (95.63 kWh/m2/a).

VF 14 - ETFE - HL Heating Demand (kWh/m2/a)

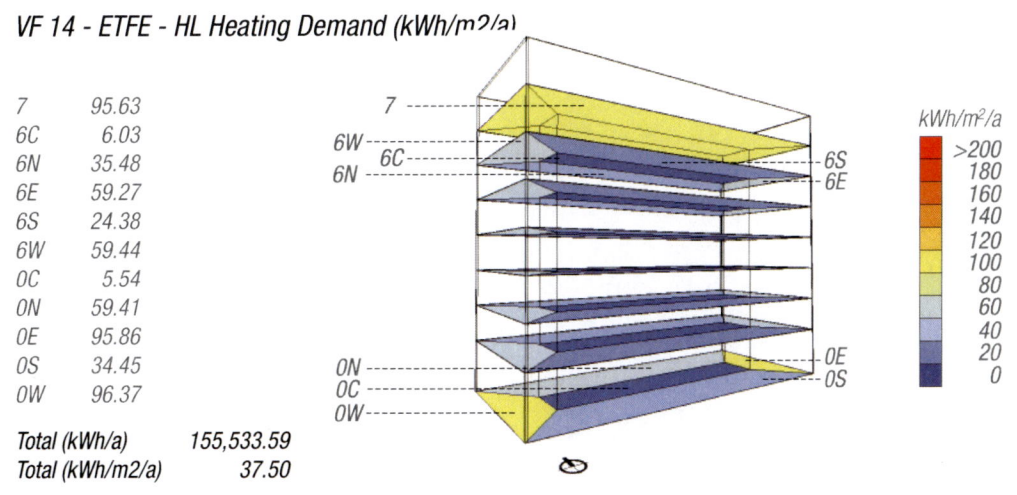

7	95.63
6C	6.03
6N	35.48
6E	59.27
6S	24.38
6W	59.44
0C	5.54
0N	59.41
0E	95.86
0S	34.45
0W	96.37

Total (kWh/a)   155,533.59
Total (kWh/m2/a)   37.50

Fig.128.   VF14 Heating Demand (kWh/m2/a) Double ETFE Facade, HL-Schedule

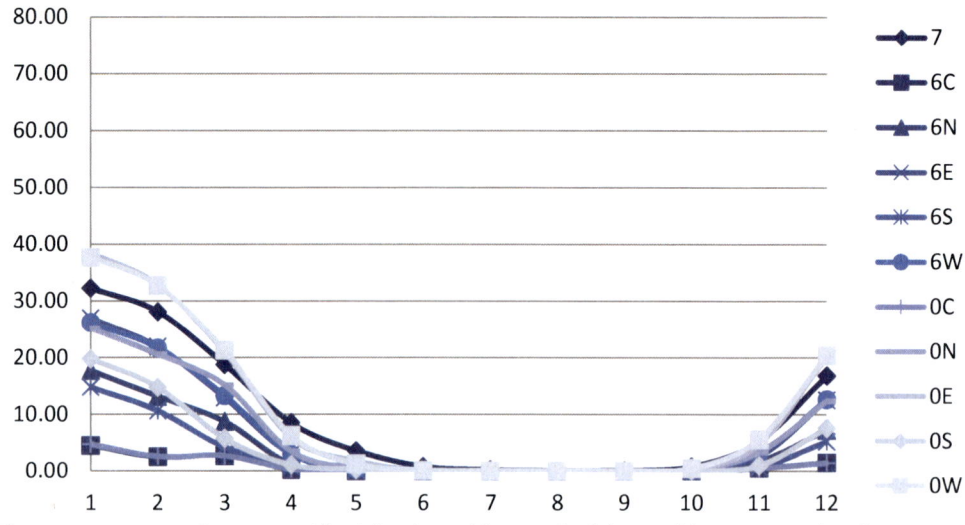

Fig.129. VF14 -Heating Demand (kWh/m2/month), HH-schedule, Double ETFE Facade, all zones

Differences in heating demand between the HH schedule and the HL schedule, especially at the beginning and the end of the year, lead to a difference of 62,226.27 kWh/a. per square meter this means a reduction from 52.51 kWh/m2/a to 37.50 kWh/m2/a which is a reduction of 28.59%. Compared to VF14 with its SG facade, the period with no or minimum heating demand has already been doubled. By the middle of

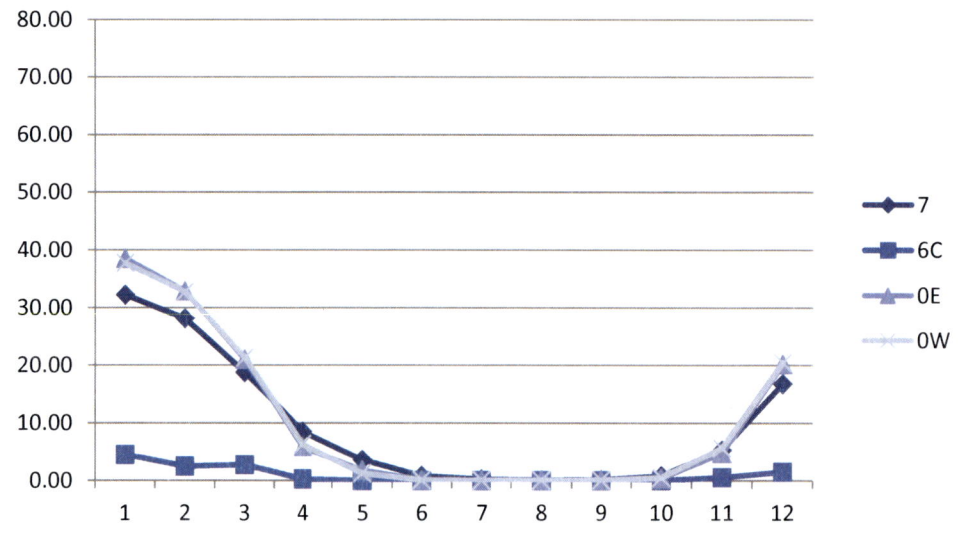

Fig.130. VF14 -Heating Demand (kWh/m2/month), HH-schedule, Double ETFE Facade, selected zones

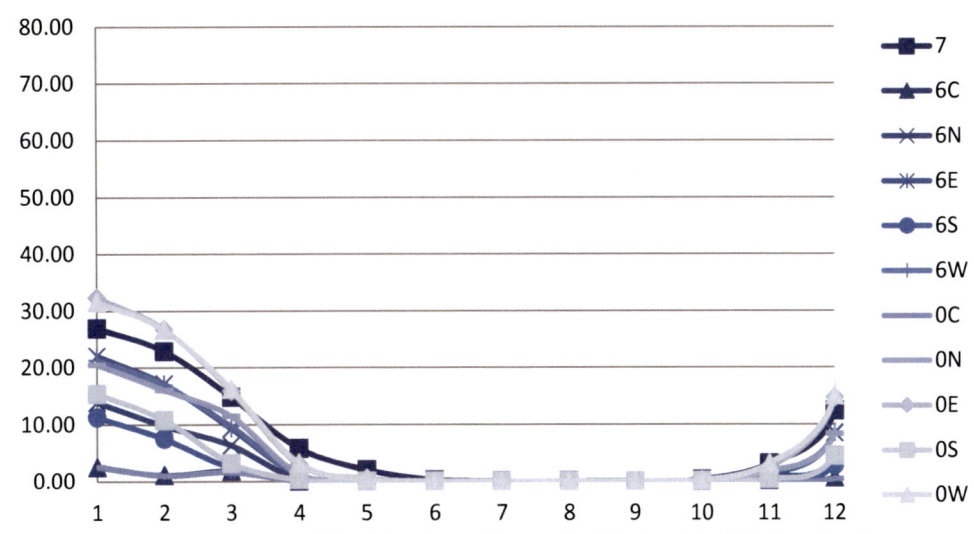

Fig.131. VF14 -Heating Demand (kWh/m2/month), HL-schedule, Double ETFE Facade, all zones

June the values come close to zero for all zones and only start to rise again at the end of September.
Heating demand at zone 7 is reduced from 122.40 kWh/m2/a to 95.63 kWh/m2/a by changing the HH schedule to the HL schedule. Heating demand for the center zones is halved.

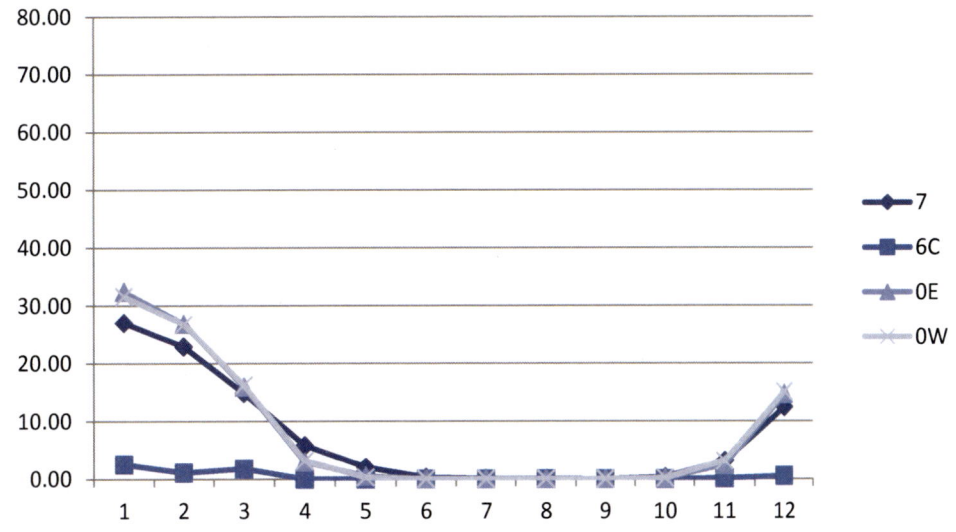

Fig.132. VF14 -Heating Demand (kWh/m2/month), HL-schedule, Double ETFE Facade, selected zones

VERTICAL FARM - SIMULATION MODEL

DOUBLE GLAZING FACADE

VF 14 - DG - HH Heating Demand (kWh/m²/a)

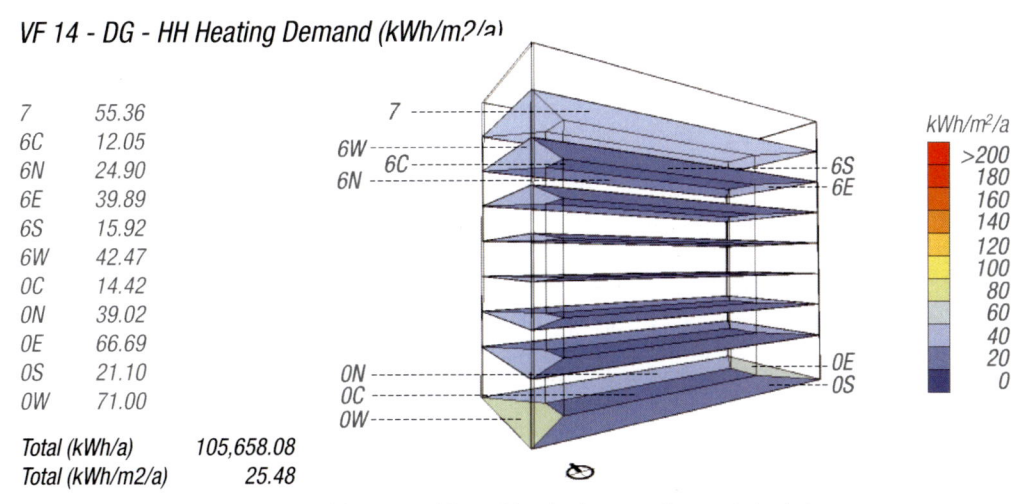

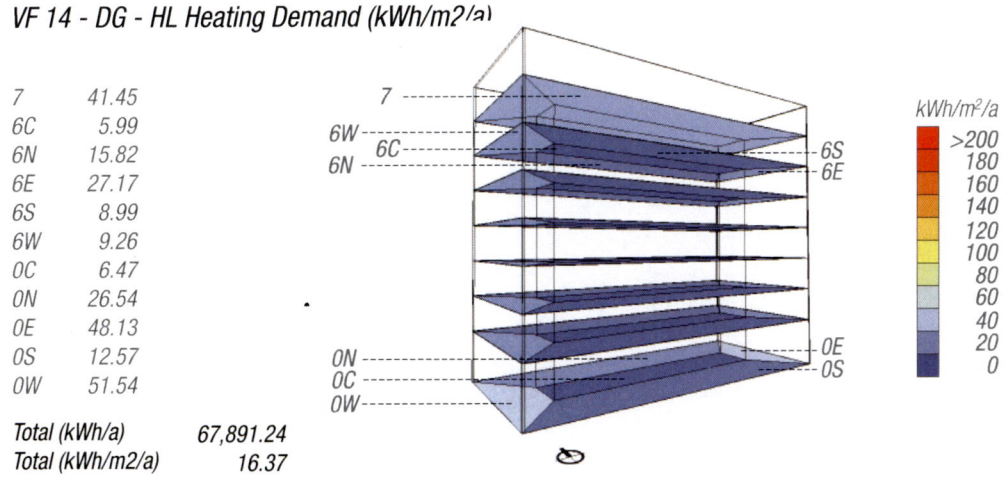

7	55.36
6C	12.05
6N	24.90
6E	39.89
6S	15.92
6W	42.47
0C	14.42
0N	39.02
0E	66.69
0S	21.10
0W	71.00
Total (kWh/a)	105,658.08
Total (kWh/m2/a)	25.48

Fig.133.   VF14 Heating Demand (kWh/m2/a) Double Glazing Facade, HH-Schedule

The double glazing facade of VF14 with a U-Value of 1.7 W/m2/K, clearly has the lowest heating demand values. More than 75% of the total annual heating demand can be reduced by changing the SG facade with a double glazing facade. In absolute terms this is a reduction from 425,230.74 kWh/a to 105,658.08 kWh/a or 102.53 kWh/m2/a to 25.48 kWh/m2/a using the higher heating schedule (HH schedule).

Simulating with the HL schedule, and comparing the extreme values of VF14 an overall reduction of 84.04% can be achieved.

VF 14 - DG - HL Heating Demand (kWh/m²/a)

7	41.45
6C	5.99
6N	15.82
6E	27.17
6S	8.99
6W	9.26
0C	6.47
0N	26.54
0E	48.13
0S	12.57
0W	51.54
Total (kWh/a)	67,891.24
Total (kWh/m2/a)	16.37

Fig.134.   VF14 Heating Demand (kWh/m2/a) Double Glazing Facade, HL-Schedule

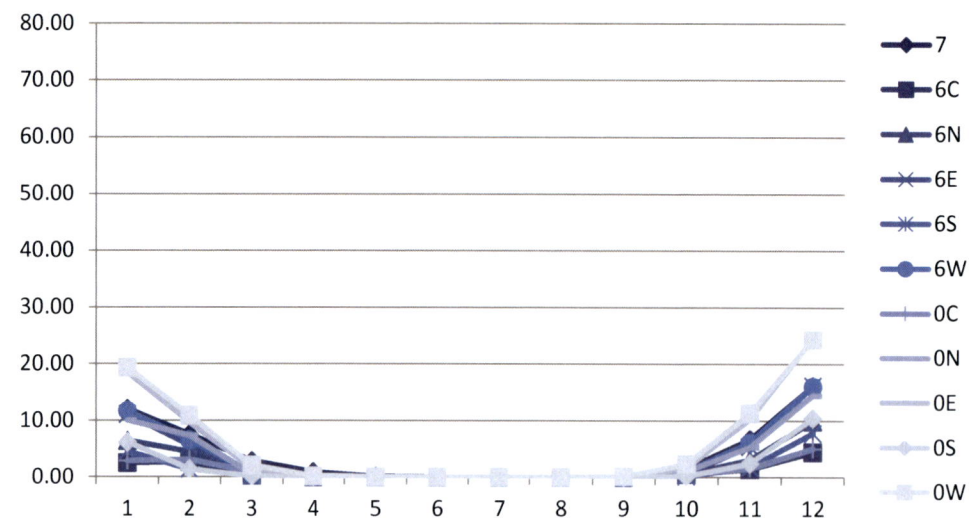

Fig.135. VF14 -Heating Demand (kWh/m2/month), HH-schedule, Double ETFE Facade, all zones

The overall heating demand in this case is of 945,657.54 kWh/a or 234.54 kWh/m2/a. The biggest reduction in terms of heating energy demand per square meter, compared to the other building envelopes, can be observed on the top level, in zone 8, where the value is reduced from 253.47 kWh/m2/a (SG) to 116.02 kWh/m2 (ETFE) to only 48.64 kWh/m2 (DG).
Values in the east and west zones of 274.19 kWh/m2/a and 275.42 kWh/m2/a in an exemplary 3.5 m high storey or 386.68 kWh/m2 and 388.77 kWh/m2/a on the ground level with a zone height of 5.00 m. Similar high values we have in the central zones where no solar heat gain is measured with a heating demand of 258.20 kWh/m2/a (7C) and 353.25 kWh/m2/a on the ground level.

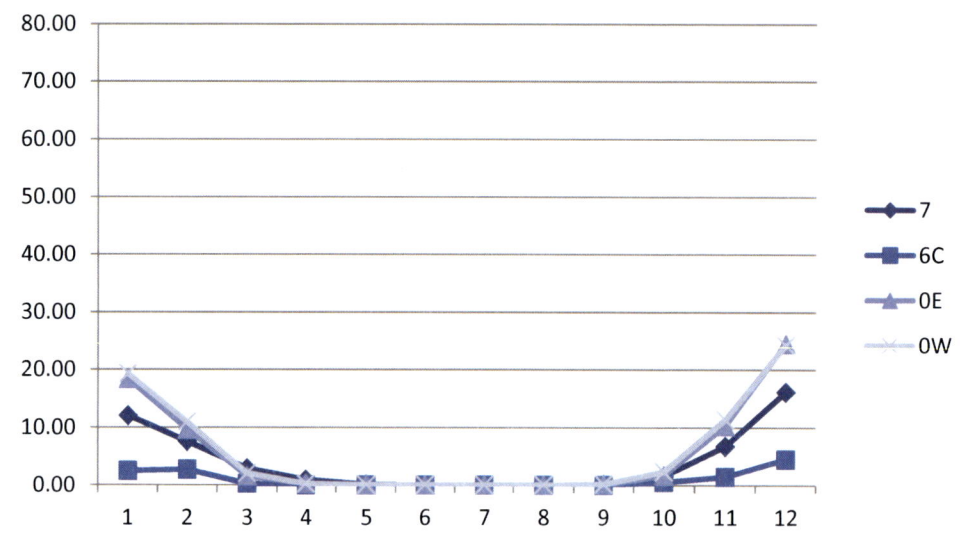

Fig.136. VF14 -Heating Demand (kWh/m2/month), HH-schedule, Double Glazing Facade, selected zones

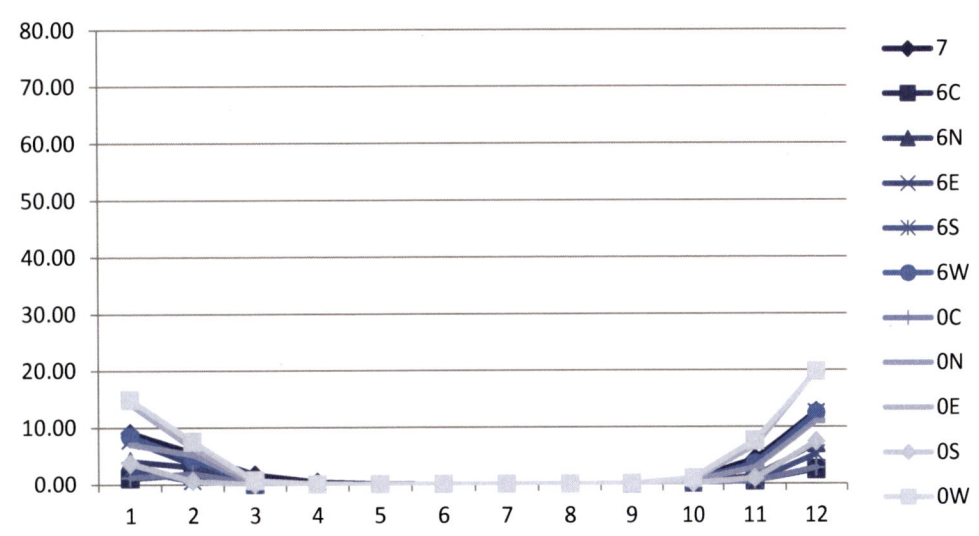

Fig.137. VF14 -Heating Demand (kWh/m2/month), HL-schedule, Double Glazing Facade, all zones

On examining the monthly results we see the same pattern for the other building envelopes, but with a reduction of more than 20% seen for the zones with the highest heating demand in winter. The months from January to March and from November and December make the biggest difference to the annual heating demand. The results with double glazing are more than 28% lower compared to a single glazing building envelope and 13.25% lower compared to double ETFE-facade.

The top level, zone 8, comprises only around 25% of the east and west oriented zones. The solar heat gain throughout the year in zone 8 is 2.3 times higher than in the east and west zones of 7E and 7W and around the double of the solar heat gain in 0E and 0W.

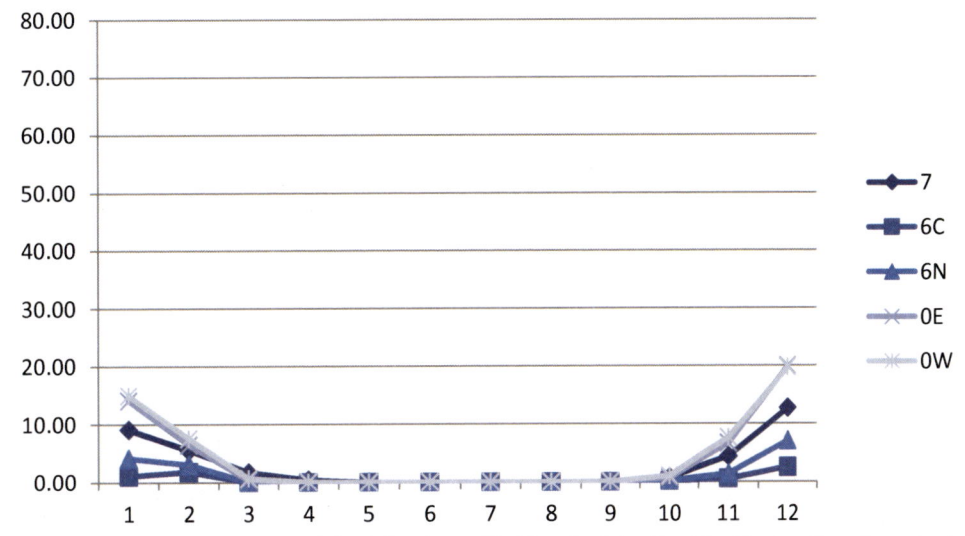

Fig.138. VF14 -Heating Demand (kWh/m2/month), HL-schedule, Double Glazing Facade, selected zones

5.7.3. VF7 - Heating demand

SINGLE GLAZING FACADE

VF 7 - SG - HH Heating Demand (kWh/m2/a)

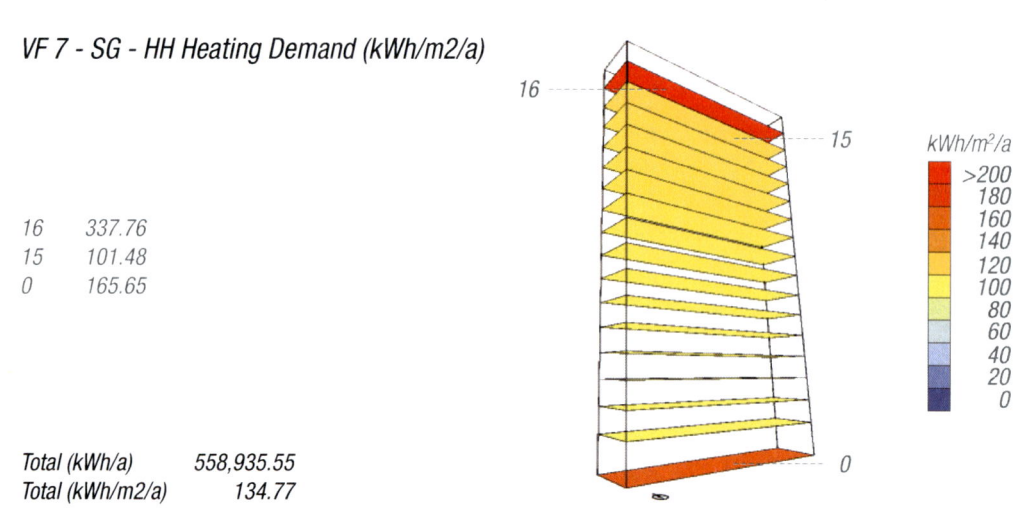

16	337.76
15	101.48
0	165.65

Total (kWh/a)   558,935.55
Total (kWh/m2/a)   134.77

Fig.139.   VF7 Heating Demand (kWh/m2/a) Single Glazing Facade, HH-Schedule

VF7 is a special example in terms of the simulation results. Its dimensions of 36.00 m x 7.20 m x 61.00 m and a building volume of 15,811.20 m3 has the highest A/V-ratio of 0.36 (compared to 0.26 of VF14 and 0.28 of VF32). This leads to the highest heating demand. The heating demand for VF7 with the SG building envelope is 558,935.55 kWh/a or 134.77 kWh/m2/a and is reduced by changing from HH schedule to HL schedule by 24.6% to 421,515.12 kWh/a or 95.66 kWh/m2/a.

VF 7 - SG - HL Heating Demand (kWh/m2/a)

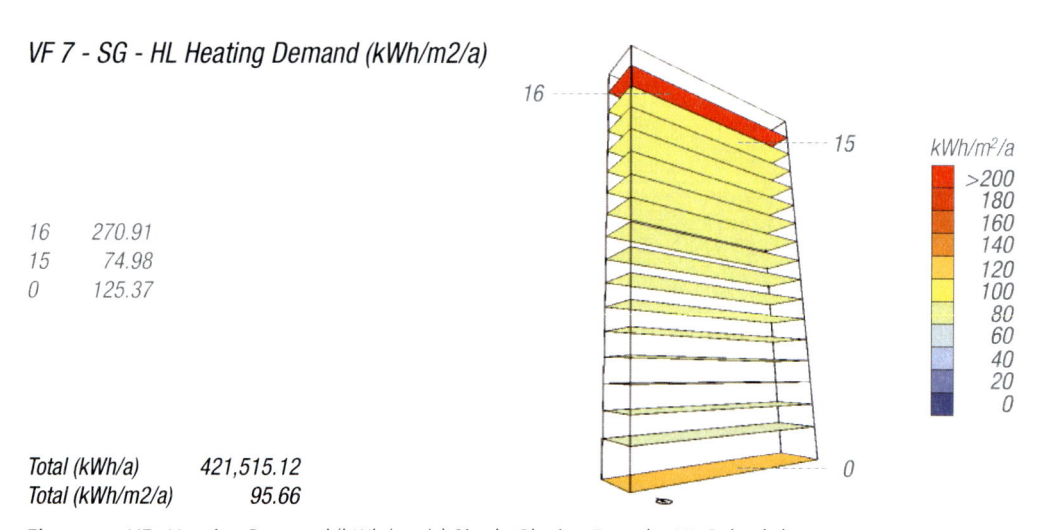

16	270.91
15	74.98
0	125.37

Total (kWh/a)   421,515.12
Total (kWh/m2/a)   95.66

Fig.140.   VF7 Heating Demand (kWh/m2/a) Single Glazing Facade, HL-Schedule

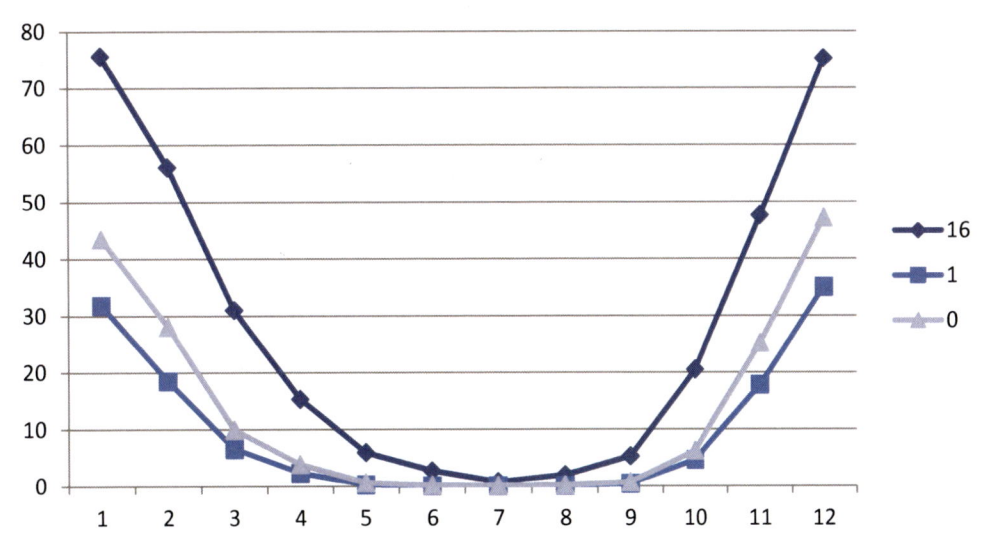

Fig.141. VF7 -Heating Demand (kWh/m2/month), HH-schedule, Single Glazing Facade, selected zones

VF7 with its reduced building depth, the highest A/V-ratio and the biggest north facing glass facade leads to the highest heat losses and the biggest heating demand. Compared to VF14 for SG facade, using the HH schedule the total amount is 31.4% higher. In addition we note on a comparison of the monthly values, that this VF has the shortest period where only little or no heating is needed. It also shows the biggest differences between the zone values, whereas heating demand on the lop level is more than three times higher compared to a typical level (HH schedule) or 3.6 times higher (HL schedule). Heating demand values for the ground floor, e.g. for the HH schedule, (level 0) are comparable to the simulation results of 0N both for VF14 (156.91 kWh/m2/a) and VF32 (165.43 kWh/m2/a).

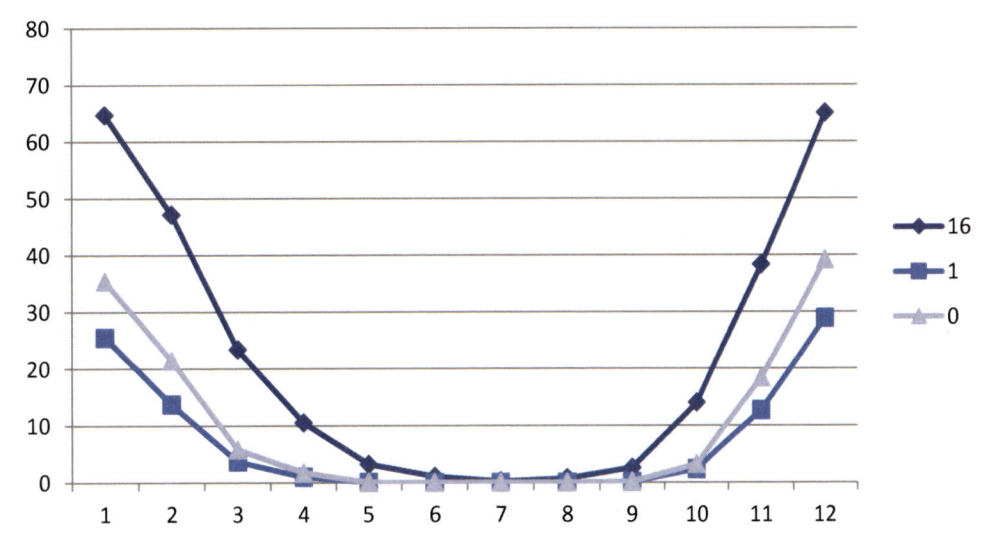

Fig.142. VF7 -Heating Demand (kWh/m2/month), HL-schedule, Single Glazing Facade, selected zones

DOUBLE ETFE FACADE

VF 7 - ETFE - HH Heating Demand (kWh/m2/

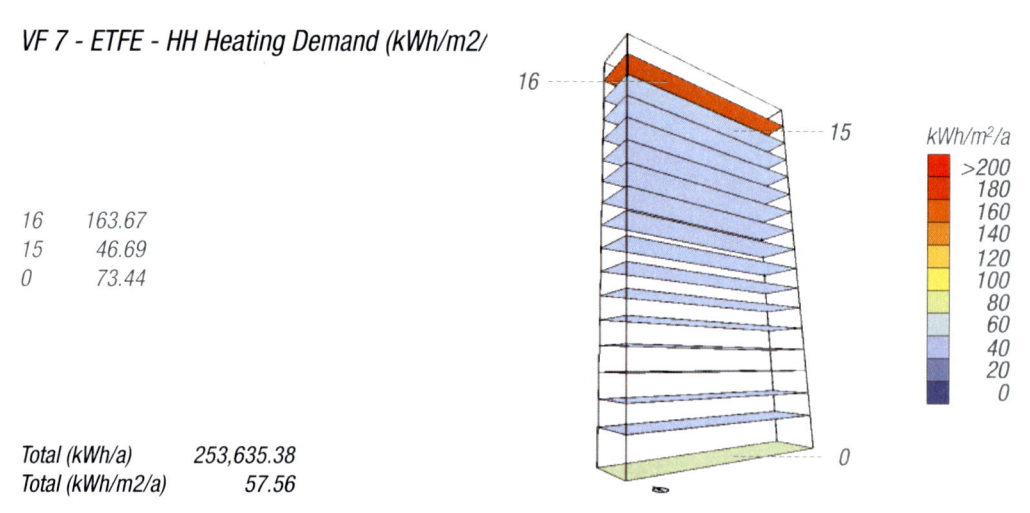

16	163.67
15	46.69
0	73.44

Total (kWh/a) 253,635.38
Total (kWh/m2/a) 57.56

Fig.143. VF7 Heating Demand (kWh/m2/a) Double ETFE Facade, HH-Schedule

Reducing the U-value of 5.88 W/m2/K of the single glazing facade to 2.9 W/m2/K by applying double ETFE-building envelope, the total annual heating demand can be reduced by 55% from 558,935.55 kWh/a to 253,635.38 kWh/a or 134.77 kWh/m2/a to 57.56 kWh/m2/a. The difference by applying the HL schedule is similar (42.65%) with a reduction from 95.66 kWh/m2/a to 40.80 kWh/m2/a.

Comparing the absolute results for total heating, this VF consumes 49% more energy than VF32 to provide the requested temperature by using HH schedule or 25% more by adapting HL schedule for simulation.

VF 7 - ETFE - HL Heating Demand (kWh/m2/

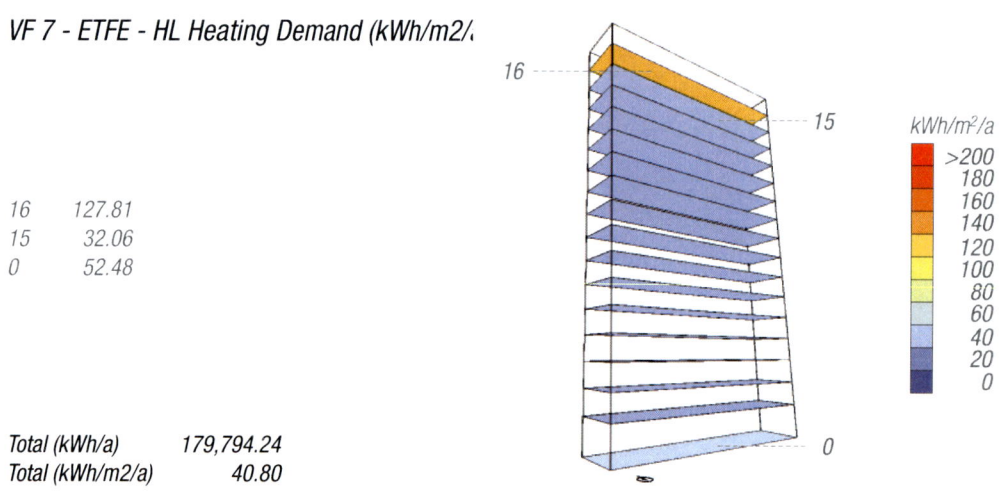

16	127.81
15	32.06
0	52.48

Total (kWh/a) 179,794.24
Total (kWh/m2/a) 40.80

Fig.144. VF7 Heating Demand (kWh/m2/a) Double ETFE Facade, HL-Schedule

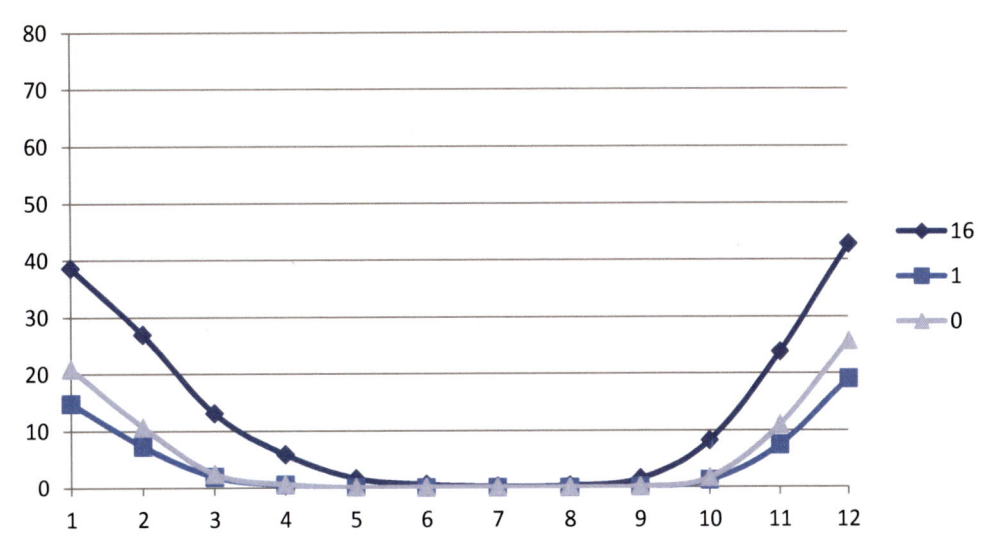

Fig.145. VF7 -Heating Demand (kWh/m2/month), HH-schedule, Double ETFE Facade, selected zones

Although the plateau with low or no heating demand during the warm season is increased, compared to SG-building envelope, heating demand during winter times is still very high. The monthly results show that the during top level (zone 16) during January energy consumption is roughly twice that of the ground level and more than 60% of zone 15.

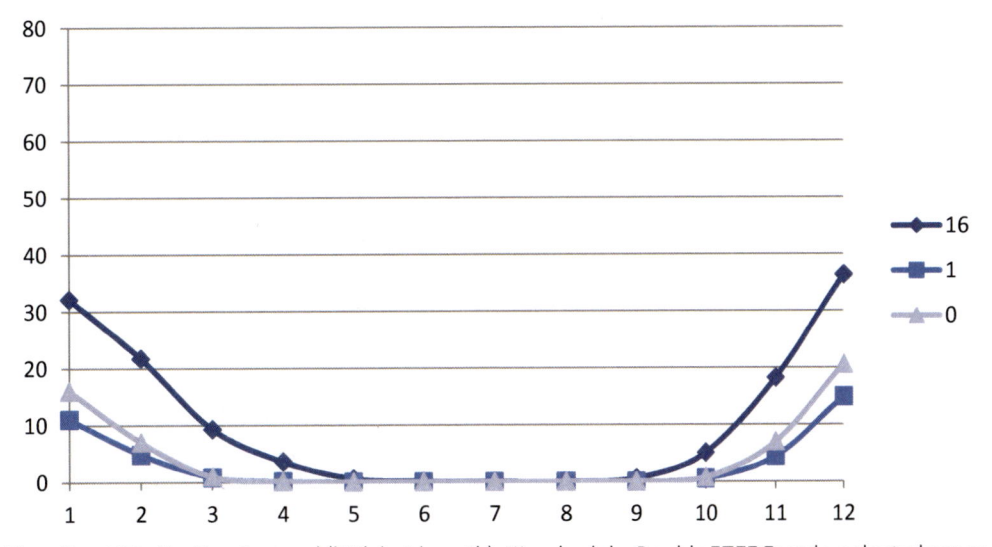

Fig.146. VF7 -Heating Demand (kWh/m2/month), HL-schedule, Double ETFE Facade, selected zones

DOUBLE GLAZING FACADE

VF 7 - DG - HH Heating Demand (kWh/m2/a)

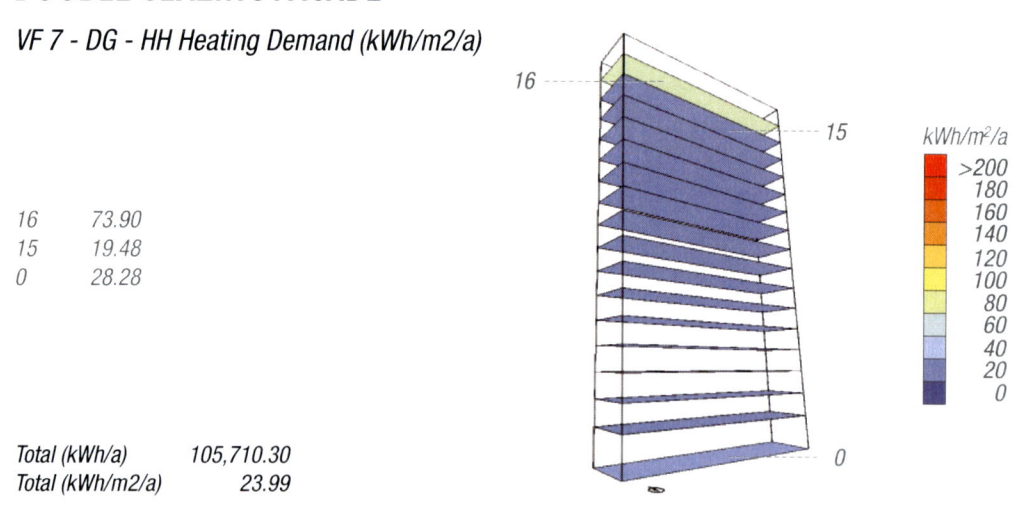

16	73.90
15	19.48
0	28.28

Total (kWh/a)   105,710.30
Total (kWh/m2/a)   23.99

Fig.147.   VF7 Heating Demand (kWh/m2/a) Double Glazing Facade, HH-Schedule

The last example in this Chapter shows the results of VF7 with the double glazing-facade. This typology, enclosed with the double glazing material offers the lowest values of the analyzed VF building types. The heating demand is comparable to VF32 DG, in absolute terms, although somewhat higher. When using the HH schedule VF7 has a per square meter heating demand of 23.99 kWh/m2/a (21.41 kWh/m2/a for VF32 DG) and 15.01 kWh/m2/a (14.08 kWh/m2/a for VF32 DG) by using the HL schedule.

The total reduction from SG HH to DG HL is the highest for VF7. In absolute numbers it is 492,777.11 kWh/a or 119.76 kWh/m2/a, which is a reduction of 88.17%.

VF 7 - DG - HL Heating Demand (kWh/m2/a)

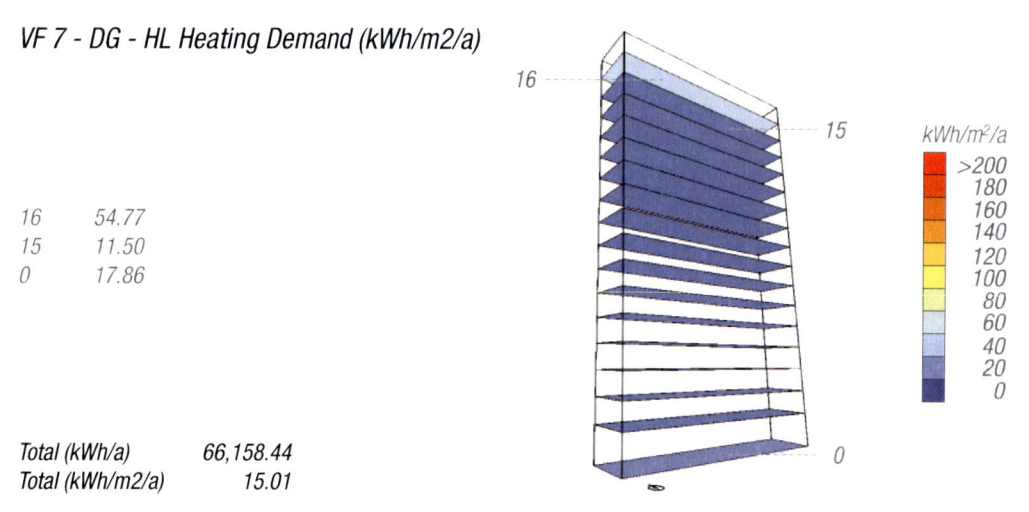

16	54.77
15	11.50
0	17.86

Total (kWh/a)   66,158.44
Total (kWh/m2/a)   15.01

Fig.148.   VF7 Heating Demand (kWh/m2/a) Double Glazing Facade, HL-Schedule

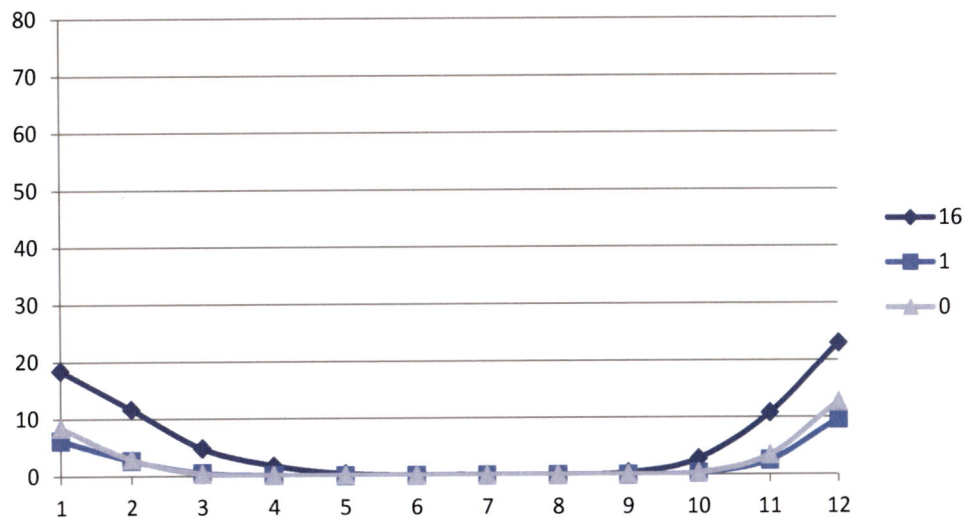

Fig.149. VF7 -Heating Demand (kWh/m2/month), HH-schedule, Double Glazing Facade, selected zones

Heating demand in winter at the most exposed zone is around 20 kWh/m2/month (HH schedule) or below. Recalling the figures achieved with SG facade this is a reduction of more than 50 kWh/m2/month.

By February every zone is already below a heating demand of 10 kWh/m2/month, from March to the end of September in most of the zones heating demand is close to 0. It is only by the end of September that the curves strongly increase again and reach the highest values for heating demand in December.

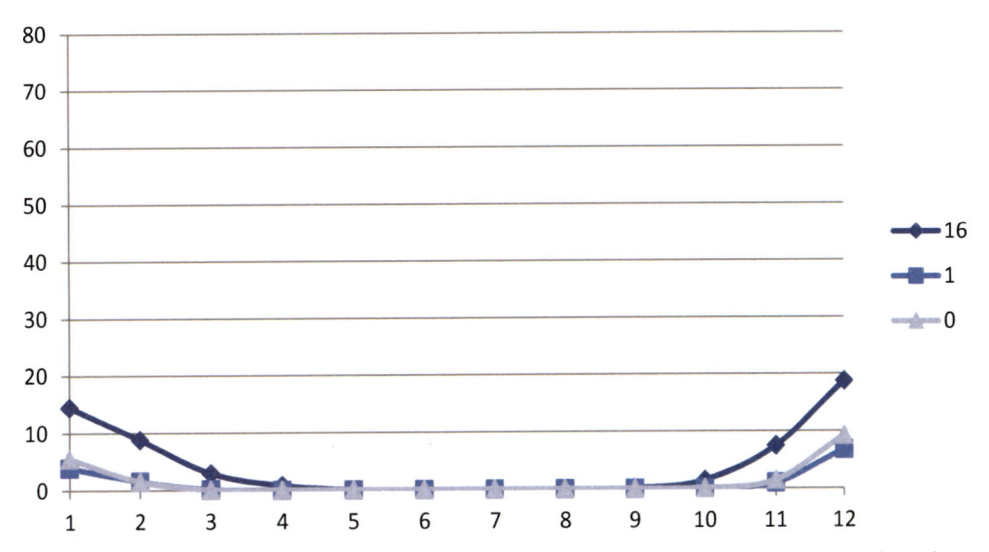

Fig.150. VF7 -Heating Demand (kWh/m2/month), HL-schedule, Double Glazing Facade, selected zones

5.8. Annual Simulation Results - Discussion

The following pages now conclude the Chapter "Vertical Farm - simulation model" with the diagrammatic representation of the total annual heating demand for the building entity and the total annual heating demand per square meter.

The results will be converted into values which represent the total primary energy supply (TPES), based on conversion factors specific for Austria, Vienna, using the actual numbers for heating and power-mix.
These results then will be compared to the actual situation, both in terms of energy consumption and land consumption, based on a variety of assumptions explained below.

For remembering the following abbreviations on the next pages will be used:

- VF7   Vertical Farm 36 m x 7.20 m x 61 m (V=15,811.20 m3), A/V-ratio: 0.36
- VF14 Vertical Farm 36 m x 14.40 m x 33 m (V=15,292.80 m3), A/V-ratio: 0.26
- VF32 Vertical Farm 36 m x 32.00 m x 12 m (V=13,824.00), A/V-ratio: 0.28
- SG   single glazing facade (U=5.88 W/m2/K, SHGC=0.80, VT=0.85)
- ETFE double ETFE facade (U=2.90 W/m2/K, SHGC=0.65, VT=0.85)
- DG   Double ETFE facade (U=1.70 W/m2/K, SHGC=0.70, VT=0.91)
- HH   Heating-High schedule from 14°C to 24°C
- HL   Heating-Low schedule from 12°C to 22°C
- L   Lighting Demand
- GH   Greenhouse
- EE   Embodied Energy

SG HH

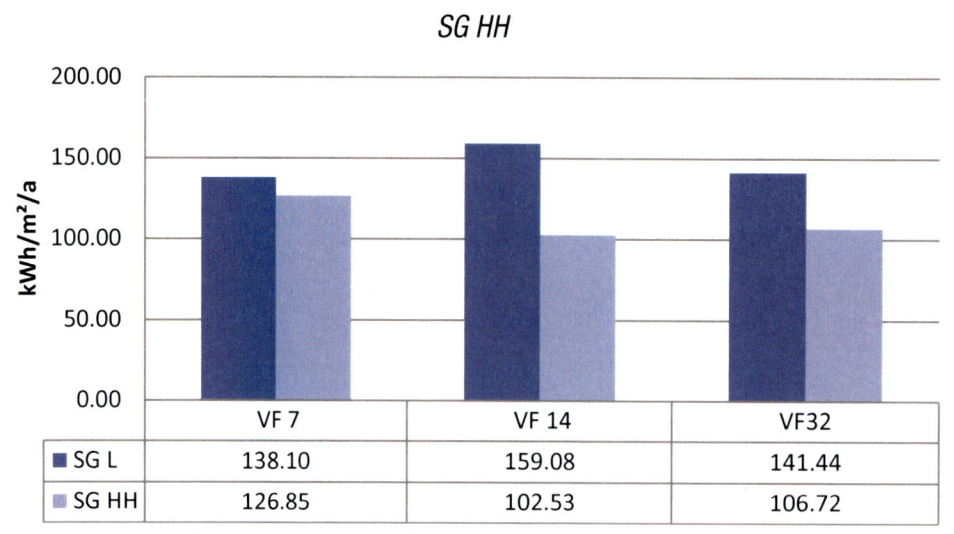

	VF 7	VF 14	VF32
■ SG L	138.10	159.08	141.44
■ SG HH	126.85	102.53	106.72

Tab.15. All VF - Heating Demand for SG and HH-Schedule (kWh/m2/a)

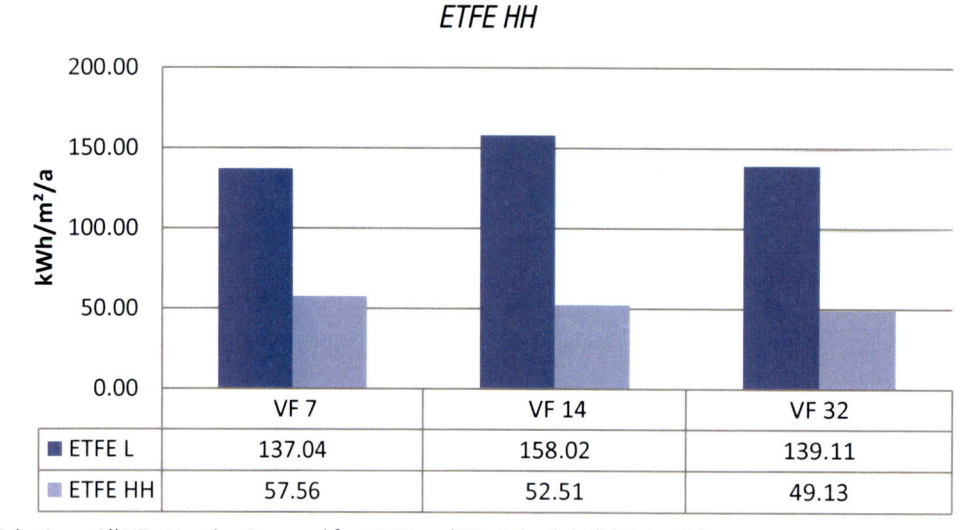

ETFE HH

	VF 7	VF 14	VF 32
■ ETFE L	137.04	158.02	139.11
■ ETFE HH	57.56	52.51	49.13

Tab.16. All VF - Heating Demand for ETFE and HH-Schedule (kWh/m2/a)

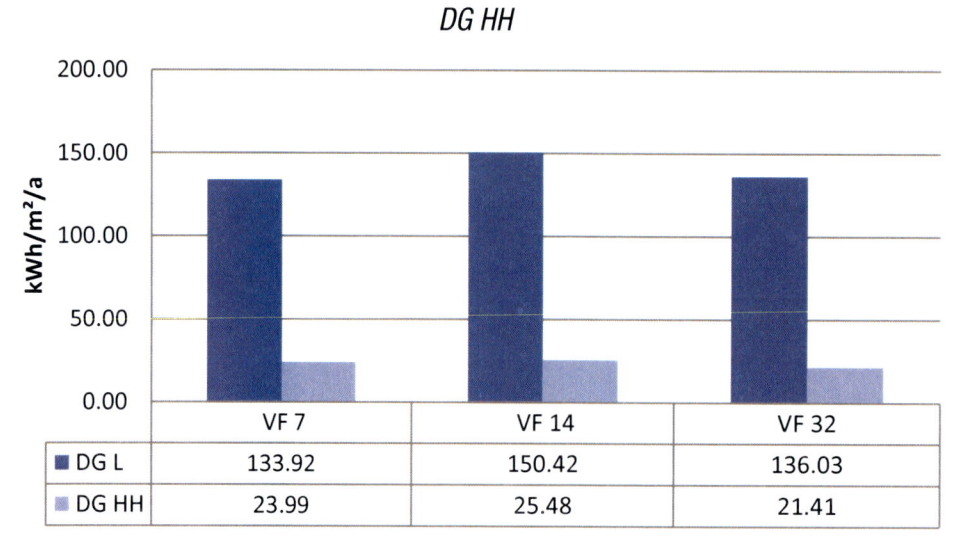

DG HH

	VF 7	VF 14	VF 32
■ DG L	133.92	150.42	136.03
■ DG HH	23.99	25.48	21.41

Tab.17. All VF - Heating Demand for DG and HH-Schedule (kWh/m2/a)

The diagrams show the comparison of all VF in terms of expectable lighting- and heating demand (using HH-schedule) per square metre.

VF32, summing up lighting- and heating demand, has the lowest results (DG HH) with a sum of 157.44 kWh/m2/a (136.03 kWh/m2/a per lighting and 21.41 kWh/m2/a for heating, closely followed by VF7 (DG HH) with 157.91 kWh/m2/a (133.92 kWh/m2/a for lighting and 23.99 kWh/m2/a for heating.

The detailed analysis provided in the pages above made visible that the difference of the results for lighting is negligible. The diagrams on the left with the total end energy values per square meter show the biggest differences.

Whereas for VF7 lighting demand can be reduced for 3% by changing SG with DG and for VF32 3.9%, the highest difference is shown by VF14 with a reduction of 5.4%. Results for ETFE are between the results of SG and DG.

A remarkable impact on reducing energy demand, however, is shown by the different U-values of the three building envelopes chosen (SG: U=5.88 W/m2/K; ETFE: U=2.90 W/m2/K, DG: U=1.70 W/m2/K).

VF7 SG shows the highest heating demand (HH-schedule) with 126.85 kWh/m2/a, followed by VF32 and VF14 (106.72 kWh/m2/a and 102.53 kWh/m2/a respectively). By changing the building envelope to ETFE, heating demand for VF7 can be reduced by over 54% or by 81.1% with a DG-facade.

Accordingly, VF14 SG shows a heating demand of 102.53 kWh/m2/a (HH-schedule). This value is minimized to 52.51 kWh/m2/a with ETFE or to 25.48 kWh/m2/a with DG, a total reduction from SG to DG of 75.15%.

The heating demand of VF32 for SG is 106.72 kWh/m2/a and gets reduced by 53.07% for ETFE to a value of 49.13 kWh/m2/a. The lowest heating demand is 21.41 kWh/m2/a for DG which is a reduction of nearly 80% compared to SG.

What we see is that lighting demand results vary much stronger between the building types, especially influenced by the building depth. Differences in heating demand are higher depending on the building envelope.

SG HL

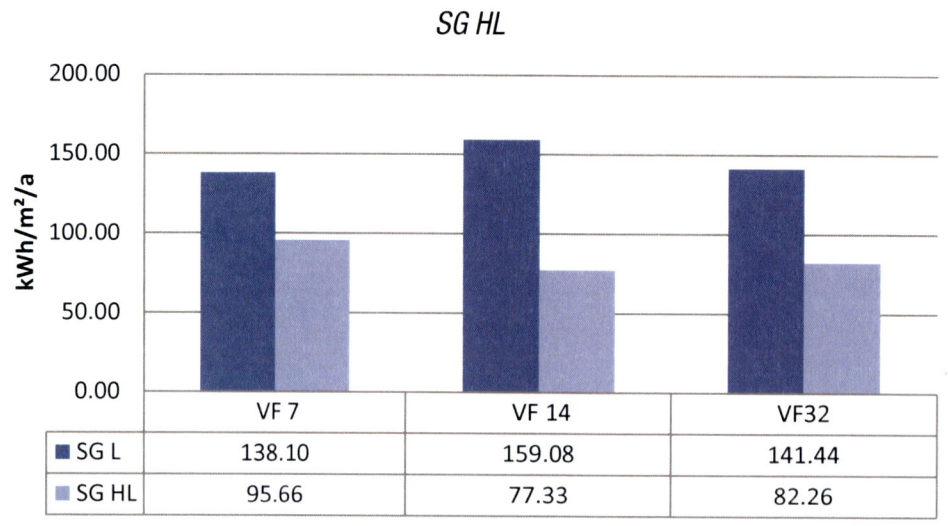

	VF 7	VF 14	VF32
SG L	138.10	159.08	141.44
SG HL	95.66	77.33	82.26

Tab.18. All VF - Heating Demand for SG and HL-Schedule (kWh/m2/a)

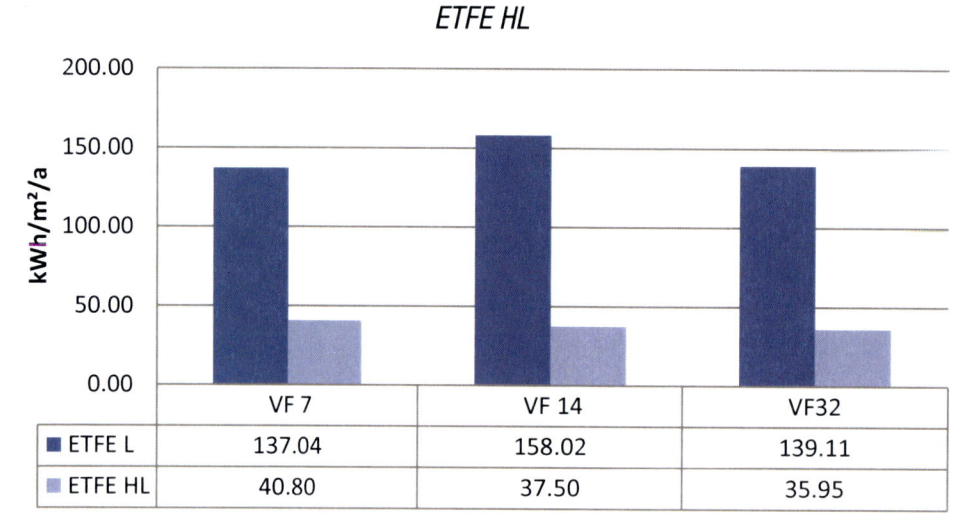

ETFE HL

	VF 7	VF 14	VF32
ETFE L	137.04	158.02	139.11
ETFE HL	40.80	37.50	35.95

Tab.19. All VF - Heating Demand for SG and HL-Schedule (kWh/m2/a)

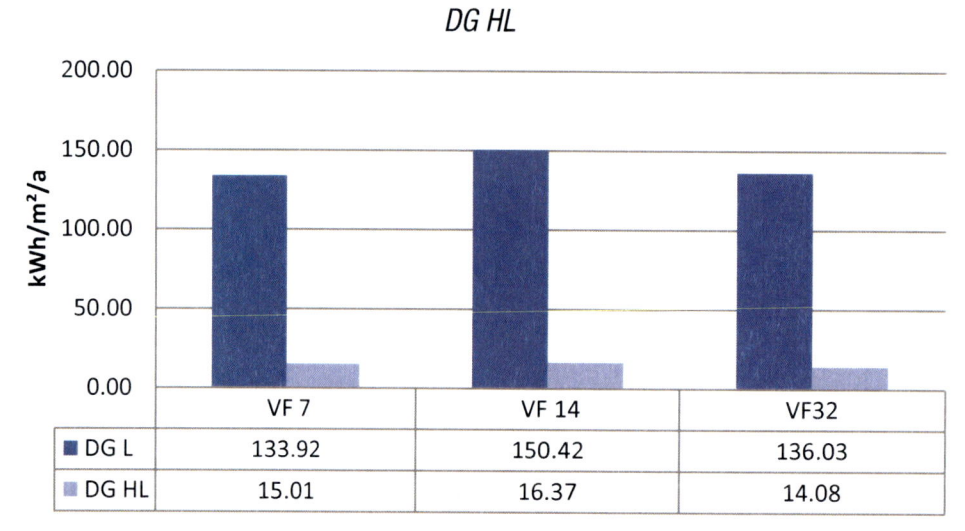

DG HL

	VF 7	VF 14	VF32
DG L	133.92	150.42	136.03
DG HL	15.01	16.37	14.08

Tab.20. All VF - Heating Demand for DG and HL-Schedule (kWh/m2/a)

The effect of reducing the temperature by 2°C (changing from HH-schedule to HL-schedule) varies from 25% (VF7 SG) to 29.12% (VF7 ETFE) to 28.44% (VF7 DG). Similar percentages are perceptible for VF14 and VF32.
Heating demand with the HL-schedule follows the same pattern: the highest heating demand is for VF7, followed by VF32 and, subsequently VF14. VF7 SG shows the highest heating demand (HH-schedule) with 95.66 kWh/m2/a, followed by VF32 and VF14 (77.33 kWh/m2/a and 82.26 kWh/m2/a respectively). With ETFE, heating demand for VF7 can be reduced by over 57.35% or by 84.31% with a DG facade.

Accordingly, VF14 SG shows a heating demand of 77.33 kWh/m2/a (HL-schedule). This value is minimized to 37.50 kWh/m2/a with ETFE or to 16.37 kWh/m2/a with DG, a total reduction from SG to DG of 78.84%.
The heating demand of VF32 for SG is 82.26 kWh/m2/a and is reduced by 56.30% for ETFE to a value of 35.95 kWh/m2/a. The lowest heating demand of all the Vertical Farm types simulated amounts to 14.08 kWh/m2/a for DG which is a reduction of nearly 82.89% compared to SG.

By changing HH-schedule with HL-schedule VF32 no longer has the lowest total values. VF7 has an overall end energy demand of 148.93 kWh/m2/a (133.92 kWh/m2/a for lighting and 15.01 kWh/m2/a for heating. VF32's total end energy demand is 150.11 kWh/m2/a (136.03 kWh/m2/a for lighting and 14.08 kWh/m2/a for heating.

This simulation results show how strongly the building type influences the simulation results. Whereas lighting demand is strongly dependent from the building type, heating demand is more influenced by the building envelope. The high differences in lighting demand of course are provoked by chosing a cultivar with high lighting requirement for optimum plant growth for crops with high ligh requirements such as *L. esculentum*.
Low lighting demand show VF32 with its compactness which led to the biggest top level surface, where a third of the cultivars gain daylight throughout the whole day-hours, only reduced by the building envelope. Maximizing all facades to all cardinal directions (nearly) equally, also positively influences the relatively low lighting demand. The fact that 1,144 m2 (0C= 572 m2 and 1C = 572m2), which is 33.10% of the cultivation area, are offset by 5m from the facade and therefor has the maximum lighting demand, still makes results compareable to VF7.
Compactness, activation of the top level for cultivation and optimizing the building orientation towards the sunpath seems to be the recommended way for following studies to optimize the building shape for Vertical Farming.
The difference of the results to VF7 only are around 2.4% (SG) to 1.5% (DG). VF7 with its minimized building depth might also be worth to be investigated more deeply for future Vertical Farm building typology studies. The building depth of 7.2m and south orientation has the lowest light requirement of all three Vertical Farm building types analyzed. Although, through its highest A/V ratio of 0.36 heating demand is the higher, this picture doesn't add up in the moment when the values, shown on this pages, get changed from end energy use to total primary energy supply (TPES), visualized on the next pages.

VERTICAL FARM - SIMULATION MODEL

	VF 7			VF14		
	SG	ETFE	DG	SG	ETFE	DG
LIGHT total (kWh/a)		603,857.60	590,100.34	659,717.57	655,334.90	623,836.53
HH total (kWh/a)		253,635.38	105,710.30	425,230.74	217,759.86	105,658.08
HL total (kWh/a)		179,794.24	66,158.44	320,719.20	155,533.59	67,891.24
LIGHT (kWh/m²/a)		137.05	133.93	159.08	158.02	150.42
HH/ (kWh/m²/a)		57.57	23.99	102.53	52.51	25.48
HL (kWh/m²/a)		40.81	15.02	77.33	37.50	16.37
LIGHTING total (kWh/a)		603,857.60	590,100.34		655,334.90	623,836.53
HEATING total [ØHH/HL] (kWh/a)		216,714.81	85,934.37		186,646.73	86,774.66
TPES LIGHTING total (kWh/a)		1,582,106.91	1,546,062.89		1,716,977.43	1,634,451.72
TPES HEATING total [ØHH/HL] (kWh/a)		266,559.22	105,699.28		229,575.47	106,732.83
TPES (kWh/a)		1,848,666.13	1,651,762.17		3,145,186.06	2,890,430.36
TPES/m² LIGHTING		359.08	350.90		414.01	394.11
TPES/m² HEATING [ØHH/HL]		60.50	23.99		55.36	25.74
TPES/m²		445.76	398.28		780.06	716.87
L CO2 (kg)		166,664.70	162,867.69		180,872.43	172,178.88
H CO2 (kg)		187,799.71	183,521.21		203,809.15	194,013.16
CO2 (t)		354.46	346.39		384.68	366.19
Cultivation Area		4,406.40	4,406.40		4,147.20	4,147.20
GH-Plants eq.		10,233.86	10,233.86		9,631.87	9,631.87
GH-Yield eq.		330,480.00	330,480.00		311,040.00	311,040.00
VF-L.esculentum Stems	4,680.00	4,680.00	4,680.00		4,368.00	4,368.00
VF-Plants		11,700.00	11,700.00		10,920.00	10,920.00
VF-Increasing Factor (Bedsize)		1.14	1.14		1.13	1.13
VF-Yield		377,825.62	377,825.62		352,637.24	352,637.24
Greenhouses eq. (3.6x36x3.2m) = 129.6 m²		67.80	67.80		63.28	63.28
Greenhouse area eq. (ha):		0.879	0.879		0.820	0.820
Soil based eq. (ha):		1.399	1.399		1.306	1.306
Plant Biomass kg (end year)		93,600.00	93,600.00		87,360.00	87,360.00
C (kg)		21,996.00	21,996.00		20,529.60	20,529.60
fruit yield		311,044.00	311,044.00		302,400.00	302,400.00
C (kg) of fruit yield		8,491.50	8,491.50		8,255.52	8,255.52
∑ C (kg) Plant and Yield		30,487.50	30,487.50		28,785.12	28,785.12
CO2 (kg)		111,889.13	111,889.13		105,641.39	105,641.39
CO2 (t)		111.89	111.89		105.64	105.64
Δ CO2		242.58	234.50		279.04	260.55
CO2/kg L.esculentum		0.62	0.60		0.74	0.69

Tab.21. All VF TPES, Area, Yield, C and CO2-comparison

5.8.1. Total Primary Energy Supply for the Vertical Farms

The final energy demand of the various Vertical Farms with their different building envelopes is converted into primary energy values here to make a comparison possible between the total primary energy supply involved here with the current situation in soil based agriculture. The next page show averaged numbers. An „ideal" VF model gets generated with averaged values (HH-schedule and HL-schedule) to be used as a simplified picture of expectable TPES-values for comparative reasons.

The Austrian policy guideline OIB RL 6 (October 2011) is used for this conversion with the conversion factor for heating defined as 1.23 and the conversion factor for power (typical electricity mix in Austria) as 2.62. The conversion factors from kWh to CO2eq. were also retrieved from the same policy guideline.

The table on this page also provides an overview of the total primary energy consumption for heating and lighting (with both the HH schedule and the HL schedule) for the entire Vertical Farm per square meter. In a second step the HH and HL results were averaged to obtain a single value for each Vertical Farm with the specific building envelopes. These figures were used to determine the CO2eq. emissions for every building.

	VF 32		Ø VF7/VF32
SG	ETFE	DG	DG
488,801.98	480,773.03	470,115.06	530,107.70
368,833.83	169,780.09	73,999.79	89,855.05
284,286.09	124,237.60	48,654.25	57,406.35
141.44	139.11	136.03	134.98
106.72	49.13	21.41	22.70
82.26	35.95	14.08	14.55
488,801.98	480,773.03	470,115.06	530,107.70
326,550.96	147,008.85	61,327.02	73,630.70
1,280,661.19	1,259,625.34	1,231,701.46	1,388,882.17
101,668.75	180,820.88	75,432.23	90,565.75
1,835,884.11	1,764,533.52	1,625,665.19	1,638,713.68
370.56	364.47	356.40	353.65
110.22	52.32	21.83	22.91
577.16	510.57	470.39	376.56
	132,693.36	129,751.76	146,309.73
152,012.42	149,520.41	146,205.78	164,863.49
286.93	282.21	275.96	311.17
3,456.00	3,456.00	3,456.00	**3,931.20**
8,026.56	8,026.56	8,026.56	9,828.00
259,200.00	259,200.00	259,200.00	294,840.00
3,978.00	3,978.00	3,978.00	4,329.00
9,945.00	9,945.00	9,945.00	10,822.50
1.24	1.24	1.24	1.19
321,151.78	321,151.78	321,151.78	386,732.51
57.63	57.63	57.63	69.40
0.747	0.747	0.747	0.899
1.189	1.189	1.189	1.43
79,560.00	79,560.00	79,560.00	86,580.00
18,696.60	18,696.60	18,696.60	20,346.30
259,200.00	259,200.00	259,200.00	386,732.51
7,076.16	7,076.16	7,076.16	10,557.80
25,772.76	25,772.76	25,772.76	30,904.10
94,586.03	94,586.03	94,586.03	113,418.04
94.59	94.59	94.59	113.42
192.34	187.63	181.37	197.76
0.59	0.58	0.56	0.51

Subsequently all of the VF7, VF14 and VF32 final results were averaged for the building envelopes ETFE and DG. The SG results were ignored: as mentioned at the beginning of this chapter this glazing material has only been chosen for use as a comparison with classical glass-covered greenhouses on the one hand and on the other the use of SG is against building law for this specific building typology.

	Ø VF7/VF32
	DG
TPES LIGHTING total (kWh/a)	1,388,882.17
TPES HEATING total [ØHH/HL] (kWh/a)	90,565.75
TPES (kWh/a)	1,638,713.68
TPES/m² LIGHTING	353.65
TPES/m² HEATING [ØHH/HL]	22.91
TPES/m²	376.56
L CO2 (kg)	146,309.73
H CO2 (kg)	164,863.49
CO2 (t)	311.17
Cultivation Area	3,931.20
GH-Plants eq.	9,828.00
GH-Yield eq.	294,840.00
VF-L.esculentum Stems	4,329.00
VF-Plants	10,822.50
VF-Increasing Factor (Bedsize)	1.19
VF-Yield	386,732.51
Greenhouses eq. (3.6x36x3.2m) = 129.6 m²	69.40
Greenhouse area eq. (ha):	0.899
Soil based eq. (ha):	1.43
Plant Biomass kg (end year)	86,580.00
C (kg)	20,346.30
fruit yield	386,732.51
C (kg) of fruit yield	10,557.80
$\sum$ C (kg) Plant and Yield	30,904.10
CO2 (kg)	113,418.04
CO2 (t)	113.42
Δ CO2	197.76
CO2/kg L.esculentum	0.51

Tab.22.   VF7/32, DG: TPES, Area, Yield, C and CO2-comparison

The cultivation area defines the number of plants (whereas 2.5 heads, or plants, at the beginning of the crop rotation are on one stem). 2.5 plants/m2 is a greenhouse average (Chapter 4.3.5 and 4.3.6) and is increased through a higher plant density. The VF-Increasing Factor leads to higher output (expected 75 kg/m2 with 2.5 plants/m2).

At the end of Tab. 21 (page before) and Tab. 22 a rough assumption of net primary production of carbon throughout the crop rotation can be found, with the weight of dry matter from a single *L. esculentum* plant assumed to be 20 kg by the end of the year.[1][2][3] The percentage of carbon in one ton dry matter is 47%.[4] The Carbon content of the fruit yield is defined as 2.73%.[5]

The averaged figure for stems (plant stems retrieved from the parametric simulation model) is 4,173 together producing 386,732.51 kg of *L. esculentum*. The dry matter resulting from all the plants is 86,580 kg. The carbon content of both plants and fruit is 30,904.10 kg which corresponds to 113.42 t CO2. The difference between the carbon dioxide produced and that captured through plant and fruit growth is 197.76 t or 0.51 kg/*L. esculentum* fruit yield per crop rotation.

---

1   Dry matter production of *L. esculentum* is difficult to compare (http://aob.oxfordjournals.org/content/75/4/369.full.pdf, retrieved 12.09.2015). Photomorphogenesis is dependent on a variety of factors. *L. esculentum*, the indeterministic tomato plant can reach a dry matter weight of up to 200% of the yielded fruit weight.
2   http://www.researchgate.net/post/How_can_organic_carbon_from_soil_be_converted_to_CO2_under_the_impact_of_burning_biomass, retrieved 12.09.2015
3   SMIL, V. 2008. Energy in Nature and Society, Cambridge, Mass., MIT Press. p.74 ff.
4   THEURL, M. C. 2008. CO2-Bilanz der Tomatenproduktion. Analyse acht verschiedener Produktionssysteme in Österreich, Spanien und Italien. Vienna: Institute of Social Ecology. p.29
    This value has set as a standard value by Mc.Groddy et al. in IPPC 2006
5   ibid.

> The resulting values for VF7/32 (DG) are: TPES for lighting 353.65 kWh/m2/a, 22.91 kWh/m2/a and a sum of 376.56 kWh/m2/a. CO2 emissions calculated with these values are 311.17 t CO2/a or 0.51 kg CO2/kg of *L. esculentum*.[1]

These final results in lighting, again, are based on the assumptions[2], that the low energy output in $\mu mol/m^{-2}/s^{-1}$ of LED (81 $\mu mol/m^{-2}/s^{-1}$ of blue light with 444 nm and red light at 660 nm[3], communicated as 300 $\mu mol/m^{-2}/s^{-1}$ 'Adjusted PAR')[4] compared to actually needed PAR-values (300 $\mu mol/m^{-2}/s^{-1}$) for *L. esculentum* are sufficient for optimum plant growth (see p. 202 ff.).

---

[1] http://www.oib.or.at/sites/default/files/aenderungen_oib-richtlinie_6_26.03.15_0.pdf, p.10, retrieved 21.10.2015: 276 g CO2 Electricity-Mix Austria per kWh and 311 g CO2 for heating per kWh

[2] Chapter 4. p.203

[3] see Appendix, Lumigrow Evaluation of Simulation Parameters

[4] DIN 5031-10:2013-12 (D) „Strahlungsphysik im optischen Bereich und Lichttechnik - Teil 10: Photobiologisch wirksame Strahlung, Größen, Kurzzeichen und Wirkungsspektren" http://www.beuth.de/de/norm/din-5031-10/195406448, retrieved 29.10.2015

5.8.2. Vertical Farming and Land use

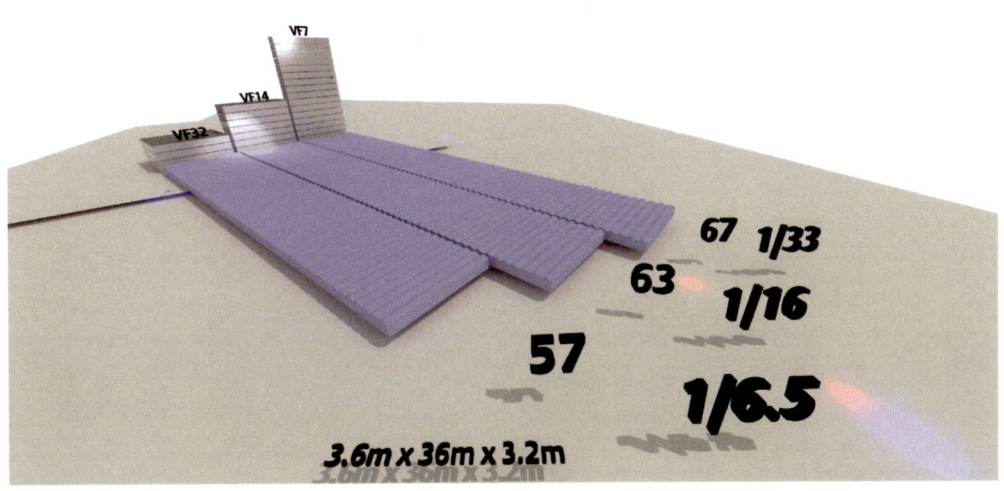

Fig.151.   Greenhouse equivalent to achieve same yield of VF

A very clear picture can be drawn in terms of land use reduction by referring to the crop yield per year. The diagram on this page shows the greenhouse equivalents theoretically needed to produce the same amount of *L. esculentum* during a whole crop rotation in diagram form. Depending on the ground floor dimensions, where the plant density of 60cm x 70cm was generated parametrically, the number of stems varies from one Vertical Farm to another. This explains the different ratios established when comparing verticalized greenhouse production to the horizontally oriented equivalent.

Dimensions of typical Venlo-Greenhouses (Reference Greenhouse for area comparison): Building depth = 3.6m, eave height = 3.2m, rooftop = 3.71m, building length: modularily expandable, for area comparative reasons the building length corresponds to the y-axis of all Vertical Farms, or the length of the North- and South-facade, 36.0m.
A=129.6 m2

2.5 plants per squaremeter for conventional greenhouse cultivation get assumed. Comparing the building footprint to the glasshouse footprint needed to reach the same fruit yield (same output of 75kg/m2 or more *L. esculentum* is assumed). The greenhouse buildings-equivalent for VF7 to achieve the same output is the highest with 67, followed by VF14 with 63 and 57 of VF32. In terms of area this is a ratio of 1/33 for VF7 (8,800 m2 / 259.2 m2), 1/16 for VF14 (8,200 m2 / 518.4 m2) and 1/6.5 for VF32 (7,500 m2 / 1,152 m2).

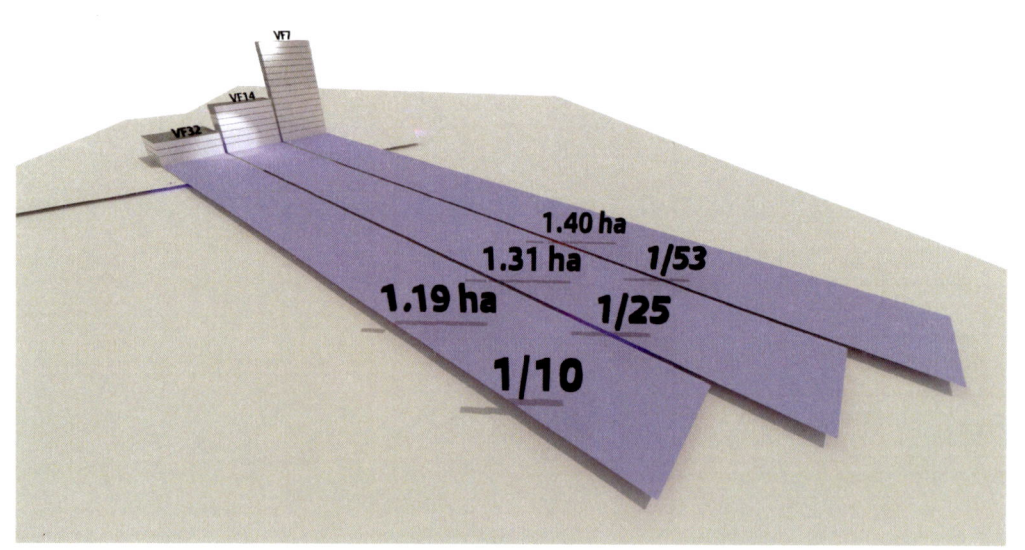

Fig.152. Soil Based equivalent to achieve same yield of VF

Comparing the building footprint with land needed for soil based agriculture (instead of land requirement for greenhouses) the three different Vertical Farm's land reduction factor increases even more. Comparing the yield output of the Vertical Farms to the output that can be achieved by traditional soil based agriculture, in this case for Austria, the reduction is due to the different plant types. Whereas in closed environments greenhouse optimized F1 hybrids (indeterministic tomatoes) are planted, outside on the fields deterministic plants are usually cultivated. The average yield on the fields in Austria is 27.2 kg/m2, whereas in greenhouses 43 kg/m2 are produced on average.[1]

For High-tech greenhouses, so for Vertical Farms, for year round crop with supporting artificial lighting and supplemental heating, yields of 75 kg/m2 are achievable. This already leads, for *L. esculentum*, to a reduction of two thirds of the needed area. Additional stacking of the cultivation area in sequence reduces the land area requirement even further.

VF32 (building footprint 1,152 m2) with its 3,978 *L. esculentum* stems produces 321,151.78 kg of fruit yield on a cultivation area of 3,456 m2 per year. The needed soil based area for this Vertical Farm type corresponds to 11,890 m2. The ratio between the building footprint and the soil based area equivalent is more than 1/10.

VF14 (building footprint 518.4 m2) with its 4,368 *L. esculentum* stems on a cultivation area of 4,147.20 m2 produces 352,637 kg of fruit yield on a cultivation area of 4,368 m2 per year. The needed soil based area for this Vertical Farm type corresponds to 13,060 m2. The building footprint therefor is 3,96 % of the corresponding area needed for soil based agriculture. The resulting ratio is 1/25.

Lastly, VF7 with a building footprint of 259.2 m2 and a cultivation area of 4,406.40 m2 achieves a yield of 377,825.62 kg/a. The soil based equivalent for this Vertical Farm is 13,990 m2. Only 1.85% of the traditional agricultural area is needed for the building footprint. The ratio of the building footprint to the soil based equivalent is 1/53.

---

[1] www.statistik-austria.at, retrieved 26.10.2014

5.8.3. Energy Land - Resumée

With some raw assumptions now we can draw a final picture of energy consumption compared to traditional greenhouse practises, theoretical High-Tech-greenhouses and conventional SBA. An additional comparative parameter gets introduced: Energy Land.[1]

L.esculentum	VF7	VF32	Conv.GH	HT-GH	SBA
Land Use	259.20	1,152.00	8,990.00	4,000.00	14,323.43
Cultivation Area	4,406.00	3,456.00	8,990.00	4,000.00	14,323.43
EMBODIED ENERGY	28.09	28.09	23.22	25.54	2.16
LIGHTING	350.90	356.40	261.60	254.92	0.00
HEATING	23.99	21.83	400.00	43.47	0.00
kWh/m2	402.98	406.32	684.82	323.93	2.16
kWh/a	1,775,544.33	1,404,226.56	6,156,497.18	1,295,719.86	30,938.60
TPES/m² Land Use	104,453.27	468,075.52	6,156,497.18	1,295,719.86	30,938.60
Energy Land (m2)	15,853	12,538	54,969	11,569	276
∑ Built up + NRG-Land	16,112.27	13,689.74	63,958.72	15,568.93	14,599.66

Tab.23. TPES VF, Greenhouse, SBA and Energy Land

To compare energy consumption between the different cultivation methods (VF, GH and SBA) energy land highlights the actual land needed to generate the requested amount of energy with renewables.

Tab. 23 on this page compares the land use, which corresponds to the building footprint of VF7, and VF32, the greenhouse-area equivalent calculated in the subchapter 5.8.1 and the soil based equivalent. Both, for greenhouses and SBA, the average yield of Austria gets taken. EE (embodied energy) values for conventional greenhouses are leant on the numbers taken from Theurl (ca. 0.1 kg of CO_2/kg of Tomato yield)[2]. A surplus of 10% for EE was calculated for HT-GH (High-Tech-Greenhouse) and an additional 10% for VF. For SBA it is the average of FAO's numbers calculated in TPES on page 91; numbers for retail, preparation and cooking got substracted. This value, then, was divided by the world's arable land (ca. 15,000,000 km2) and can be read as the world's average energy consumption for the food supply chain from the farm to the retail-gate.

---

1 CODY, B. 2014. The Role of Technology in Sustainable Architecture, Cloud-Cuckoo-Land, International Journal of Architectural Theory. cloud-cuckoo.net/fileadmin/issues_en/issue_33/article_cody.pdf, retrieved 27.10.2015
2 THEURL, M. C. 2008. CO 2-Bilanz der Tomatenproduktion. Analyse acht verschiedener Produktionssysteme in Österreich, Spanien und Italien. Vienna: Institute of Social Ecology. p. 80

Energy land got calculated by dividing TPES for VF, GH or SBA by 10% of the annual total horizontal radiation of Vienna (1,119.32 kWh/m2/a) and is an assumption of the effective power output of mono- or polycrystalline PV-cells, knowing that this efficiency potential is located on the low side of its potentials.

Lighting demand is retrieved from VF7 and VF32 DG results, for conventional GH and HT-GH the value for VF32 SG 2 was used and VF32 DG 2 respectively. Heating demand for conventional GH is oriented on the statistics of the upmentioned study.[1]

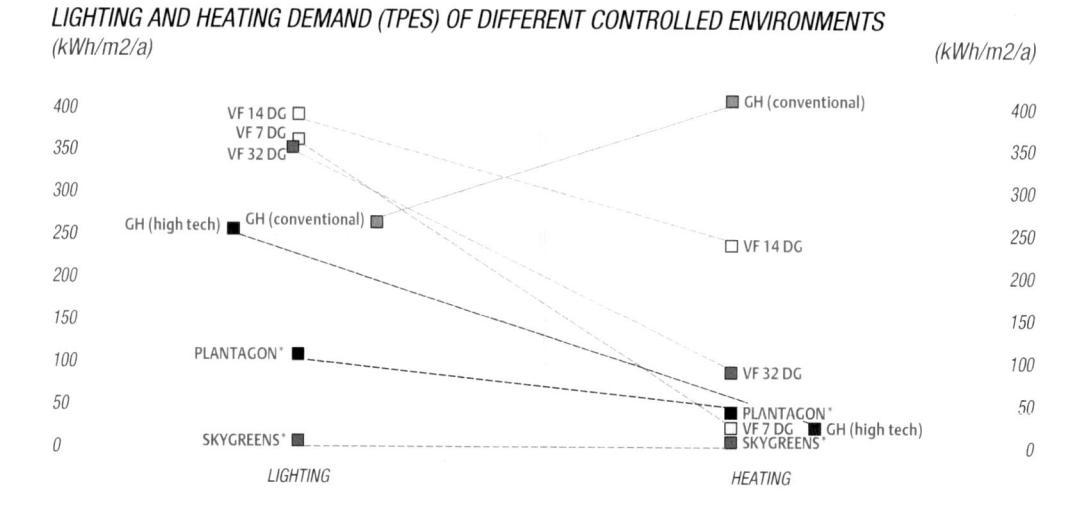

LIGHTING AND HEATING DEMAND (TPES) OF DIFFERENT CONTROLLED ENVIRONMENTS
(kWh/m2/a)   (kWh/m2/a)

* = assumptions

Fig.153.   TPES for lighting and heating of different controlled environments

---

[1]   THEURL, M. C. 2008. CO 2-Bilanz der Tomatenproduktion. Analyse acht verschiedener Produktionssysteme in Österreich, Spanien und Italien. Vienna: Institute of Social Ecology. p.32

The results clearly show that Vertical Farming, developed as a stacked greenhouse with intermediate levels for *L. esculentum* with conventional power mix and oil heating, although through its strong reduction in land use, in terms of energy still can't compete with traditional SBA. Encounting energy land needed to cover the energy demand of the stacked greenhouse, is still higher than the actual energy-intense practise of SBA.

Theoretical Crop 1	VF7	VF32	Conv.GH	HT-GH	SBA
Land Use	259.20	1,152.00	8,990.00	4,000.00	14,323.43
Cultivation Area	4,406.00	3,456.00	8,990.00	4,000.00	14,323.43
EMBODIED ENERGY	28.09	28.09	23.22	25.54	2.16
LIGHTING	245.63	249.48	183.12	178.45	0.00
HEATING	19.19	15.28	280.00	30.43	0.00
kWh/m2	292.92	292.85	486.34	234.41	2.16
kWh/a	1,290,585.61	1,012,086.45	4,372,172.37	937,651.91	30,938.60
TPES/m² Land Use	75,923.69	337,362.15	4,372,172.37	937,651.91	30,938.60
Energy Land (m2)	11,523	9,036	39,037	8,372	276
∑ Built up + NRG-Land	11,782.29	10,188.49	48,027.25	12,371.89	14,599.66

Tab.24. Energy consumption - VF for Theoretical Crop 1

The break even, counted with the values shown in the table on this page, would be by some 365 kWh/m2 TPES within Vertical Farming. Then the sum of the built up land and the Energy land would be equal.

This figure, understood as a raw sketch of comparison, shows the lighting and heating demand (both TPES) based on the upmentioned sources and assumptions. We see that lighting demand by far is the highest for the simulated VF (both VF7 and VF32) of this thesis, followed by conventional greenhouses and High-Tech greenhouses (GH high tech), both using the same lighting schedule than VF with the building envelope DG. Lighting and heating demand within Figure 154 (Plantagon) are the numbers communicated at the Urban Agriculture Summit, 2013 in Linköping, Sweden.

Theoretical Crop 2	VF7	VF32	Conv.GH	HT-GH	SBA
Land Use	259.20	1,152.00	8,990.00	4,000.00	14,323.43
Cultivation Area	4,406.00	3,456.00	8,990.00	4,000.00	14,323.43
EMBODIED ENERGY	28.09	28.09	23.22	25.54	2.16
LIGHTING	35.09	35.64	26.16	25.49	0.00
HEATING	21.59	19.64	360.00	39.12	0.00
kWh/m2	84.77	83.38	409.38	90.15	2.16
kWh/a	373,517.80	288,152.02	3,680,322.74	360,606.20	30,938.60
TPES/m² Land Use	21,973.63	96,050.67	3,680,322.74	360,606.20	30,938.60
Energy Land (m2)	3,335	2,573	32,860	3,220	276
∑ Built up + NRG-Land	3,594.18	3,724.79	41,850.02	7,219.70	14,599.66

Tab.25. Energy consumption - VF for Theoretical Crop 2

They got converted with the same TPES-factor mentioned above. If these numbers about lighting demand really are on a low side, then this obviously must be argued with the substitution of intermediate cultivation levels with a cultivation method which increases light penetration of the vertical greenhouse, the inclination in the z-axis of the building envelope, must have, at least to some percentages, an influence in lighting demand to arrive on this very low value and the high difference in light-needs between *L. esculentum* and Pak Choi, the cultivar chosen by Plantagon.

On the other side, looking on the heating demand, by far the highest value is for conventional greenhouses in Easter Austria, followed by VF14 and VF32 (both DG). Plantagon, again, has a very low heating demand, compareable with VF7 and a High-Tech-Greenhouse.

On the very bottom Skygreens must be situated. Looking on the specific relatively constant temperature in Singapore and the vertical circulation system wich enables equal light distribution, although no numbers cold have been retreived, the assumtion on the actual TPES is most likely arranged on the very low side.

LIGHTING AND HEATING DEMAND (TPES) FOR L.ESCULTENTUM AND TWO THEORETICAL CROPS
(kWh/m2/a) *(kWh/m2/a)*

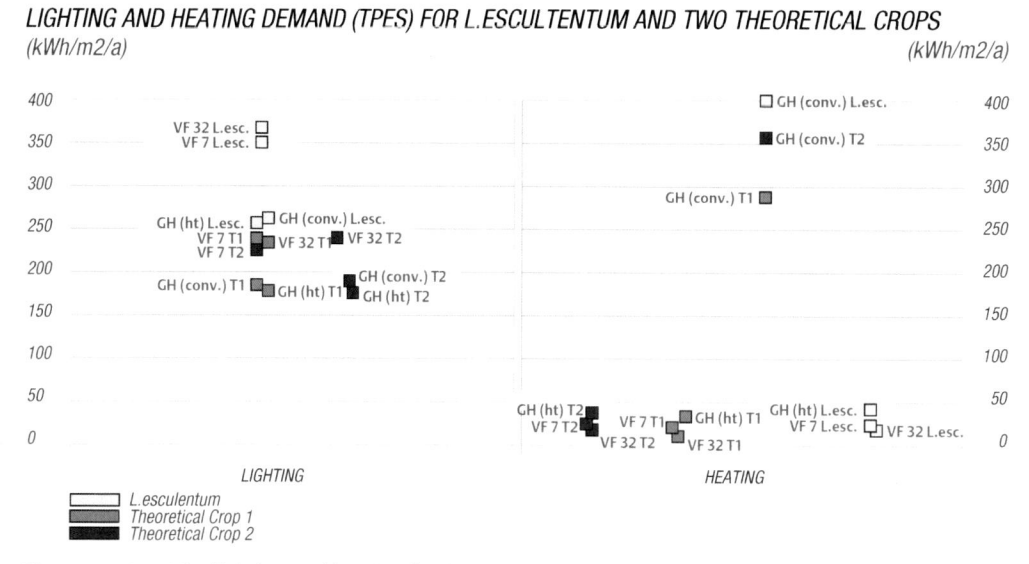

Fig.154. TPES for lighting and heating for theoretical crops

Assuming values for „Theoretical Crop 1" with the same land use and yield needing 30% less light and 20% less heat, VF7 and VF32 would reduce the land use by some 4,000 m2.

In addition, reducing lighting demand by 90% and heating demand by 10%, this ratio would again increase, whereas only 26% of the total land (SBA plus energy land) would be needed.

„Theoretical Crop 2" would be a crop with very low light requirement (10% of *L. esculentum*) and 10% less heating demand (e.g. a tropical starchy root). The sum of built up land and energy land now is by some 24% of SBA.

The theoretical crop-tables show the dependency of VF-TPES from the crop needs. Figure 158 clearly shows that lighting demand for VF is still the highest compared to GH-practises, although the gap between the vertical and horizontal cultivation practise is way smaller. A relatively low heating demand with a DG-building envelope, leads to negligible differences between the different cultivation methods.

6. Conclusio

6.1. From Modernism to Sustainism

Little imagination is needed to perceive that numerous essential conditions of human life are based on limited resources: food or space for living, matter and energy, space and time. But this fact alone is not enough to explain the phenomenon of scarcity or shortage. Shortage can entail the potential to enable social decisions and regulations when considered in the context of social perception.[1]

And it would appear that awareness of scarcity is also social awareness - above all in some of the most highly developed countries. The twentieth century and the birth of modernist ideas and values, with all their consequences for architecture, urbanism and economic interdependencies, has become modernist culture. At present "we can see a new era emerging, one that embraces more sustainable ways of living and an interconnected world. (...) [N]e w approaches to both local and global issues (...)"[2] simultaneously emerging on many cultural levels, overcoming modernist and postmodernist ideas - a perspective that "promises a networked, globalized, sustainable future." ."[3]

This dissertation has shown the complete dependency of today's world agriculture on hydrocarbon energy. With all those many actions during the course of the twentieth century, which have led to an increase in productivity of soil based agriculture, when expressed euphemistically, have caused an ease of the pressure on the necessity for agricultural land expansion.
But there has been a high price to pay: an estimated 176 EJ of TPES (32% of global energy production) is needed every year to keep this practice up and running effectively. The need for new agricultural land is growing exponentially in the contemporary world, primarily as a result of world population growth and changes in diets that are taking place in countries with emerging economies. Water scarcity and land erosion and other environmental impacts of conventional agriculture, problems which have not been delved into in this work, are additional challenges that add their weight in making a necessity of land expansion for crop production. But urban expansion has reached a point where it has become evident "that the ecosystems on which the city depends have a limited capacity."[4]

---

1 LUHMANN, N. 1994. Die Wirtschaft der Gesellschaft, Frankfurt/Main, Suhrkamp., p. 177 ff.
2 SCHWARZ, M. & ELFFERS, J. 2010. Sustainism is the New Modernism, New York, D. A. P. p.3
3 ibid.
4 GAUSA, M., GUALLART, V., MÜLLER, W., SORIANO, F., PORRAS, F. & MORALES, J. 2003. The Metapolis Dictionary of Advanced Architecture, Barcelona, Actar. p. 580

CONCLUSIO

The definition of scarcity, though, is always dependent on the perspective, or "scale"[1]. Chapter 2 also has highlighted that potential does exist:

- The biocapacity of the earth is entirely adequate for the supply of future generations. But land conversion is needed to supply the additional food needs, and if all suitable land were converted for agriculture, the result would be an intolerable CO_2 burden for the planet.
- There is enormous potential, however, to increase productivity on existing agricultural land in many regions of the world, but this too will increase our dependency on hydrocarbon energy sources.
- Changes in diets, above all reducing the consumption of meat and dairy products, can reduce the per capita footprint. In order to achieve this, rigorous political decisions must be made – supported and carried by broad social acceptance.
- Food losses and food waste must be reduced, although it is not possible for all food that is lost from the food supply chain to be eliminated from the statistics, since our globally expanded food supply chain makes losses and wastes immanent.

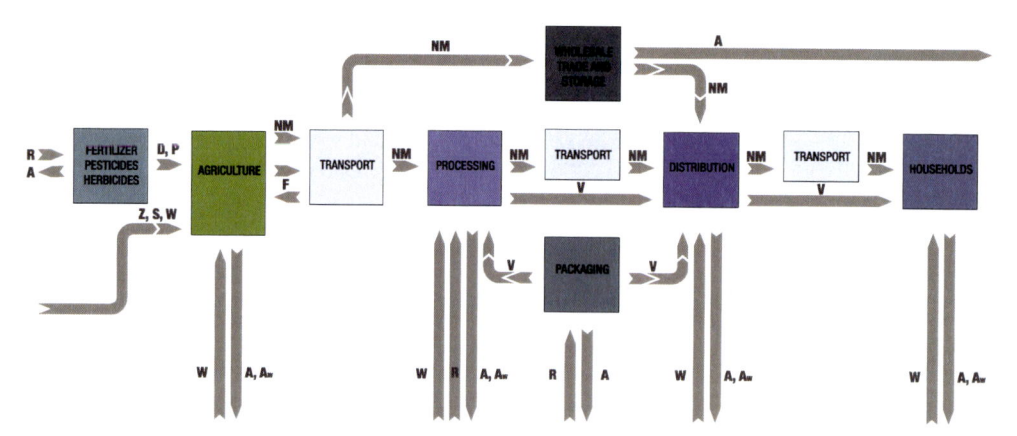

Fig.155. Systemborders of the food sector (A=Waste, Aw=Wastewater, D=Fertilizers, F=Feed, P=Pesticides, NM= Food Products, R=Resources, S=Seeds, V=Packaging Material, W=Water, Z=Growth) - based on FAIST, 2006.

- Food used for energy production (e.g. biofuels) could be reduced to bring it back into the range of consumable calories for humans. But the trends clearly show movement in a different direction.

As a global average, we need some 1,700 m2 to produce our daily food to cover on average 2,700 kcal/day by consuming over 16,000 kcal of hydrocarbon energy to produce, distribute and prepare our food. For every calorie we consume, we need six calories of hydrocarbon energy to produce it.

---

[1] german: „Maßstab" also could be translated by using „standard", „point of view"

CONCLUSIO

The Neolithic Revolution implemented a nucleus of agricultural land and built-up land. Settlements, emergence of societies, creation of cities and rising of civilizations were first made possible as a result of this development. This practice was first interrupted some decades ago. A food production- and cultivation method, which has been in discussion for the past decade, would re-establish the practice of food being produced where it is consumed – that is the use of the Vertical Farm.[1][2]

The results of this work have shown the limits of Vertical Farming. Crops with a high lighting and heating demand within a stacked greenhouse with intermediate floors simply cannot compete with the already inefficient energy consuming food sector. Bomford's estimations and calculations[3] clearly highlight the risks that Vertical Farms, instead of easing cities from the dependency on hydrocarbon energy for food supply, rather aggravate the situation.
But this thesis also outlines a way to make Vertical Farms, as structural elements of a city, competitive with energy numbers compared to the today's world agriculture practice.

The principle of the Vertical Farming and the value and meaning it contains – its raison d'être – is primarily dependent on five factors:

TO WHAT EXTENT VERTICAL FARMING REDUCES THE NEED FOR ADDITIONAL LAND CONVERSION

VF7, 14 and 32 each produces the same amount of crops what more than 60 greenhouses would yield. This reduces land use already by a factor of approximately 15. Comparing the yield to soil based agriculture, a reduction of up to 50 times can be achieved.
In Chapter 2, we have seen that since the 1980s roughly 80-90% of the additional land for agriculture came from forests. Today, this is still an ongoing practice. A ha of forest every year binds more than 11 t of CO_2[4] and it would release up to 737 t of CO_2[5] which, by converting it into agricultural land, would be released by current slash-and-burn practices. Yield of VF7, for example, could keep untouched some 1.3 ha with a building footprint of roughly 260 m2.

---

1 DESPOMMIER, D. 2010. The Vertical Farm, New York, St. Martin's Press.
  First releases of findings at the Institute of Microbiology and Public Health, headed by Dickson Despommier, Columbia University, were published in the early years of the millenium.
2 DANIEL PODMIRSEG, 2008. The Vertical Farm Project for London - SPUROPE 2050, Diploma, Akademy of Fine Arts, Vienna, Prof. Markus Schäfer
3 https://energyfarms.wordpress.com/2010/12/02/energy-and-vertical-farms/, retrieved 20.10.2015
4 SMIL, V. 2008. Energy in Nature and Society, Cambridge, Mass., MIT Press. p.69
5 http://pubs.iied.org/pdfs/16023IIED.pdf. p.10

CONCLUSIO

CONSIDERING THE TOTAL ENERGY BALANCE, IF THE SELECTED CROP WERE PRODUCED VERTICALLY, THIS WOULD EASE THE CURRENT SITUATION WITH REGARD TO HYDROCARBON ENERGY DEPENDENCE

With its high demand on light, *L.esculentum* challenges the typological development of a Vertical Farm to an extent that it is very difficult to achieve an energy consumption comparable to practices in soil based agriculture, as long it is connected to a conventional energy grid. It is the crop that determines energy consumption. Crops with lower light and temperature requirements can drastically reduce TPES of a Vertical Farm, e.g. lettuce, strawberries or tropical starchy roots typically require a low amount of light for photosynthesis.

IF THE BUILDING TYPOLOGY AND PRODUCTION METHOD IS OPTIMIZED FOR LIGHT PENETRATION

Lighting demand of Vertical Farms is the key energy consumer. What results in Chapter 5 have shown is that the skyscraper, or a similar building typology with intermediate floors is most likely inadequate to reach an energy efficiency that would make cities more independent of hydrocarbon energy for food production, regardless of all the positive impacts through the spatial implosion of the food supply chain.
Solar altitude, climate data and light availability are key strategic design components to develop the Vertical Farm building typology with a positive impact for the urban system.

IF THE VERTICAL FARM WERE IMPLEMENTED, SEEN AS AN URBAN OPERATION OR A STRUCTURAL ELEMENT, IT WOULD ENABLE AN ECOLOGICAL SYMBIOSIS BETWEEN AGRICULTURE, SOCIETY AND ARCHITECTURE

The Neolithic Revolution implemented a nucleus of spaces for food production and cities. This development has drastically changed within the last decades. The city has to be understood as a system, and the Vertical Farm as an integrative structural element of it. Energy and material flows within the future city have to be redrawn from linear to circular.

THE VERTICAL FARM WITH A LINEAR PROGRAMMATIC ENLARGEMENT WOULD ENABLE SYNERGY POTENTIALS IN ADDITION TO FOOD PRODUCTION

by adding items inside the systemic borders of the food sector using synergy potentials on an energetic level (compare Fig. 156 and Fig. 157). Besides energy consumption, DEFRA[1], highlight negative effects of the global food supply chain and its impacts on cities.

---

[1] WATKISS, P., SMITH, A., TWEEDLE, G., MCKINNON, A., BROWNE, M., HUNT, A., TRELEVEN, C., NASH, C. & CROSS, S. 2005. The Validity of Food Miles as an Indicator of Sustainable Development: Final report produced for DEFRA. AEA Technology.

6.2. Vertical Farming and Energy Consumption

Three Vertical Farms with similar volumes have each been simulated with three different building envelopes. Daylight availability and heating demand were at the center of the interest. Simulation results for *L. esculentum*, though, show that the simulated Vertical Farms for L. esculentum cannot compete with energy intensive soil based agriculture-practices. Total primary energy supply and energy land needed to supply Vertical Farms with renewable energy are higher compared to today's traditional agricultural practice.

Economic pressure on maximizing yields within agriculture (and as a consequence, in greenhouse and Vertical Farm production) automatically leads to an intensification in plant density. This, in addition, reduces daylight penetration of plants moving away from the facade to the inside of the production level. To find the balance between the optimum density of crops and the necessary daylight availability is an interesting topic for future research. Maximizing yield and optimizing the building type for daylight gain "becomes the most important design consideration for architects, and ultimately will be the primary criteria from which the efficacy of a design will be determined."[1]

Results of three VF volumes with different A/V ratios show a TPES need ranging from 398.28 kWh/m2/a to 842.50 kWh/m2/a. Whereas the influence of different VT-values for each VF-type are negligible, heating demand strongly differs between the different U-values of each building envelope.
Lighting demand of the different types varies up to 25%, clearly showing the necessity of the potential of precise typological studies for a verticalized cultivation practice. Opaque levels between the cultivation spaces must be substituted.[2]

By considering these values we see that vertical production is more energy intense than the actual practise in world agriculture. Around 1.50 GWh/a (400 kWh/m2/a TPES) are needed for annual production of *L. esculentum*. The actual world average of energy supply for the food sector per square meter of agricultural land is 11.73 MJ/m2/a or 3.25 kWh/m2/a. Subtracting the energy for retail, preparation and cooking, this number is reduced to 7.80 MJ/m2 or 2.16 kWh/m2/a.[3] Energy intensive cultivation practices focused on vegetable production is some 6.5 MJ/m2/a[4] (1.81 kWh/m2/a) up to the farm gate, or 27 MJ/m2/a (7.50 kWh/m2/a), when considering the whole food supply chain.[5]

---

1 GRAFF, G. 2011. Skyfarming. Master of Architecture, Waterloo, Ontario, Canada. p.76
2 http://plantagon.com/urban-agriculture/vertical-greenhouse, retrieved 29.10.2015
3 see Chapter 2.4.4, Landuse, Energy Consumption and Biocapacity
4 FOOD AND AGRICULTURE ORGANIZATION OF THE UNITED NATIONS 2011.Energy-Smart Food for People and Climate, Issue Paper, Rome:FAO, p. 13
5 We assuming the ratio that 24% of TPES of the food sector is related to the energy consumption until the farmgate. See Chapter 2.

CONCLUSIO

The aim therefor for future Form Follows Energy[1]-studies for Vertical Farms is to optimize the building shape to reduce TPES, and secondly, to read urban food production as a structural entity of the system "city", i.e. the city as an ecosystem. "The one characteristic they all [Ecosystems, Ed.] share is that primary productivity (the total mass of plants produced over a year in a given geographically defined region) is limited by the total amount of energy received and processed."[2]

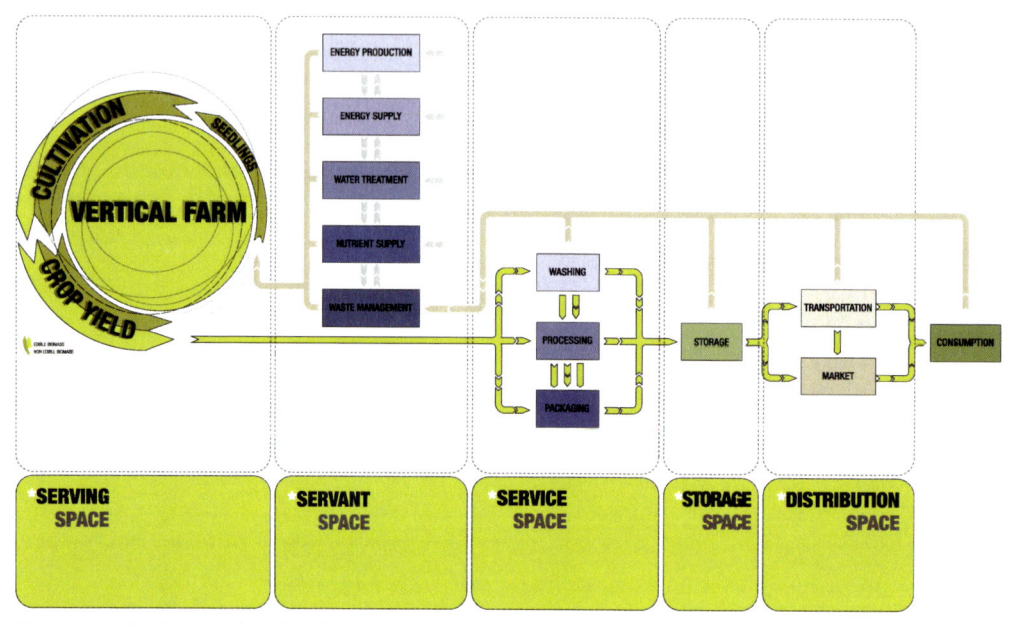

Fig.156.   Ideal systemborder of a Vertical Farm

Research projects such as "Hyper-Building-City"[3], where building typologies for vertical structures (spatially, temporally and digitally densified) entailing all necessary infrastructural elements of a society, including industry and agricultural use, visualize the potential of future cities, when synergy potentials between functions really become activated and systemic borders are rearranged to reduce the spatial dimension between the links of the FBS.

---

1   CODY, B. 2012. „Form follows Energy - Beziehungen zwischen Form und Energie in der Architektur und Urban Design, DBZ Deutsche BauZeitschrift, Bauverlag BV GmbH, Gütersloh, p.48 ff.
2   DESPOMMIER, D. 2010. The Vertical Farm, New York, St. Martin's Press. p.19
3   CODY, B. 2014. Form Follows Energy - Die Zukunft der Energie-Performance, energy2121, Bilder zur Energiezukunft, Klima- und Energiefonds, Vienna, omninum, p. 121 ff.

6.3. Vertical Farming and Land use

The effect of reducing land use for agricultural production based on the above mentioned simulation models and considering the assumptions of other Vertical Farms[1] results in the emergence of a clear picture: Land use can be reduced to 30%, when comparing the cultivation area of the production entity to the alternatively needed area for traditional soil based agriculture, or referring to the three simulated building types, by 12.08%, 4.53% and 2.26%, when comparing the building footprint to the soil based area equivalent.[2]

Compared to traditional greenhouse practices, VF 32's ratio is 1/6.5, for VF14 1/16 and VF7 1/33 for land use.
This strong image is relevant for two reasons: Land in urban areas is more expensive and the virtual area made available by Vertical Farms can be used in various ways. On this issue it is necessary to point out that Vertical Farming need not necessarily be seen as the complete substitution of all current food production practices. But it will bring relief to the current situation in which natural land is being converted into agricultural land.

On a small, (peri) urban scale, agricultural area made available by implementing Vertical Farms, can be used for alternative practices such as the intensification of socially highly desirable areas for urban farming (still soil-based) activation of area for practices with yields that tend to be lower, such as permaculture restoring natural land by abandoning agricultural land with eroded soils (a consequence of conventional agricultural practices) and capturing CO_2.

---

[1] Chapter 3
[2] Chapter 5

6.4. Final Considerations

Vertical Farming has a notable impact on urbanism at many levels. For example, the ground level strongly reshapes modernist cities, developed on the drawing table and optimized for private transport.[1] On a human scale, market and trade areas, and communal spaces can be implemented again. Of great importance at a social level is the acceptance that Vertical Farms achieve within a dense city.

Although refusal was noticeable during the public presentations and discussions organized by the author, there is also strong interest in public perception: There is no doubt that "[t]he public is no longer convinced that big agriculture has its best interest at heart, and this has resulted in a deep mistrust of mass-produced food."[2] To make this statement plausible we should add that the mistrust primarily comes from the fact that agricultural production is not visible and therefore not perceptible for the urban population. In terms of acceptance for Vertical Farms within a city, it must assert high aspirations in terms of architectural aesthetics extending beyond mere functionality to increase the potential for identification, as well as to provide an urban ground floor level that is carefully planned as public space with integrated market and trade areas.

The Vertical Farm might contribute to make cities more resilient and to reduce environmental impacts, if energy- and material flows are interconnected with the urban system. Research on agricultural sciences, financed by public funding, which has been drastically reduced within the past few decades, must be re-established.[3] Public private partnerships[4] are needed to continue the process of establishing the experimental Vertical Farms, which were started in 2009.[5]

---

[1] Daniel Podmirseg, IMDP, Exhibition, 2012. „Dio Nero - Aesthetics of Disease - Rebuilding local social and economic interdependencies, with Lucas Kulnig and Maria Huber. Venice
[2] DESPOMMIER, D. 2010. The Vertical Farm, New York, St. Martin's Press. p.125
[3] Wiederaufbau der öffentlichen Agrarforschung, Conclusion of the lecture „Ist die Herkulesaufgabe lösbar?" held by Prof. Dr. Harald von Witzke, on „Getreidehandelstag" on June 17th and 18tz 2014, Warberg, Germany
[4] DESPOMMIER, D. 2010. The Vertical Farm, New York, St. Martin's Press. p.252 ff.
[5] Suwon Vertical Farm, Chapter 3

CONCLUSIO

This doctoral thesis aims to contribute to the debate by summarizing the status quo of the situation in world agriculture today and to release impulses for future investigations of crop production using a verticalized cultivation method. The sustainability and energy efficiency discussions in architecture are mostly reduced to the narrow confinements of living, office and retail spaces. When considering that the per capita area for Austria (similar to Germany) is 45m2 living, 10m2 office[1] and 2 m2 retail[2] space, , it should be recalled that the area needed to meet the daily energy requirement for a person is 2,300 m2, a contrast which clearly shows how great the potential is for adding momentum to this discussion.

Results clearly show that the building type, mainly related to natural light gain, the building envelope with respect to heating demand reduction, and the specific physiological needs of the cultivated crop require very careful investigation. These three main drivers determine whether Vertical Farming has a positive impact on the energy demand of urban agglomerations.

Simulation results clearly show how high the expected TPES is when captured by Vertical Farms adopting Skyscraper typologies, and enveloped for completely different purposes. The design of a new typology for Vertical Farming not only needs to enable light penetration within the vertical cultivation space, but use potentials for the production of renewable energy in addition.

From the perspective of an architect, "[t]he end of cheap oil for architecture carries the possibility to move away from the representation of abstract, non-spatial processes and identities back to the presentation of current, local relationships."[3] As plant needs are internalized and sensitivity for the potential of Vertical Farming is considered to make cities more resilient, accepting "trends that are transforming our living urban (and rural) space on a massive and unstoppable scale" , a fascinating opportunity arises: the development of a new architectural typology, the typology of the Vertical Farm.

---

1 https://www.wko.at/Content.Node/branchen/k/sparte_iuc/Immobilien--und-Vermoegenstreuhaender/tabellenband_wohnen_2013_079206.pdf, retrieved 13.09.2015
2 https://www.dghyp.de/fileadmin/media/dg_hyp_deutsch/downloads/broschueren_marktberichte/marktberichte/Immomarkt_Deutschland_2011_DRUCK.pdf, p.3 ff. retrieved 13.09.2015
3 Prof. Markus Schäfer for DANIEL PODMIRSEG, 2008. The Vertical Farm Project for London - SPUROPE 2050, Diploma, Academy of Fine Arts, Vienna, Guest-Prof. Markus Schäfer

CONCLUSIO

To conclude, this dissertation presents findings primarily on energy consumption. It also seeks to discuss Vertical Farming in a broad context by opening the following perspectives on the subject:

- Cities and water: as urban populations increase, megacities are today already suffering severe water shortages in terms of both quality and quantity. Vertical Farming and other greenhouse practices, with their system-immanent water control practices, may well contribute to minimizing this problem. World agriculture is responsible for more than 70% of annual global water withdrawal. Cultivation methods such as hydroponics in closed environments drastically reduce water consumption for crop production.

- City and land use: with every m2 the urban area increases, agricultural land inevitably increases ten times. Central urban areas are expensive. In most Asian megacities, the settlement density is much higher than in old established European cities. These circumstances are influential on two levels: the light gain for Vertical Farms in an area of high density might decrease by such an amount that the power demand required may well increase drastically. On an urbanistic level, investigations could well prove interesting on whether Vertical Farms in peri-urban zones, still connected to the existing traffic network, might have a positive effect for a typologically different redensification of the low-density peri-urban belt compared to the city center.

- City, land use and carbon: Results of Chapter 5 show average soil based area equivalents for achieving the same yield of 13,000 m2 or 1.3 ha. It might be interesting to follow the concept of "carbon sink" to achieve more precise data for evaluating the advantage of reducing the conversion of natural land (over all forests) into agricultural land. The aforementioned area potentially stores from 1.45 to 3.2 t C/ha[1]. This value corresponds to a CO2eq. from 5.3 to 11.7 t CO2eq.

- Cities and carbon: private and public transport, industry and the urban dwellers themselves release enormous quantities of CO2, which is essential for photosynthesis. The extent to which CO2-capturing could be possible, e.g. by enabling CO2-cycles between different urban uses and plant growth, should be investigated in greater depth. The concept of Vertical Farming developed by Plantagon and Sweco in Linköping must be quoted here. CO2 from the biodigester is delivered to the production greenhouse, as well as excess heat. The concept of "industrial symbiosis" adds weight to this idea.[2]

- Cities and employment: "Vertical Harvest" in Jackson, Wyoming[3] follows the concept of working together intensively with the community, also ensuring meaningful employment opportunities for disabled people. Vertical Farming will

---

1 Raw assumption based on numbers retrieved from SMIL, V. 2008. Energy in Nature and Society, Cambridge, Mass., MIT Press. p.69
2 http://plantagon.com/urban-agriculture/industrial-symbiosis, retrieved 03.11.2015
3 http://verticalharvestjackson.com/our-team/, retrieved 03.11.2015

CONCLUSIO

undoubtedly be more labor intensive than soil based practices. "Large dimensioned farm machinery will not be an option"[1] To what extent Vertical Farming could contribute to the widespread issue of underemployment in urban areas must be investigated.

- Cities and politics: developing and developed countries and their cities in particular are vastly dependent on agricultural production areas outside their national borders. Food politics leads not infrequently, to geopolitical conflicts on both a domestic and an international scale. If cities reduce their dependency on agricultural land outside their national borders, the question arises of achieving a stabilization of (inter-)national relationships.
- Cities and real estate: urban voids are expensive. Additional feasibility studies are necessary to determine what conditions are necessary to make Vertical Farming a profitable business proposition. It is estimated that only 20% of the money spent on food finds its way to the farmers. The differences in the food price this imbalance represents have their roots in the long food chain. Vertical Farms could sell their produce directly in the city areas where they are located. The question of whether this will result in a balance with the investment costs out is worth examining in further studies.

World population will continue to grow, and changes in diets are most likely to be expected. Agricultural land most likely will continue to expand and change natural habitat into productive land. World agriculture is dependent on hydrocarbon energy sources and world transport is run almost entirely on oil.

Although Vertical Farming per se is not the solution to these problems, we have seen that this production method comprises the potential to relieve the pressures of the current situation. The right design strategies defined by architects entail the potential to drastically reduce the energy demand. Furthermore, if the implementation of the building is understood as an urban operation, i.e. as an integration of a structural element into a system, Vertical Farms could contribute to change urban material and energy flows from linear to circular.

---

1 DESPOMMIER, D. 2010. The Vertical Farm, New York, St. Martin's Press. p.171 ff.

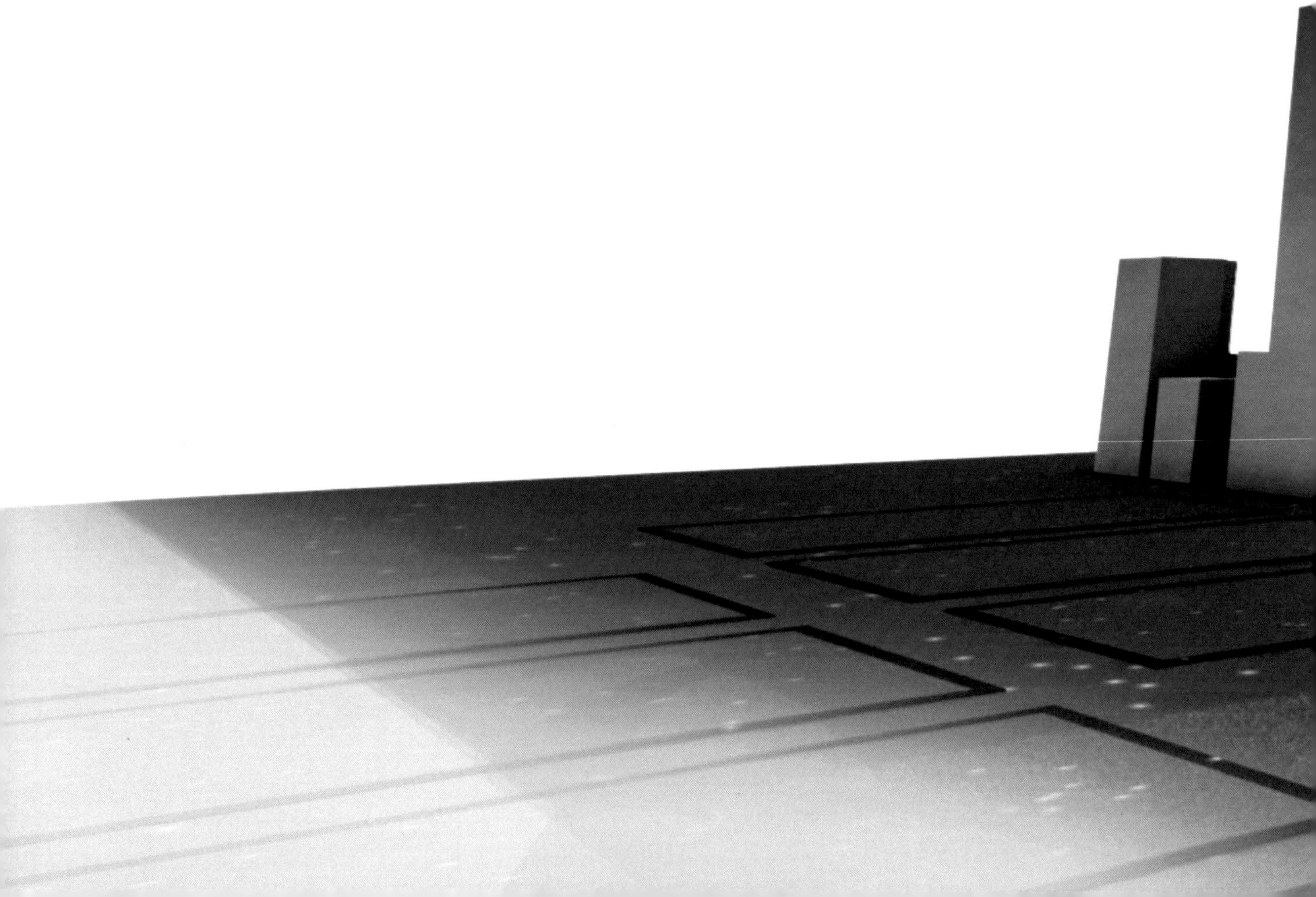

© 2012 – „Aesthetics of Disease" IMDP - DANIEL PODMIRSEG

V. List of Figures

Fig.1.	Arable land per person and food items consumption ratio (blue) and the related area needed (green) exemplary for Germany	49
Fig.2.	Land use per person for food production and related ecological food footprint	51
Fig.3.	Subcontinental divisions of FBS-groups used by Kastner et al.	53
Fig.4.	World land mass and land use	59
Fig.5.	World land mass and soil conditions for agricultural production	62
Fig.6.	World crop production and calories directly consumed by people and losses from food chain	65
Fig.7.	Increase in Energy Subsidies in the 20th century	71
Fig.8.	Energy conversion coefficient change since 1945 using the example of wheat	72
Fig.9.	Increase in productivity per area from 1900 to 2000	72
Fig.10.	Ratio of energy consumption within the food sector (left) and ratio of CO_2eq. emissions	77
Fig.11.	Energy consumption within the food sector comparing USA (left) and Africa	85
Fig.12.	Energy consumption of the food sector until farmgate	87
Fig.13.	Food losses and wastes related to the food supply chain, per capita and Food production quantity per world regions	88
Fig.14.	World energy consumption, energy consumption of the food sector, food footprint per person and world biocapacity	91
Fig.15.	End energy consumption of the food sector (ratio and amount)	93
Fig.16.	End energy consumption within the food sector	96
Fig.17.	Total primary energy consumption (est.) within the food sector	97
Fig.18.	Single horizontal layer	112
Fig.19.	Suwon Vertical Farm, Multiple stacked horizontal layers, ©Heinrich Holtgreve, Ostkreuz	113
Fig.20.	Paignton Zoo: Rendering of the production entity	115
Fig.21.	Paignton Zoo, Devon: Photo of the production volume, ©Verticrop/ Paignton Zoo	115
Fig.22.	Skygreens: Rendering of the production entity	118
Fig.23.	SkyGreens, Singapore: Production Volume, ©Sky Greens Pte Ltd	119

LIST OF FIGURES

Fig.24.	Vertical Harvest: Rendering of the production entity...............	121
Fig.25.	Vertical Harvest, Jackson, WY: Vertically rotating cultivation conveyor, © Vertical Harvest Jackson Hole.................	121
Fig.26.	Plantagon Linköping: Rendering of the production entity............	123
Fig.27.	Plantagon, Linköping: Production helix, © Plantagon, Sweco..........	124
Fig.28.	Suwon Vertical Farm, South Korea, ©Heinrich Holtgreve, Ostkreuz......	133
Fig.29.	Suwon Vertical Farm, South Korea - Cultivation Room, ©Heinrich Holtgreve, Ostkreuz..................	133
Fig.30.	Total daily radiation and 24 h mean temperature, Suwon, South Korea....	134
Fig.31.	Paignton Zoo, Devon: Photo of the production volume, ©Verticrop/ Paignton Zoo..................	137
Fig.32.	Paignton Zoo, Devon: Photo of the greenhouse envelope, ©Verticrop/ Paignton Zoo..................	137
Fig.33.	Total daily radiation and 24 h mean temperature, Devon, United Kingdom.	138
Fig.34.	Paignton Zoo, Devon: Areas and Dimensions...................	139
Fig.35.	Paignton Zoo: land use reduction, production volume and A/V-ratio.....	140
Fig.36.	Vertical Harvest, Jackson, Wyoming - Rendering of vertical rotating production © Vertical Harvest Jackson Hole..................	143
Fig.37.	Vertical Harvest, Jackson, Wyoming © Vertical Harvest Jackson Hole....	143
Fig.38.	Total daily radiation and 24 h mean temperature, Jackson, WY, USA.....	144
Fig.39.	Vertical Farm Vertical Harvest, Jackson: Functions...............	145
Fig.40.	Vertical Farm Vertical Harvest, Jackson: Areas and Dimensions........	146
Fig.41.	Vertical Harvest: land use reduction, production volume and A/V-ratio...	147
Fig.42.	Vertical Harvest: land use reduction, production volume and A/V-ratio...	148
Fig.43.	Skygreens, Singapore, Vertical production, ©Sky Greens Pte Ltd.......	149
Fig.44.	Skygreens, Singapore, ©Sky Greens Pte Ltd...................	149
Fig.45.	Total daily radiation and 24 h mean temperature, Singapore..........	150
Fig.46.	SkyGreens, Singapore: Areas and Dimensions..................	151
Fig.47.	Skygreens Singapore: land use reduction, production volume and A/V-ratio	152
Fig.48.	Plantagon, Vertical Farm, Linköping, Sweden, 3D conveyor belt, © Plantagon, Sweco..................	155
Fig.49.	Plantagon, Vertical Farm, Linköping, Sweden, © Plantagon, Sweco......	155
Fig.50.	Total daily radiation and 24 h mean temperature, Linköpin, Sweden.....	156

LIST OF FIGURES

Fig.51.	Vertical Farm Plantagon, Linköping, Areas and Dimensions	157
Fig.52.	Plantagon, Linköping: land use reduction, production volume and A/V-ratio	159
Fig.53.	Plantagon, Linköping: land use reduction, production volume and A/V-ratio, large scale.	160
Fig.54.	Electromagnetic radiation with ratio of light and PAR.	166
Fig.55.	Visible light vs. PAR and its related sensitivity curves for the human eye and plants	172
Fig.56.	World production and yield of tomatoes.	178
Fig.57.	World production of tomatoes.	180
Fig.58.	World's ten largest producers of tomatoes	181
Fig.59.	Land use of Austria and Vieanna and Tomato-production	183
Fig.60.	L. esculentum: PPFD need change along the crop rotation.	190
Fig.61.	Light spectrum - Human Eye Sensitivity	202
Fig.62.	Light spectrum - Plants Photosynthesis Sensitivity	203
Fig.63.	Light spectrum - Grow Light LED.	204
Fig.64.	LumiGrow Pro 325 and 650 LED light fixture.	205
Fig.65.	Lighting schedule used for simulation model.	210
Fig.66.	Daily heating schedules used for simulation model	210
Fig.67.	Total solar radiation, ratio of PAR and needed radiation for L. esculentum in Vienna	212
Fig.68.	Total daily radiation and 24 h mean temperature, Vienna, Austria	221
Fig.69.	PPFD availability in Vienna and PPFD needed along the crop rotation	222
Fig.70.	Annual Temperature curves in Vienna	222
Fig.71.	Watt (PAR) availability in Vienna.	223
Fig.72.	Daylength in Vienna and Photoperiod for Photosynthesis within the Vertical Farm.	223
Fig.73.	Three Simulation Models - VF32, VF14, VF7.	225
Fig.74.	Vertical Farm Types - Dimensions.	226
Fig.75.	VF32 - Cultivation levels with LED-fixtures, screenshot perspective	228
Fig.76.	VF32 - Cultivation levels with LED-fixtures.	229
Fig.77.	VF14 - Cultivation levels with LED-fixtures.	229
Fig.78.	VF7 - Cultivation levels with LED-fixtures	231

LIST OF FIGURES

Fig.79.	Vertical Farm Resulting Cultivation Area	230
Fig.80.	Vertical Farm Types and Zoning of Cultivation levels	231
Fig.81.	Vertical Farm Types and Zoning Dimensions	233
Fig.82.	Typical winter's day, Radiation availability, DLI and LED coverage on the top level of VF32	250
Fig.83.	Typical equinox day, Radiation availability, DLI and LED coverage on the top level of VF32	251
Fig.84.	Typical summer's day, Radiation availability, DLI and LED coverage on the top level of VF32	251
Fig.85.	VF32 Annual Daylight Availability - Single Glazing Facade	255
Fig.86.	VF32 Annual Daylight Availability - Double ETFE Facade	255
Fig.87.	VF32 Annual Daylight Availability - Double Glazing Facade	255
Fig.88.	VF32 - Lighting Demand (kWh/m2/month), Single Glazing Facade, all zones	256
Fig.89.	VF32 - Lighting Demand (kWh/m2/month), Double ETFE Facade, all zones	256
Fig.90.	VF32 - Lighting Demand (kWh/m2/month), Double Glazing Facade, all zones	257
Fig.91.	VF14 Annual Daylight Availability - Single Glazing Facade	259
Fig.92.	VF14 Annual Daylight Availability - Double ETFE Facade	259
Fig.93.	VF14 Annual Daylight Availability - Double Glazing Facade	259
Fig.94.	VF14 - Lighting Demand (kWh/m2/month), Single Glazing Facade, all zones	260
Fig.95.	VF32 - Lighting Demand (kWh/m2/month), Single Glazing Facade, all zones	260
Fig.96.	VF7 Annual Daylight Availability - Single Glazing Facade	263
Fig.97.	VF7 Annual Daylight Availability - Double ETFE Facade	263
Fig.98.	VF7 Annual Daylight Availability - Double Glazing Facade	263
Fig.99.	VF7 - Lighting Demand (kWh/m2/month), Single Glazing Facade, all zones	264
Fig.100.	VF7 - Lighting Demand (kWh/m2/month), Double ETFE Facade, all zones	264
Fig.101.	VF7 - Lighting Demand (kWh/m2/month), Double Glazing Facade, all zones	265
Fig.102.	Comparison of monthly mean lighting demand for VF7, 14, 32 to a theoretical black box	265
Fig.103.	VF32 Heating Demand (kWh/m2/a) Single Glazing Facade, HH-Schedule	267
Fig.104.	VF32 Heating Demand (kWh/m2/a) Single Glazing Facade, HL-Schedule	267
Fig.105.	VF32 - Heating Demand (kWh/m2/month), HH-schedule, Single Glazing Facade, all zones	268

LIST OF FIGURES

Fig.106.	VF32 - Heating Demand (kWh/m2/month), HH-schedule, Single Glazing Facade, selected zones	268
Fig.107.	VF32 - Heating Demand (kWh/m2/month), HL-schedule, Single Glazing Facade, all zones	269
Fig.108.	VF32 - Heating Demand (kWh/m2/month), HL-schedule, Single Glazing Facade, selected zones	269
Fig.109.	VF32 Heating Demand (kWh/m2/a) Double ETFE Facade, HH-Schedule	271
Fig.110.	VF32 Heating Demand (kWh/m2/a) Double ETFE Facade, HL-Schedule	271
Fig.111.	VF32 - Heating Demand (kWh/m2/month), HH-schedule, Double ETFE Facade, all zones	272
Fig.112.	VF32 - Heating Demand (kWh/m2/month), HH-schedule, Double ETFE Facade, selected zones	272
Fig.113.	VF32 - Heating Demand (kWh/m2/month), HL-schedule, Double ETFE Facade, all zones.	273
Fig.114.	VF32 - Heating Demand (kWh/m2/month), HL-schedule, Double ETFE Facade, selected zones.	273
Fig.115.	VF32 Heating Demand (kWh/m2/a) Double Glazing Facade, HH-Schedule . .	275
Fig.116.	VF32 Heating Demand (kWh/m2/a) Double Glazing Facade, HL-Schedule. . .	275
Fig.117.	VF32 - Heating Demand (kWh/m2/month), HH-schedule, Double Glazing Facade, all zones	276
Fig.118.	VF32 - Heating Demand (kWh/m2/month), HH-schedule, Double Glazing Facade, selected zones	276
Fig.119.	VF32 - Heating Demand (kWh/m2/month), HL-schedule, Double Glazing Facade, all zones	277
Fig.120.	VF32 - Heating Demand (kWh/m2/month), HL-schedule, Double Glazing Facade, selected zones	277
Fig.121.	VF14 Heating Demand (kWh/m2/a) Single Glazing Facade, HH-Schedule . . .	279
Fig.122.	VF14 Heating Demand (kWh/m2/a) Single Glazing Facade, HL-Schedule. . . .	279
Fig.123.	VF14 -Heating Demand (kWh/m2/month), HH-schedule, Single Glazing Facade, all zones	280
Fig.124.	VF14 -Heating Demand (kWh/m2/month), HH-schedule, Single Glazing Facade, selected zones	280
Fig.125.	VF14 -Heating Demand (kWh/m2/month), HL-schedule, Single Glazing Facade, all zones	281

LIST OF FIGURES

Fig.126. VF14 -Heating Demand (kWh/m2/month), HL-schedule, Single Glazing Facade, selected zones 281

Fig.127. VF14 Heating Demand (kWh/m2/a) Double ETFE Facade, HH-Schedule 283

Fig.128. VF14 Heating Demand (kWh/m2/a) Double ETFE Facade, HL-Schedule 283

Fig.129. VF14 -Heating Demand (kWh/m2/month), HH-schedule, Double ETFE Facade, all zones. 284

Fig.130. VF14 -Heating Demand (kWh/m2/month), HH-schedule, Double ETFE Facade, selected zones. 284

Fig.131. VF14 -Heating Demand (kWh/m2/month), HL-schedule, Double ETFE Facade, all zones. 285

Fig.132. VF14 -Heating Demand (kWh/m2/month), HL-schedule, Double ETFE Facade, selected zones. 285

Fig.133. VF14 Heating Demand (kWh/m2/a) Double Glazing Facade, HH-Schedule . . 287

Fig.134. VF14 Heating Demand (kWh/m2/a) Double Glazing Facade, HL-Schedule . . . 287

Fig.135. VF14 -Heating Demand (kWh/m2/month), HH-schedule, Double ETFE Facade, all zones. 288

Fig.136. VF14 -Heating Demand (kWh/m2/month), HH-schedule, Double Glazing Facade, selected zones 288

Fig.137. VF14 -Heating Demand (kWh/m2/month), HL-schedule, Double Glazing Facade, all zones 289

Fig.138. VF14 -Heating Demand (kWh/m2/month), HL-schedule, Double Glazing Facade, selected zones 289

Fig.139. VF7 Heating Demand (kWh/m2/a) Single Glazing Facade, HH-Schedule. . . . 290

Fig.140. VF7 Heating Demand (kWh/m2/a) Single Glazing Facade, HL-Schedule 290

Fig.141. VF7 -Heating Demand (kWh/m2/month), HH-schedule, Single Glazing Facade, selected zones 291

Fig.142. VF7 -Heating Demand (kWh/m2/month), HL-schedule, Single Glazing Facade, selected zones. 291

Fig.143. VF7 Heating Demand (kWh/m2/a) Double ETFE Facade, HH-Schedule. 292

Fig.144. VF7 Heating Demand (kWh/m2/a) Double ETFE Facade, HL-Schedule 292

Fig.145. VF7 -Heating Demand (kWh/m2/month), HH-schedule, Double ETFE Facade, selected zones. 293

Fig.146. VF7 -Heating Demand (kWh/m2/month), HL-schedule, Double ETFE Facade, selected zones. 293

Fig.147. VF7 Heating Demand (kWh/m2/a) Double Glazing Facade, HH-Schedule . . . 294

LIST OF FIGURES

Fig.148.	VF7 Heating Demand (kWh/m2/a) Double Glazing Facade, HL-Schedule	294
Fig.149.	VF7 -Heating Demand (kWh/m2/month), HH-schedule, Double Glazing Facade, selected zones	295
Fig.150.	VF7 -Heating Demand (kWh/m2/month), HL-schedule, Double Glazing Facade, selected zones	295
Fig.151.	Greenhouse equivalent to achieve same yield of VF	307
Fig.152.	Soil Based equivalent to achieve same yield of VF	308
Fig.153.	TPES for lighting and heating of different controlled environments	310
Fig.154.	TPES for lighting and heating for theoretical crops	312
Fig.155.	Systemborders of the food sector (A=Waste, Aw=Wastewater, D=Fertilizers, F=Feed, P=Pesticides, NM= Food Products, R=Resources, S=Seeds, V=Packaging Material, W=Water, Z=Growth)	316
Fig.156.	Ideal systemborder of a Vertical Farm	320

VI. List of Tables

Tab.1.	World food energy content requirement estimation	44
Tab.2.	FBS comparision of different countries (gramms supplied per person/day) for USA, Italy, Germany, Austria and China (left to right)	52
Tab.3.	World FBS 2011	54
Tab.4.	FBS (kCal and area) food supply per person comparison (Kastner et al.)	56
Tab.5.	Agricultural land availability and forest land and its distributions over the continents	60
Tab.6.	GAEZ - Potential for Agricultural production	64
Tab.7.	Mass, kCal and proteins lost from FSC	66
Tab.8.	Changes in food diets related to GDP worldwide and, exemplary China	67
Tab.9.	World food transport	86
Tab.10.	Light Spectrum - Wavelength, Frequency and Photon Energy	174
Tab.11.	Lumigrow 325 PRO - PPFD, DLI values depending on fixture height	206
Tab.12.	CSV-schedule for lighting, Top Level, VF32, for the 21st december	252
Tab.13.	CSV-schedule for lighting, Top Level, VF32, for the 21st march	253
Tab.14.	CSV-schedule for lighting, Top Level, VF32, for the 21st june	253
Tab.15.	All VF - Heating Demand for SG and HH-Schedule (kWh/m2/a)	298
Tab.16.	All VF - Heating Demand for ETFE and HH-Schedule (kWh/m2/a)	298
Tab.17.	All VF - Heating Demand for DG and HH-Schedule (kWh/m2/a)	298
Tab.18.	All VF - Heating Demand for SG and HL-Schedule (kWh/m2/a)	300
Tab.19.	All VF - Heating Demand for SG and HL-Schedule (kWh/m2/a)	300
Tab.20.	All VF - Heating Demand for DG and HL-Schedule (kWh/m2/a)	300
Tab.21.	All VF TPES, Area, Yield, C and CO2-comparison	302
Tab.22.	VF7/32, DG: TPES, Area, Yield, C and CO2-comparison	304
Tab.23.	TPES VF, Greenhouse, SBA and Energy Land	309
Tab.24.	Energy consumption - VF for Theoretical Crop 1	311
Tab.25.	Energy consumption - VF for Theoretical Crop 2	311
Tab.26.	Vienna - Daylength, available solar radiation, light- and temperature need for *L. esculentum*	380
Tab.27.	All VF - Total Heating Demand for SG and HH-Schedule (kWh/a)	382
Tab.28.	All VF - Total Heating Demand for ETFE and HH-Schedule (kWh/a)	382

LIST OF TABLES

Tab.29.	All VF - Total Heating Demand for DG and HH-Schedule (kWh/a)	382
Tab.30.	All VF - Total Heating Demand for SG and HL-Schedule (kWh/a)	383
Tab.31.	All VF - Total Heating Demand for ETFE and HL-Schedule (kWh/a)	383
Tab.32.	All VF - Total Heating Demand for DG and HL-Schedule (kWh/a)	383
Tab.33.	VF7 - Total Heating Demand	384
Tab.34.	VF32 - Total Heating Demand	384
Tab.35.	VF7 - Total LightingDemand	385
Tab.36.	VF32 - Total Lighting Demand	385
Tab.37.	VF14 - Total Heating Demand	386
Tab.38.	VF14 - Total Lighting Demand	387
Tab.39.	All VF - End Energy Demand for representative Zones	388
Tab.40.	All VF - End Energy Demand for representative Levels	388
Tab.41.	All VF - End Energy Demand for representative Zones	389
Tab.42.	All VF - End Energy Demand for representative Levels	389

VII. List of Abbreviations

ASHRAE	America Society of Heating, Refrigerating, and Air-Conditionioning Engineers
ASPO	Association for the Studies of Peak Oil and Gas
A/V	Area divided by volume - ratio
BMR	Basal Metabolic Rate
CIE	Commission Internationale de l'Eclairage
DG	Double Glazing Facade
DGE	Deutsche Gesellschaft für Ernährung
DLI	Daylight Integral
EC	Electric Conductivity
EE	Embodied Energy
EJ	Exajoule
ETFE	Double-ETFE-layered Facade
F1	Filial 1 hybrid
FAO	Food and Agriculture Organization of the United Nations
FBS	Food Balance Sheets

LIST OF ABBREVIATIONS

FSC	Food Supply Chain
GAEZ	Global Agro-Ecological Zones
GDP	Gross Domestic Product
GFA	Gross Floor Area
GH	Greenhouse
GHG	Greenhouse Gas
GW	Gigawatt
ha	hectare
HID	High Intensity Discharge
HPS	High Pressure Sodium Lamps
HVAC	Building services, heating, ventilation and air conditioning
IEA	International Energy Agency
IFPRI	International Food Policy Research Institute
IIASA	International Institute for Applied Systems Analysis
J	Joule
K	Kelvin
kW	Kilowatt
LED	Light Emitting Diode

LIST OF ABBREVIATIONS

LUC	Land-use change
MJ	Megajoule
MW	Megawatt
Mt	Metric tonne (1000 kg)
NFA	Net Floor Area
nm	Nanometer
PAL	Physical Activity Level
PAR	Photosynthetically Active Radiation
PJ	Petajoule
Pcal	Petacalorie
PPFD	Photosynthetic Photon Flux Density
RDA	Rural Development Administration (South Korea)
SG	Single Glazing Facade
SI	International System of Units
THz	Terahertz
TPES	Total Primary Energy Supply
UNCED	United Nations Conference on Environment and Development
UNDP	United Nations Development Programme

LIST OF ABBREVIATIONS

UNFCCC	United Nations Framework Convention on Climate Change
UNO	United Nations Organizations
UV	Ultra Violet
VF	Vertical Farm
VF7	Simulation Model Vertical Farm 36m x 7.2m x 61m
VF14	Simulation Model Vertical Farm 36m x 14.4m x 33m
VF32	Simulation Model Vertical Farm 36m x 32m x 12m
W	Watt
WPAR	Watt within the range of 400 - 700 nm
WHO	World Health Organization
OECD	Organisation for Economic Cooperation and Development
ZINEG	Zukunftsinitiative Niedrigenergiegewächshaus

VIII. Bibliography

ABEL, C. 2011. The Vertical Garden City: Towards a New Urban Topology. Available: http://www.planetizen.com/node/50957 [Accessed 14.04.2014].

ADAMS, S. R., COCKSHULL, K. E. & CAVE, C. R. J. 2001. Effect of Temperature on the Growth and Development of Tomato Fruits. Annals of Botany ; 88, 869-877.

ALEXANDRATOS, N. & BRUINSMA, J. 2012. World Agriculture Towards 2030/2050. The 2012 Revision. [Rome]: Agricultural Development Economics Division, FAO.

ANDERSSON, N. E. & NIELSON, O. F. 2000. Energy Consumption, Light Distribution and Plant Growth in Greenhouse partly insulated with non-transparent Material. European Journal of Horticultural Science / Gartenbauwissenschaft ; 65 (5), 190-194.

ANKRI, D. S. 2010. Urban Kibbutz: Integrating Vertical Farming and Collective Living in Jerusalem, Israel. Master of Architecture, University of Maryland, College Park, Faculty of the Graduate School.

ATANAS, G.D. 2005. Integrierte Produktion von Tomaten (Lycopersicon esculentum Mill.) im Gewächshaus unter besonderer Berücksichtigung der integrierten Bekämpfung der Weissen Fliege), Dissertation, Humboldt-Universität Berlin.

BANERNJEE, C. 2012. Market Analysis for Terrestrial Application of Advanced Bio-Regenerative Modules: Prospects for Vertical Farming. Masterarbeit, Rheinische Friederichs-Wilhems-Universität, Hohe Landwirtschaftliche Fakultät.

BAUDOIN W. et al. 2013. Food and Agriculture Organization of the United Nations, Plant Production and Protection Division. Good Agricultural Practices for greenhouse vegetable crops - Principles for Mediterranean climate areas. Rome.

BAYLEY, J. E., FREDIANI, K. & YU, M. 2011. Sustainable Food Production Using High Density Vertical Growing (Verticrop). ISHS Acta Horticulturae ; 12, 95-104.

BLACK, M. & KING, J. 2009. The Atlas of Water, London, Earthscan.

BOGENSBERGER, M., GOJIC, D., HASLER, M., STRITTMATTER, A. & WALLMÜLLER, F. (eds.) 2010. Joint Action in Architecture, Graz: Haus der Architektur.

BOMFORD, M. 2010. Solar Greenhouses, Chinese-style. Available:http://energyfarms.wordpress.com/2010/04/05/solar-greenhouses-chinese-style/.

BOMFORD, M. 2011. Beyond Food Miles. Available: http://energyfarms.wordpress.com/2011/03/09/beyond-food-miles.

BRAUNGART, M. & MCDONOUGH, W. (eds.) 2009. Die nächste industrielle Revolution, Hamburg: Europäische Verlagsanstalt.

BROHM, D., DOMURATH, N. & SCHROEDER, F.-G. Urban Agriculture - a Challenge and a Chance. Available: http://www.researchgate.net/profile/Daniel_Brohm.

BRUINSMA, J. 2009. The Resource Outlook to 2050. By how much do land, water and crop yields need to increase by 2050? [Paper presented at the FAO Expert Meeting, 24-26 June 2009, Rome on „How to Feed the World in 2050".] Available: ftp://ftp.fao.org/agl/aglw/docs/ResourceOutlookto2050.pdf.

BUNDESMINISTERIUM FÜR VERKEHR, BAU UND STADTENTWICKLUNG, BERLIN & BUNDESAMT FÜR BAU-, STADT- UND RAUMFORSCHUNG IM BUNDESAMT FÜR BAUWESEN UND RAUMORDNUNG, BONN 2009. Nutzung städtischer Freiflächen für erneuerbare Energien. [ein Projekt des Forschungsprogramms „Experimenteller Wohnungs- und Städtebau" (ExWoSt) des Bundesministeriums für Verkehr, Bau und Stadtentwicklung (BMVBS) und des Bundesamtes für Bauwesen und Raumordnung (BBR)]. Berlin: BMVBS.

CAMPBELL, C. J. 1996. The Oil Depletion Protocol. Available: http://richardheinberg.com/odp/theprotocol.

CAMPBELL, C. J., LIESENBORGHS, F., SCHINDLER, J. & ZITTEL, W. 2008. Ölwechsel!, München, Deutscher Taschenbuch Verlag.

CASSIDY S. EMILY, WEST C. PAUL, GERBER S JAMES and JOLEY A JONATHAN, 2013. Redefining agricultural yields: from tonnes to people nourished per hectare.

Environmental Research Letters, IOP PUblishing. Available: http://iopscience.iop.org/article/10.1088/1748-9326/8/3/034015/pdf;jsessionid=20DC604FACFE4B90495112F30F7FB9F4.c1

CODY, B. 2009. Bauphysik (VO). ige, Institut für Gebäude und Energie, Graz University of Technology.

CODY, B. 2014. The Role of Technology in Sustainable Architecture. Wolkenkuckucksheim, Internationale Zeitschrift zur Theorie der Architektur ; vol. 19, issue 33, 239-249.

CODY, B. 2012. „Form follows Energy - Beziehungen zwischen Form und Energie in der Architektur und Urban Design, DBZ Deutsche BauZeitschrift, Bauverlag BV GmbH, Gütersloh.

CODY, B. 2014. Form Follows Energy - Die Zukunft der Energie-Performance, energy2121, Bilder zur Energiezukunft, Klima- und Energiefonds, Vienna, omninum

DARRAGH, A. 2009. The Facts of Light, Co. Down, Northern Ireland, Booklink.

DAXBECK, H., KISLIAKOVA, A. & OBERNOSTERER, R. 2001. Der ökologische Fußabdruck der Stadt Wien. Wien: Ressourcen Management Agentur (RMA).

DEKAY, M. & BROWN, G. Z. 2014. Sun, Wind and Light, Hoboken, Wiley.

DELGADO, C., ROSEGRANT, M., STEINFELD, H., EHUI, S. & COURBOIS, C. 1999. Livestock to 2020. The Next Food Revolution, Washington, International Food Policy Research Institute.

DESPOMMIER, D. 2010. The Vertical Farm, New York, St. Martin's Press.

DUECK, T., JANSE, J., LI, T., KEMPKES, F. & EVELEENS, B. 2012. Influence of Diffuse Glass on the Growth and Production of Tomato. VII International Symposium on Light in Horticultural Systems, 75-82.

EISELE, J. & KLOFT, E. (eds.) 2002. Hochhaus-Atlas, München: Callwey.

EMMOTT, S. 2013. Zehn Milliarden, Berlin, Suhrkamp.

ENERGIEAGENTUR WESTSTEIERMARK 2003. Energiebedarf der Gartenbaubetriebe - Auswertung der Gartenbaubetriebe. Zusammenfassung. Basierend auf der Projektstudie Grundlagen für den Einsatz von erneuerbaren Energieträgern wie Biomasse und Solarenergie zur Beheizung und Trocknung im Gartenbau (Endbericht 31. Oktober 2003). Stainz: EAW.

FAIST, M. C. J. R. 2000. Ressourceneffizienz in der Aktivität Ernähren. Dissertation, ETH Zürich.

FEENSTRA, G. W. 1997. Local Food Systems and Sustainable Communities. American Journal of Alternative Agriculture ; 12/1 (1997), 28-36.

FISCH, M. N., WILKEN, T. & STÄHR, C. 2012. EnergiePLUS, Leonberg, Fisch.

FISCHER, G., VELTHUIZEN, H. V. & NACHTERGAELE, F. O. 2000. Global Agro-Ecological Zones Assessment: Methodology and Results. International Institute for Applied Systems Analysis, FAO.

FOLEY, J. A. 2011. Solutions for a Cultivated Planet. Available: http://www.readcube.com/articles/10.1038/nature10452.

FOOD AND AGRICULTURE ORGANISATION OF THE UNITED NATIONS 2001. Food Balance Sheets. A handbook. Rome: FAO.

FOOD AND AGRICULTURE ORGANIZATION OF THE UNITED NATIONS 2011.Energy-Smart Food for People and Climate, Issue Paper, Rome:FAO

FOOD AND AGRICULTURE ORGANISATION OF THE UNITED NATIONS (ed.) 2012. Improving Food Systems for Sustainable Diets in a Green Economy, Rome: FAO.

FRANK, M. 2011. An Introduction to Urban Agriculture. Past, Present, and Future. Dovetail Partners Inc.

FRANKEN, M. 2013. Bericht aus der Zukunft, München, Oekom.

FREDIANI, K. 2010. Feeding Time at the Zoo. The Horticulturist, April 2010.

FUENTES, C. & CARLSSON-KANYAMA, A. (eds.) 2006. Environmental Information in the Food Supply System, Stockholm: FOI - Swedish Defence Research Agency.

GAUSA, M., GUALLART, V., MÜLLER, W., SORIANO, F., PORRAS, F. & MORALES, J. 2003. The Metapolis Dictionary of Advanced Architecture, Barcelona, Actar.

GE, S., SMITH, R. G., JACOVIDES, C. P., KRAMER, M. G. & CARRUTHERS, R. I. 2011. Dynamics of Photosynthetic Photon Flux Density (PPFD) and Estimates in Coastal Northern California. Theoretical and Applied Climatology, 105, no. 1-2, 107-118.

GESANG, B. 2011. Klimaethik, Berlin, Suhrkamp.

GIACOMELLI, G. 1998. Components of Radiation Defined: Definition of Units, Measuring Radiation Transmission, Sensors. CCEA, Center for Controlled Environment Agriculture, rutgers university, Cook College.

GOODLAND, R. & ANHANG, J. 2009. Livestock and Climate Change. What if the Key Actors in Climate Change are... Cows, Pigs and Chickens? World Watch, November/December, 10-19.

GRAFF, G. 2011. Skyfarming. Master of Architecture, University of Waterloo, Ontario, Canada.

GUIL-GUERRERO, J.-L. & REBOLLOSO-FUENTES, M. M. 2009. Nutrient composition and antioxidant activity of eight Tomato (Lycopersicon Esculentum) varieties. Journal of Food Composition and Analysis; 2, 123-129.

HABERL, H., ERB, K.-H., KRAUSMANN, F., BERECZ, S., LUDWICZEK, N., MARTÍNEZ-ALIER, J., MUSEL, A. & SCHAFFARTZIK, A. 2009. Using embodied HANPP to analyze teleconnections in the global land system: Conceptual considerations. Geografisk Tidsskrift-Danish Journal of Geography 109 (2), 119-130.

HANAN, J. J. 1998. Greenhouses, Advanced Technology for protected horticulture. Boca Raton, CRC Press.

HANFORD, A. J. 2004. Advanced Life Support. Baseline Values and Assumptions Document. Houston: National Aeronautics and Space Administration.

HAUSLADEN, G., LIEDL, P. & DE SALDANHA, M. 2012. Klimagerecht Bauen, Basel, Birkhäuser.

HAYNER, M., RUOFF, J. & THIEL, D. 2011. Faustformel Gebäudetechnik, München, Deutsche Verlags-Anstalt.

HEGGER, M., FUCHS, M., STARK, T. & ZEUMER, M. 2008. Energie-Atlas, Basel, Birkhäuser.

HENDRICKS, P. 2012. Life Cycle Assessment of Greenhouse Tomato (Solanum lycopersicum L.). University of Guelph.

HERZOG, T., KRIPPNER, R. & LANG, W. 2004. Fassaden-Atlas, Basel, Birkhäuser.

HIX, J. 2005. The Glasshouse, London, Phaidon.

JANSEN, H., BACHTHALER, E., FÖLSTER, E. & SCHARF, H.-C. 1998. Gärtnerischer Pflanzenbau. Grundlagen des Anbaues unter Glas und Kunststoffen, Stuttgart, Ulmer.

JONES, J. B. 2007. Tomato Plant Culture, Boca Raton, Fla. [u.a.], CRC Press.

KASTNER, T., IBARROLA RIVAS, M. J., KOCH, W. & NONHEBEL, S. 2012. Global Changes in Diets and the Consequences for Land Requirements for Food. Available: http://www.pnas.org/content/early/2012/04/10/1117054109.full.pdf+html.

KECKL, G. 2011. Die globalisierte Kuh, Augsburg, Ölbaum-Verlag.

KITZES, J., PELLER, A., GOLDFINGER, S. & WACKERNAGEL, M. 2007. Current Methods for Calculating National Ecological Foodprint Accounts. Science for Environment & Sustainable Society; Vol. 4 No. 1, 1-9.

KLANTEN, R., EHMANN, S. & BOLHÖFER, K. (eds.) 2011. My Green City, Berlin: Gestalten.

KLOOSTER, T. (ed.) 2009. Smart Surfaces, Basel: Birkhäuser.KOERBER, K. V., KRETSCHMER, J., PRINZ, S. & DASCH, E. 2009. Globale Nahrungssicherung für eine wachsende Weltbevölkerung - Flächenbedarf und Klimarelevanz sich wandelnder Ernährungsgewohnheiten. Journal für Verbraucherschutz und Lebensmittelsicherheit; 4, 174-189.

KOPP, G. & LEAN, J. L. 2011. A new, lower value of total solar irradiance: Evidence and climate significance. Geophysical Research Letters.

KRASNY, E. (ed.) 2012. Hands-On Urbanism 1850-2012, Hong Kong: MCCM Creations.

KRAUSMANN, F. 2000. Landnutzung, Energie und industrielle Modernisierung. Eine historische Perspektive mit Blick in die Zukunft. Natur und Kultur; 1/2, 44-61.

KRICHMAYR, K. 2012. Fleischesser verschlingen mehr Land. Der Standard, Wien, 18.4.2012, p.17.

KRISTINSSON, J. 2006. The Energy-producing Greenhouse. PLEA2006 - The 23rd Conference on Passive and Low Energy Architecture, Geneva, Switzerland, 6-8 September 2008.

LANDWIRTSCHAFTSKAMMER NORDRHEIN-WESTFALEN, VERSUCHSZENTRUM GARTENBAU STRAELEN / AUWEILER, CHRISTOPH, A. & REINTGES, T. 2013. Prüfung mit LED an Rispentomaten 2013.

LARCHER, W. 2003. Physiological Plant Ecology, Foruth Edition. Springer-Verlag Berlin, Heidelberg, New York.

LAUGHLIN, R. B. 2013. Der Letzte macht das Licht aus, München, Piper.

LEE, S. (ed.) 2011. Aesthetics of Sustainable Architecture, Rotterdam: 010 Publishers.

LEIBUNDGUT, H. 2011. LowEx Building Design, Zürich, vdf.

LINA AHLSTRÖM, M. Z. 2012. Integrating a Greenhouse in an Urban Area - Exploring how Urban Industrial Vertical Agriculture can be Integrated in Göteborg, Sweden. Master of Science, Chalmers University of Technology.

LIM CJ, ED LIU. 2010. Smartcities + Eco-warriors. Oxfordshire (first published), New York. Routledge.

LOCSMANDY, D. 2012. Nachhaltige Tomatenproduktion durch Einsatz hocheffizienter KWK. Vösendorf: IG Energieautarkie.

LOVELOCK, J. 2007. Gaias Rache, Berlin, List.

LSTIBUREK, J. W. 2008. Why Green Can Be Wash. Available: http://www.pinp.org/pdf/acf222b.pdf.

LUHMANN, N. 1994. Die Wirtschaft der Gesellschaft, Frankfurt/Main, Suhrkamp.

LUHMANN, N. 2011. Einführung in die Systemtheorie, Heidelberg, Carl-Auer-Verlag.

LUITEL, B. P., ADHIKARI, P. B., YOON, C.-S. & KANG, W.-H. 2011. Yield and Fruit Quality of Tomato (Lycopersicon esculentum Mill.) Cultivars Established at Different Planting Bed Size and Growing Substrates. Horticulture, Environment, and Biotechnology ; 53 (2), 102-107.

LUNDY, R. 2006. In Praise of Tomatoes, New York, Lark.

LYOTARD, J.-F. 2012. Das postmoderne Wissen, Wien, Passagen-Verlag.

MAAS, W. (ed.) 2010. Green Dream, Rotterdam: NAi Publishers.

MARSHALL, N. L. 1990. The Potential for Commercial Food-producing Greenhouses in the Northeast : a Review of the Literature. Available: http://www.thegreencenter.net/rr/rr005.htm.

MAZOYER, M. & ROUDART, L. 2006. A History of World Agriculture, London, Earthscan.

MCCUE, G. A. 1952. The History of the use of the Tomato : An annotated bibliography. Annals of the Missouri Botanical Garden, 289-348.

MC.KINNEY, M. et al. 2012 „Environmental Science: Systems and Solutions", Burlington, Logan Yonavjak Jones & Bartlett Publishers.

MIAZZO, F., MINKJAN, M. & CITIES (eds.) 2013. Farming the City, Amsterdam: Trancity Valiz.

MILLSTONE, E. & LANG, T. 2008. The Atlas of Food, London, Earthscan.

MITCHELL, C. A. 1994. Biogenerative life-support systems. The American Society for Clinical Nutrition, Inc. ; 60 (5), 820-824.

MONTEITH, J. L. & UNSWORTH, M. H. 2013. Principles of environmental physics, Burlington, Elsevier Science.

MOORE, R. 2012. Liz Diller: ‚We thought we would have been fired a long time ago'. Available: http://www.theguardian.com/artanddesign/2012/dec/30/diller-scofidio-renfro-high-line.

MORTENSEN, L. M. 2014. The Effect of Photosynthetic Active Radiation and Temperature on Growth and Flowering of Ten Flowering Pot Plant Species. Available: http://www.scirp.org/journal/PaperInformation.aspx?paperID=46931#.VOdf7ix_RbI.

MOUGEOT, L. J. A. (ed.) 2005. Agropolis, Ottawa, London: International Development Research Centre, Earthscan.

MÜLLER, W. & VOGEL, G. 1997. dtv-Atlas zur Baukunst (Bd. 1), München, Deutscher Taschenbuch Verlag.

MÜLLER, W. & VOGEL, G. 1997. dtv-Atlas zur Baukunst (Bd. 2), München, Deutscher Taschenbuch Verlag.

NEDERHOFF, E. M. & VEGTER, J. G. 1994. Photosynthesis of Stands of Tomato, Cucumber and Sweet Pepper Measured in Greenhouses under Various Co_2-concentrations. Annals of Botany ; 73, 252-361.

NELSON, J. A. & BUGBEE, B. 2014. Economic Analysis of Greenhouse Lighting: Light Emitting Diodes vs. High Intensity Discharge Fixtures. PLOS ONE ; 9/6, 1-10. Available: http://journals.plos.org/plosone/article?id=10.1371/journal.pone.0099010 [Accessed 15.01.2015]

NELSON, N. 2009. Planning the productive city. Available: http://www.nelsonelson.com/DSA-Nelson-renewable-city-report.pdf.

NOLEPPA, S. & WITZKE, H. V. 2012. Tonnen für die Tonne. Berlin: WWF.

OEKOM E.V. - VEREIN FÜR ÖKOLOGISCHE KOMMUNIKATION (ed.) 2011. Post-Oil City, München: Oekom-Verl.

PAARLBERG, R. 2010. Food Politics, Oxford, Oxford University Press.

PEVSNER, N., HONOUR, H. & FLEMING, J. 1992. Lexikon der Weltarchitektur, München, Prestel.

PIMENTEL, D. et al. 2008. Food, Energy and Society, third edition, CRC Press Boca Raton

POLLAN, M. 2008. Farmer in Chief. Available: http://www.nytimes.com/2008/10/12/magazine/12policy-t.html?pagewanted=all&_r=0.

POLLAN, M. 2008. In Defense of Food, London, Penguin Books.

POPKIN, B. 2009. The World Is Fat, New York, NY, Avery.

REITANO, E., DEL GIACCO, E., TOURÉ, S., GIN, G. & RAMIREZ, I. 2006. Socioeconomic and Political Implications of Vertical Farming. New York: Columbia University, Medial Ecology.

REYNOLDS, D. B. 2011. Energy Civilization, Fairbanks, Alaska, Alaska Chena Publisher.

RORABAUGH, P. A. 2014. Introduction to Hydroponics and Controlled Environment Agriculture. Tucson: University of Arizona, Controlled Environment Agriculture Center.

RUBY, I., RUBY, A. & JANSON, N. (eds.) 2014. The Economy of Sustainable Construction, Berlin: Ruby Press.

RUNKLE, E. 2006. Do you know what your DLI is? Available: http://www.hrt.msu.edu/energy/Notebook/pdf/Sec1/Do_you_know_what_your_DLI_is_by_Runkle.pdf.

SÄCHSISCHE LANDESANSTALT FÜR LANDWIRTSCHAFT, FACHBEREICH GARTENBAU 2004. Gewächshaustomaten. Hinweise zum umweltgerechten Anbau. Managementunterlage. [Dresden]: Sächsische Landesanstalt für Landwirtschaft, Fachbereich Gartenbau.

SÄCHSISCHE LANDESANSTALT FÜR LANDWIRTSCHAFT, F. G. 204. Salate im Gewächshaus. Hinweise zum umweltgerechten Anbau. Managementunterlage. Dresden: Sächsische Landesanstalt für Landwirtschaft, Fachbereich Gartenbau.

SADLER, P., FURFARO, R., GIACOMELLI, G. & PATTERSON, L. 2008. Prototype BLSS Lunar Habitat. Available: http://papers.sae.org/2008-01-2186/.

SADLER, P., GIACOMELLI, G., FURFARO, R., PATTERSON, R. & KACIRA, M. 2009. Prototype BLSS Lunar Greenhouse. Available: http://papers.sae.org/2009-01-2484/.

SCHITTICH, C. & INSTITUT FÜR INTERNATIONALE ARCHITEKTUR-DOKUMENTATION GMBH, MÜNCHEN (eds.) 2006. Im Detail: Gebäudehüllen, Basel: Birkhäuser.

SCHOPFER, P. & BRENNICKE, A. 2010. Pflanzenphysiologie, Heidelberg, Spektrum Akademischer Verlag.

SCHWARZ, M. & ELFFERS, J. 2010. Sustainism is the New Modernism, New York, D. A. P.

SCHWARZENBACH, R., MÜLLER, L., RENTSCH, C. & LANZ, K. (eds.) 2011. Mensch Klima!, Baden: Lars Müller Publishers.

SMIL, V. 2008. Energy in Nature and Society, Cambridge, Mass., MIT Press.

SNYDER, R. G. [2010]. Greenhouse Tomato Handbook. Mississippi: Mississippi State University Extension Service.

SPINNER, H. 1974. Pluralismus als Erkenntnismodell, Frankfurt/Main, Suhrkamp.

STEEL, C. 2009. Hungry City, London, Vintage.

STURM, A. & EGLI, N. 2006. Prognoseskizze Energie 2050 - Zusatzuntersuchung zur Studie „Energieperspektiven 2050 der Umweltorganisationen". Welche weiteren Bedingungen und Schritte sind erforderlich, um bis 2050 die Ziele von 2000 Watt Gesamtenergieverbrauch und maximal 500 Watt Fossilenergieverbrauch pro Kopf zu erreichen? Basel: Ellipson AG.

TANTAU, H.-J. 2009. Low Energy Greenhouse and the Hot-Box Approach. Den Haag.

THEURL, M. C. 2008. CO 2-Bilanz der Tomatenproduktion. Analyse acht verschiedener Produktionssysteme in Österreich, Spanien und Italien. Vienna: Institute of Social Ecology.

UNION OF CONCERNED SCIENTISTS. 2013. The Rise of Superweeds - and What to Do About It. Available: http://www.ucsusa.org/assets/documents/food_and_agriculture/rise-of-superweeds.pdf.

VALE, R. & VALE, B. 2009. Time To Eat The Dog?, London, Thames & Hudson.

VILLIERS, D. S. de., WIEN, H. C., REID, J. E. & ALBRIGHT, L. D. 2009. Energy use and yields in tomato production: field, high tunnel and greenhouse compared for the northern tier of the USA (Upstate New York). ISHS Acta Horticulturae ; 893.

VOICAN, V., LĂCĂTUS, V. & TĂNĂSESCU, M. 1995. Growth and development of tomato plants related to climatic conditions from some areas of Romania. ISHS Acta Horticulturae ; 412, 355-365.

WAGENINGEN UR GLASTUINBOUW 2008. Diffuses Licht verstärkt das Pflanzenwachstum. GHT GewächsHausTechnik Magazine ; 1 (2008) 2, 12-14.

WATKISS, P., SMITH, A., TWEEDLE, G., MCKINNON, A., BROWNE, M., HUNT, A., TRELEVEN, C., NASH, C. & CROSS, S. 2005. The Validity of Food Miles as an Indicator of Sustainable Development: Final report produced for DEFRA. AEA Technology.

WHITE, J. 2010. Sky-field: a Vertical Farming Solution for Urban New York. Master of Architecture, Dissertation, Roger Williams University. School of Architecture, Bristol.

WIEGMANN, K., EBERLE, U., FRITSCHE, U. R. & HÜNECKE, K. 2005. Umweltauswirkungen von Ernährung - Stoffstromanalysen und Szenarien. Diskussionspapier Nr. 7, Darmstadt, Hamburg: Öko-Institut e. V.

WITZKE, H. V. 2008. Weltagrarmärkte: Einige zentrale Änderungen der Rahmenbedingungen und deren Implikationen für die Landwirtschaft. Available: http://www.agrar.huberlin.de/de/institut/departments/daoe/ihe/Veroeff/Laendlicher_Raum_2008.pdf.

WITZKE, H. V. 2009. World Food Security. Available: http://www.hffa.info/index.php/press/archive/2009/20090903-world-food.html.

WITZKE, H. V. 2010. Die dritte Grüne Revolution, Augsburg, Ölbaum-Verlag.

WITZKE, H. V. 2010. Towards the Third Green Revolution, Augsburg, Ölbaum-Verlag.

WITZKE, H. V. 2011. Ananas aus dem Allgäu?, Augsburg, Ölbaum-Verlag.

WITZKE, H. V. 2011. Bananas from Bavaria?, Augsburg, Ölbaum-Verlag.

WITZKE, H. V., NOLEPPA, S. & ZHIRKOVA, I. 2011. Fleisch frisst Land. Berlin: WWF.

WOODS, J., WILLIAMS, A., HUGHES, JOHN K., BLACK, M., MURPHY, R. 2010, Energy and the Food System, Philosophical Transactions of the Royal Society. 2991-3005. Available: http://rstb.royalsocietypublishing.org/content/royptb/365/1554/2991.full.pdf, retrieved 14.10.2015

YARA INTERNATIONAL ASA. Important questions on fertilizer and the environment. Available: http://www.yara.com/doc/3734_Important_Questions_on_Fertilizer_and_the_Environment.pdf.

IX. Nomenclatura

AVOGRADO'S NUMBER: „number of units in one mole of any substance (defined as its molecular weight in grams), equal to $6.02214129 \times 10^{23}$. The units may be electrons, atoms, ions, or molecules, depending on the nature of the substance and the character of the reaction (if any). See also Avogadro's law."[1]

BASAL METABOLIC RATE: „Basal metabolic rate (BMR), index of the general level of activity of an individual's body metabolism, determined by measuring his oxygen intake in the basal state—i.e., during absolute rest, but not sleep, 14 to 18 hours after eating. The higher the amount of oxygen consumed in a certain time interval, the more active is the oxidative process of the body and the higher is the rate of body metabolism. The BMR has been used in measuring the general metabolic state during therapy. It was formerly widely used to assess thyroid function, since the thyroid hormones are prime regulators of tissue oxidation and metabolism; but, since the advent of radioactive-isotope tests and thyroid-hormone studies, BMR measurements have fallen into disuse. (...) Energy is needed not only when a person is physically active but even when the body is lying motionless. Depending on an individual's level of physical activity, between 50 and 80 percent of the energy expended each day is devoted to basic metabolic processes (basal metabolism), which enable the body to stay warm, breathe, pump blood, and conduct numerous physiological and biosynthetic activities, including synthesis of new tissue in growing children and in pregnant and lactating women. Digestion and subsequent processing of food by the body also uses energy and produces heat. This phenomenon, known as the thermic effect of food (or diet-induced thermogenesis), accounts for about 10 percent of daily energy expenditure, varying somewhat with the composition of the diet and prior dietary practices. Adaptive thermogenesis, another small but important component of energy expenditure, reflects alterations in metabolism due to changes in ambient temperature, hormone production, emotional stress, or other factors. Finally, the most variable component in energy expenditure is physical activity, which includes exercise and other voluntary activities as well as involuntary activities such as fidgeting, shivering, and maintaining posture. Physical activity accounts for 20 to 40 percent of the total energy expenditure, even less in a very sedentary person and more in someone who is extremely active."[2]

---

1 http://www.britannica.com/EBchecked/topic/45889/Avogadros-number, retrieved 11.09.2014
2 http://www.britannica.com/EBchecked/topic/54585/basal-metabolic-rate, retrieved 06.12.2014

DAYLIGHT INTEGRAL: „(…)the daylight integral, is the cumulative amount of photosynthetic light that is received each day. The DLI is measured as the number of moles of light (mol) per square meter (m2) per day (d1), or mol/m2/d. The DLI can have a profound effect on root and shoot growth of seedling plugs, root development of cutting and finish plant quality attributes such as stem thickness, plant branching and flower number."[1] DLI is measured by the cumulative amount of rain or light received during a 24-h-period. It is dependent on the time of the year (sun's angle), location, latitude and cloud cover and the daylength (photoperiod).

ETFE „stands for Ethylene Tetrafluoroethylene, a transparent polymer that is used instead of glass and plastic in some modern buildings. Compared to glass, ETFE: tansmits more light, insulates better, costs (…) less to install, is only 1/100 the weight of glass."[2]

EVAPOTRANSPIRATION: „Loss of water from the soil both by evaporation from the soil surface and by transpiration from the leaves of the plants growing on it. Factors that affect the rate of evapotranspiration include the amount of solar radiation, atmospheric vapor pressure, temperature, wind, and soil moisture. Evapotranspiration accounts for most of the water lost from the soil during the growth of a crop. Estimation of evapotranspiration rates is thus important in planning irrigation schemes."[3]

F1 - HYBRID VARIETIES: „The development of hybrid varieties differs from hybridization in that no attempt is made to produce a pure-breeding population; only the F1 hybrid plants are sought. The F1 hybrid of crosses between different genotypes is often much more vigorous than its parents. This hybrid vigour, or heterosis, can be manifested in many ways, including increased rate of growth, greater uniformity, earlier flowering, and increased yield, the last being of greatest importance in agriculture."[4] LED or LIGHT EMITTING DIODES: „in full light-emitting diode, in electronics, a semiconductor device that emits infrared or visible light when charged with an electric current. Visible LEDs are used in many electronic devices as indicator lamps, in automobiles as rear-window and brake lights, and on billboards and signs as alphanumeric displays or even full-colour posters. Infrared LEDs are employed in autofocus cameras and television remote controls and also as light sources in fibre-optic telecommunication systems. (…) By varying the precise composition of the semiconductor,

1   RUNKLE, E. 2006. Do you know what your DLI is? Available: http://www.hrt.msu.edu/energy/Notebook/pdf/Sec1/Do_you_know_what_your_DLI_is_by_Runkle.pdf.
2   http://architecture.about.com/od/construction/g/ETFE.htm
3   http://www.britannica.com/search?query=evapotranspiration; retrieved 14.09.2014
4   http://www.britannica.com/EBchecked/topic/463294/plant-breeding/67745/Hybrid-varieties; retrieved 14.09.2014

the wavelength (and therefore the colour) of the emitted light can be changed. LED emission is generally in the visible part of the spectrum [from 400 to 700 nm (A/N)] or in the near infrared [from 700 to 2.000 nm (A/N)]. The brightness of the light observed from an LED depends on the power emitted by the LED and on the relative sensitivity of the eye at the emitted wavelength. Maximum sensitivity occurs at 0.555 micrometre, which is in the yellow-orange and green region. The applied voltage in most LEDs is quite low, in the region of 2.0 volts; the current depends on the application and ranges from a few milliamperes to several hundred milliamperes. (...)"[1]

GREENHOUSE: „also called glasshouse, building designed for the protection of tender or out-of-season plants against excessive cold or heat. In the 17th century greenhouses were ordinary brick or timber shelters with a normal proportion of window space and some means of heating. As glass became cheaper and as more sophisticated forms of heating became available, the greenhouse evolved into a roofed and walled structure built of glass with a minimal wooden or metal skeleton. By the middle of the 19th century, the greenhouse had developed from a mere refuge from a hostile climate into a controlled environment, adapted to the needs of particular plants. A huge increase in the availability of exotic plants in the 19th century led to a vast increase in glasshouse culture in England.[2]

HAITZ'S LAW: In 2000 Dr. Roland Haitz on the converence „Strategies in Light" presented his observation that every ten years costs per Lumen decreases by factor 10, while the amount of light LEDs are producing increases by the factor 20 (for a certain wavelength). Meanwhile this observation is defined as Haitz's law.[3][4]

HYDROPONICS: „also called Aquaculture, Nutriculture, Soilless Culture, or Tank Farming, the cultivation of plants in nutrient-enriched water, with or without the mechanical support of an inert medium such as sand or gravel. Plants have long been grown with their roots immersed in solutions of water and fertilizer for scientific studies of their nutrition. Early commercial hydroponics (from Greek hydro, "water," and ponos, "labour") adopted this method of culture. Because of the difficulties in supporting the plants in a normal upright growing position and aerating the solution, however, this method was supplanted by gravel culture, in which gravel supports the plants in a watertight bed or bench. Various kinds of gravel and other materials have been used successful-

1 http://www.britannica.com/technology/LED, retrieved 13.03.2014
2 http://www.britannica.com/topic/greenhouse, retrieved 09.08.2014
3 http://www.nature.com/nphoton/journal/v1/n1/full/nphoton.2006.78.html
4 http://www.ledsmagazine.com/articles/2015/08/in-tribute-recognizing-the-life-and-work-of-led-pioneer-roland-haitz.html

ly, including fused shale and clay and granite chips. Fertilizer solution is pumped through periodically, the frequency and concentration depending on the plant and on ambient conditions such as light and temperature. The solution drains into a tank, and pumping is usually automatic. The solution is composed of different fertilizer-grade chemical compounds containing varying amounts of nitrogen, phosphorus, and potassium—the major elements necessary for plant growth—and various trace, or minor, elements such as sulfur, magnesium, and calcium. The solution can be used indefinitely; periodic tests indicate the need for additional chemicals or water. The chemical ingredients usually may be mixed dry and stored. As the plants grow, concentration of the solution and frequency of pumping are increased. A wide variety of vegetables and florist crops can be grown satisfactorily in gravel. The principal advantage is the saving of labour by automatic watering and fertilizing. The disadvantages are high installation costs and the need to test the solution frequently. Yields are about the same as for soil-grown crops."[1]

LIGHT COMPENSATION POINT: There is a break-even where the plant is producing as much sugar as it needs for respiration. This point is defined as light compensation point. As light increases (and water is available) carbon production also increases. The plant therefor exceeds its carbon production, the surplus is transformed into glucose. Principally, exceeding the light compensation point, is the main goal in food production. By ncreasing brightness and intensity within PAR, of photosynthesis-rate also increases, „but only up to a certain point, beyond which increasing the brightness of light has little or no effect on the rate of photosyntesis. (...) The light intensity at which the net amount of oxygen produced is exactly zero, is called the compensation point for light."[2] At this point the consumption of oxygen by the plant due to cellular respiration is equal to the rate at which oxygen is produced by photosynthesis.[3]

LIGHT SATURATION POINT: On the other side of the „photosynthesis activating point" we could call the light compensation point an other point is essential in plant cultivation: the light saturation point. „The saturation point describes the amount of light that is beyond the capability of the chloroplast to absorb. Photosynthesis still occurs, but the amount of light has exceeded the amount of pigments that are available for absorption."[4] This saturation point is different to every plant. „Different plants have different saturation points, determined by the number of pigments in their chlorophyll cells. Plants that typically

1 http://www.britannica.com/EBchecked/topic/279000/hydroponics, retrieved 13.03.2015
2 http://Tomatosphere.org/teachers/guide/grades-8-10/plants-and-light, retrieved 12.09.2014
3 ibid.
4 http://www.ehow.com/about_6535863_definition-plant-light-saturation.html, retrieved 12.09.2014

grow in shaded areas have lower saturation points, while those that grow in areas more exposed to light have higher saturation points.

LYCOPERSICON ESCULENTUM (MILL.): flowering plant of the nightshade family (Solanaceae), cultivated extensively for its edible fruits. Labelled as a vegetable for nutritional purposes, tomatoes are a good source of vitamin C and the phytochemical lycopene. The fruits are commonly eaten raw in salads, served as a cooked vegetable, used as an ingredient of various prepared dishes, and pickled. Additionally, a large percentage of the world's tomato crop is used for processing; products include canned tomatoes, tomato juice, ketchup, puree, paste, and "sun-dried" tomatoes or dehydrated pulp.[1]

MOL: „is a unit of measurement used in physics and chemistry to express amounts of elements, defined as the amount of any substance that contains as many elementary entities (e.g., atoms, molecules, ions, electrons) as there are atoms in 12 grams of pure carbon-12 (12C), the isotope of carbon with relative atomic mass of exactly 12 by definition. This corresponds to the Avogadro constant, which has a value of $6.02214129(27) \cdot 10^{23}$ elementary entities of the substance."[2] A mole of photons, therefor consists in 602 trillion light particles.

PAR, PHOTOSYNTHETICALLY ACTIVE RADIATION: „is the waveband 400 to 700 nm, which are the limits of wavelengths that are of primary importance for plant photosynthesis. The PPFD, Photosynthetic Photon Flux Density is the number of photons in the PAR waveband that are incident on a surface in a give time period ($\mu mol/m^{-2}/s^{-1}$). The quantum sensor will measure this value. A very clear sky value will approch approx. $2000 \mu mol/m^{-2}/s^{-1}$ PAR."[3] The PPFD-number for a clear sunny sky differs up to 15% regarding differend studies, from 1700 $\mu mol/m^{-2}/s^{-1}$ (also used by Gene Giacomelli) to 2000 $\mu mol/m^{-2}/s^{-1}$. This work uses 1800 $\mu mol/m^{-2}/s^{-1}$ for a clear sunny day. Most conversion calculators from horticulturalist and grow lamp manufacturers use the factor 0,018 to convert lux to $\mu mol/m^{-2}/s^{-1}$ and the factor 0.219 from Photons to W (sunlight) or 4.57 from W (sunlight to Photons.[4] „The term PAR and its units (...) [are] an important concept to understand and use.(...) Clearly, the human eye cannot even begin to respond to many of the wavelengths before 500 nm and beyond 600

---

1 http://www.britannica.com/plant/tomato, retrieved 26.09.2015
2 International Bureau of Weights and Measures (2006), The International System of Units (SI) (8th ed.), pp. 114–15, ISBN 92-822-2213-6
3 GIACOMELLI, G. 1998. Components of Radiation Defined: Definition of Units, Measuring Radiation Transmission, Sensors. CCEA, Center for Controlled Environment Agriculture, rutgers university, Cook College.
4 ibid.

nm. The plant leaf response, however, extends beyond the PAR waveband of 400 - 700 nm. (...)"[1]

PEAK OIL, PEAK OIL THEORY: „a contention that conventional sources of crude oil, as of the early 21st century, either have already reached or are about to reach their maximum production capacity worldwide and will diminish significantly in volume by the middle of the century. "Conventional" oil sources are easily accessible deposits produced by traditional onshore and offshore wells, from which oil is removed via natural pressure, mechanical walking beam pumps, or well-known secondary measures such as injecting water or gas into the well in order to force oil to the surface. The peak oil theory does not apply to so-called unconventional oil sources, which include oil sands, oil shales, oil extracted after fracking "tight rock" formations, and oil found in deepwater wells far offshore—in short, any deposit of oil that requires substantial investment and labour to exploit." Proponents of peak oil theory do not necessarily claim that conventional oil sources will run out immediately and create acute shortages, resulting in a global energy crisis. Instead, the theory holds that, with the production of easily extractable oil peaking and inevitably declining (even in formerly bounteous regions such as Saudi Arabia), crude-oil prices are likely to remain high and even rise further over time, especially if future global oil demand continues to rise along with the growth of emerging economies such as China and India. Although peak oil theory may not portend prohibitively expensive gasoline any time soon, it does suggest that the days of inexpensive fuel, as were seen for more than a decade after the collapse of OPEC cartel prices in the mid-1980s, will probably never return. [2]

PHOTOPERIODISM: „is another attribute of plants that may be changed or manipulated in the microclimate. The length of a day is a photoperiod, and the responses of the plant development to a photoperiod are called photoperiodism. Response to the photoperiod is different for different plants; long-day plants flower only under day lengths longer than 14 hours; in short-day plants, flowering is induced by photoperiods of less than 10 hours; day-neutral plants form buds under any period of illumination. There are exceptions and variations in photoperiodic response; also, it is argued that the truly critical factor is actually the amount of exposure to darkness rather than to daylight. Temperature is intimately related to photoperiodism, tending to modify reactions to daylength. Photopcriodism is one determining factor in natural distribution of plants throughout the world. The phenomenon has many practical applica-

---

1   GIACOMELLI, G. 1998. Components of Radiation Defined: Definition of Units, Measuring Radiation Transmission, Sensors. CCEA, Center for Controlled Environment Agriculture, rutgers university, Cook College.
2   http://www.britannica.com/topic/peak-oil-theory, retrieved 26.09.2014; also see M.King Hubbert, Colin J. Campbell, ASPO (Association for the Studies of Peak Oil and Gas".

tions. Selection of a plant or a variety for a given locality requires knowledge of its interaction with the photoclimate. Artificial illumination is used to control flowering seasons and to increase production of greenhouse crops. In plant breeding, such stimulation of flowering has greatly reduced the time span from germination to maturity, shortening the time necessary to develop new varieties. In sowing field crops, photoperiodism can be used to select the date of sowing to produce optimum harvest size. Crop yield is reduced both by planting in a season that will cause plants to flower early and by planting at a time that will cause very late flowering. In Sri Lanka (formerly Ceylon), certain rice varieties with a vegetative period of five to six months may extend their life to more than a year when planted in the wrong season, causing almost complete loss of yield. Cowpeas in Nigeria will flower early and produce many seeds only when planted in daylengths of 12 hours or less.[1]

PLANCK'S CONSTANT: (symbol h), „fundamental physical constant characteristic of the mathematical formulations of quantum mechanics, which describes the behaviour of particles and waves on the atomic scale, including the particle aspect of light. The German physicist Max Planck introduced the constant in 1900 in his accurate formulation of the distribution of the radiation emitted by a blackbody, or perfect absorber of radiant energy (see Planck's radiation law). The significance of Planck's constant in this context is that radiation, such as light, is emitted, transmitted, and absorbed in discrete energy packets, or quanta, determined by the frequency of the radiation and the value of Planck's constant. The energy E of each quantum, or each photon, equals Planck's constant h times the radiation frequency symbolized by the Greek letter nu, ν, or simply $E = h\nu$. A modified form of Planck's constant called h-bar ($\hbar$), or the reduced Planck's constant, in which $\hbar$ equals h divided by 2π, is the quantization of angular momentum. For example, the angular momentum of an electron bound to an atomic nucleus is quantized and can only be a multiple of h-bar. The dimension of Planck's constant is the product of energy multiplied by time, a quantity called action. Planck's constant is often defined, therefore, as the elementary quantum of action. Its value in metre-kilogram-second units is $6.62606957 \times 10^{-34}$ joule·second, with a standard uncertainty of $0.00000029 \times 10^{-34}$ joule·second."[2]

PPFD: Photosynthetic Photon Flux Density is the number of photons within the PAR-waveband that is incident on a surface in a give[n] time period (λmol m2/s). The quantum sensor will measure this value. When considered as a photon it may be expressed in energy terms, Watts per square meter (W/m2), or as the

---

[1] http://www.britannica.com/EBchecked/topic/9620/agricultural-technology/67813/The-degree-day, retrieved 16.03.2015
[2] http://www.britannica.com/EBchecked/topic/559095/planck constant, retrieved, 11.09.2014

number of photons (moles of photons) µmol/m^{-2}/s^{-1}. Wavelength (λ, A/N) has units of meters, typically nanometers (nm) [...] or micrometers (µm). Frequency (f, A/N) has units of cycle per second. Together they are related as parameters of a photon of light by the constant c, the speed of light (299.792.458 m/s, A/N). The frequency of the photon is equal to the speed of light divided by wavelength of the photon. The energy of a wavelength of light is equal to Planck's constant (h = 6,626·10-34 Js, A/N) multiplied by the speed of light and divided by the wavelength. From this relationship, an important fact is determined. For radiation (light), as its wavelength increases, its energy decreases, and as the wavelength decreases, the energy increases. Thus short wave blue light has more energy than longer wave red light."[1]

RHIZOSPHERE: is the area (...) [volume, Ed.] around the root (soil in near contact with roots) which is rich in nutrients and directly influenced by the secretions of plant root exudates and microorganisms.[2]

SOLANUM / SOLANACEA or NIGHTSHADE: genus of about 2,300 species of flowering plants in the nightshade family (Solanaceae). The term nightshade is often associated with poisonous species, though the genus also contains a number of economically important food crops, including tomato (Solanum lycopersicum), potato (S. tuberosum), and eggplant (S. melongena). Nightshades are annuals or perennials and range in size from small herbs to small trees. The alternate leaves can be simple or pinnately compound and usually feature glandular or nonglandular trichomes (plant hairs). The leaves and stems are sometimes armed with prickles. The flowers have five petals that are often fused. The flowers usually are white, yellow, or purple and are borne in clusters. The fruit is a berry. The species usually called nightshade in North America and the United Kingdom is S. dulcamara, also known as bittersweet and woody nightshade. Its foliage and egg-shaped red berries are poisonous, the active principle being solanine, which can cause convulsions and death if taken in large doses. The black nightshade (S. nigrum) is also generally considered poisonous, but its fully ripened fruit and foliage are cooked and eaten in some areas. A number of plants outside the genus Solanum are also known as nightshades. The aptly named deadly nightshade, or belladonna (Atropa belladonna), is a tall bushy herb of the same family and the source of several alkaloid drugs. Enchanter's nightshade is a name applied to plants of the genus Circaea (family Onagraceae). Malabar nightshade, also known as Mala-

---

1   GIACOMELLI, G. 1998. Components of Radiation Defined: Definition of Units, Measuring Radiation Transmission, Sensors. CCEA, Center for Controlled Environment Agriculture, rutgers university, Cook College., p.2
2   http://www.researchgate.net/post/What_is_the_best_definition_of_rhizosphere, retrieved 27.09.2015

bar spinach, refers to twining herbaceous vines of the genus Basella (family Basellaceae).

SPEED OF LIGHT: „speed at which light waves propagate through different materials. In particular, the value for the speed of light in a vacuum is now defined as exactly 299,792,458 metres per second. The speed of light is considered a fundamental constant of nature. Its significance is far broader than its role in describing a property of electromagnetic waves. It serves as the single limiting velocity in the universe, being an upper bound to the propagation speed of signals and to the speeds of all material particles. In the famous relativity equation, $E = mc^2$, the speed of light (c) serves as a constant of proportionality linking the formerly disparate concepts of mass (m) and energy (E)."[1]

TOMATO: see *LYCOPERSICON ESCULENTUM* (MILL.)

TPES or TOTAL PRIMARY ENERGY SUPPLY: „equals production plus imports minus exports minus international bunkers plus or minus stock changes. The International Energy Agency (IEA) energy balance methodology is based on the calorific content of the energy commodities and a common unit of account. The unit of account adopted is the tonne of oil equivalent (toe) which is defined as 10⁷ kilocalories (41.868 gigajoules). This quantity of energy is, within a few per cent, equal to the net heat content of one tonne of crude oil. The difference between the "net" and the "gross" calorific value for each fuel is the latent heat of vaporisation of the water produced during combustion of the fuel. For coal and oil, net calorific value is about 5% less than gross, for most forms of natural and manufactured gas the difference is 9-10%, while for electricity there is no difference. The IEA balances are calculated using the physical energy content method to calculate the primary energy equivalent."[2]

VERTICAL FARMING: Vertical Farming is defined as a highly industrialized year round cultivation method for food production, adaptable for multiple crop types, where the verticalized building typology, its programme and functions primarily focus on optimum plant growth. The building is seen as a structural element of the urban ecosystem. In addition to food production, the Vertical Farm must incorporate elements of the food sector which, at present, are spatially detached from each other on a global scale, something which has a severe impact on energy consumption and the environment.

---

[1] http://www.britannica.com/EBchecked/topic/559095/speed-of-light; retrieved 11.09.2014
[2] http://www.oecd-ilibrary.org/sites/factbook-2013-en/06/01/01/index.html?containerIt emId=%2Fcontent%2Fserial%2F18147364&itemId=%2Fcontent%2Fchapter%2Ffactbo ok-2013-41-en&mimeType=text%2Fhtml

CONCEPTS AND DEFINITIONS - RETRIEVED FROM FAO:

FOOD BALANCE SHEETS: „(FBS) provide essential information on a country's food system through three components: 1. Domestic food supply of the food commodities in terms of production, imports, and stock changes, 2. Domestic food utilization which includes feed, seed, processing, waste, export, and other uses and 3.per capita values for the supply of all food commodities (in kilograms per person per year) and the calories, protein, and fat content. Annual food balance sheets show the trends in the overall national food supply, disclose changes that may have taken place in the types of food consumed, and reveal the extent to which the food supply of the country is adequate in relation to nutritional requirements. Food balance sheets provide other relevant statistics that can be used in designing and targeting policies to reduce hunger in countries. The import dependency ratio for food, that compares the quantities of food available for human consumption with those imported, indicates the extent to which a country depends upon imports to feed itself. The amount of food crops used for feeding livestock in relation to total crop production indicates the degree to which primary food resources are used to produce animal feed which is useful information for analyzing livestock policies or patterns of agriculture. Data on per caput food supplies are an important element for projecting food demand, together with such other elements as income elasticity coefficients, projections of private consumption expenditure and population."[1]

PRODUCTION: „For primary commodities, production should relate to the total domestic production whether inside or outside the agricultural sector, i.e. including non-commercial production and production in kitchen gardens. Unless otherwise indicated, production is reported at the farm level for primary crops (i.e. excluding harvesting losses for crops) and livestock items and in terms of live weight (i.e. the actual ex-water weight of the catch at the time of capture) for primary fish items. Production of processed commodities relates to the total output of the commodity at the manufacture level (i.e. it comprises output from domestic and imported raw materials of originating products). Reporting units are chosen accordingly, e.g. cereals are reported in terms of grains and paddy rice. As a general rule, all data on meat are expressed in terms of carcass weight. Usually the data on production relate to that which takes place during the reference period. However, production of certain crops may relate to the harvest of the year preceding the utilization period if harvesting takes place late in the year. In such instances, the production of a given year largely moves into consumption in the subsequent year. In the sample Form II of the food balance sheet, located at the end of this document, a distinction is made

---

1 http://www.fao.org/economic/ess/fbs/en/, retrieved 12.08.2014

between „output" and „input". The production of primary as well as of derived products is reported under „output". For derived commodities, the amounts of the originating commodity that are required for obtaining the output of the derived product are indicated under „input", and are expressed in terms of the originating commodity."[1]

CHANGES IN STOCKS: „In principle, this comprises changes in stocks occurring during the reference period at all levels from production to the retail stage, i.e. it comprises changes in government stocks, in stocks with manufacturers, importers, exporters, other wholesale and retail merchants, transport and storage enterprises, and in stocks on farms. In practice, though, the information available often relates only to stocks held by governments, and even this is, for a variety of reasons, not available for a number of countries and important commodities. It is because of this that food balance sheets are usually prepared as an average for several years as this is believed to reduce the degree of inaccuracy contributed by the absence of information on stocks. Increases in stocks of a commodity reduce the availability for domestic utilization. They are therefore indicated by the - sign and decreases in stocks by the + sign since they increase the available supply. In the absence of information on opening and closing stocks, changes in stocks are also used for shifting production from the calendar year in which it is harvested to the year in which it enters domestic utilization or is exported."[2]

GROSS IMPORTS: „In principle, this covers all movements of the commodity in question into the country as well as of commodities derived therefrom and not separately included in the food balance sheet. It, therefore, includes commercial trade, food aid granted on specific terms, donated quantities, and estimates of unrecorded trade. As a general rule, figures are reported in terms of net weight, i.e. excluding the weight of the container."[3]

SUPPLY: „There are various possible ways to define „supply" and, in fact, various concepts are in use. The elements involved are production, imports, exports and changes in stocks (increases or decreases). There is no doubt that production, imports, and decreases in stocks are genuine supply elements. Exports and increases in stocks might, however, be considered to be utilization elements. Accordingly, the following possibilities exist for defining „supply". (a) Production + imports + decrease in stocks = total supply. (b) Production + imports + changes in stocks (decrease or increase) = supply available for ex-

---

[1]   http://www.fao.org/economic/ess/fbs/en/, retrieved 12.08.2014
[2]   ibid.
[3]   ibid.

port and domestic utilization. (c) Production + imports - exports + changes in stocks (decrease or increase) = supply for domestic utilization."[1]

GROSS EXPORTS: „In principle, this covers all movements of the commodity in question out of the country during the reference period. The conditions specified for gross imports, under 3. above, apply also to exports by analogy. A number of commodities are processed into food and feed items. Therefore, there is a need to identify the components of the processed material exported in order to arrive at a correct picture of supplies for food and feed in a given time-reference period."[2]

FEED: "This comprises amounts of the commodity in question and of edible commodities derived therefrom not shown separately in the food balance sheet (e.g. dried cassava, but excluding by-products, such as bran and oilcakes) that are fed to livestock during the reference period, whether domestically produced or imported."[3]

SEED: "In principle, this comprises all amounts of the commodity in question used during the reference period for reproductive purposes, such as seed, sugar cane planted, eggs for hatching and fish for bait, whether domestically produced or imported. Whenever official data are not available, seed figures can be estimated either as a percentage of production (e.g. eggs for hatching) or by multiplying a seed rate with the area under the crop of the subsequent year. In those cases where part of the crop is harvested green (e.g. cereals for direct feed or silage, green peas, green beans) an adjustment must be made for this area. Usually, the average amount of seed needed per hectare planted in any given country, does not greatly vary from year to year."[4]

FOOD MANUFACTURE: „The amounts of the commodity in question used during the reference period for manufacture of processed commodities for which separate entries are provided in the food balance sheet either in the same or in another food group (e.g. sugar, fats and oils, alcoholic beverages) are shown under the column Food Manufacture. Quantities of the commodity in question used for manufacture for non-food purposes, e.g. oil for soap, are shown under the element Other Uses. The processed products do not always appear in the same food group. While oilseeds are shown under the aggregate Oilcrops, the respective oil is shown under the Vegetable Oils group; similarly, skim milk is in the Milk group, while butter is shown under the aggregate Animal Fats. Barley,

1 ibid. http://www.fao.org/economic/ess/fbs/en/, retrieved 12.08.2014
2 http://www.fao.org/economic/ess/fbs/en/, retrieved 12.08.2014
3 ibid.
4 ibid.

maize, millet and sorghum are in the Cereals group, while beer made from these cereals is shown under the Alcoholic Beverages group. The same principle applies for grapes and wine."[1]

WASTE: „This comprises the amounts of the commodity in question and of the commodities derived therefrom not further pursued in the food balance sheets, lost at all stages between the level at which production is recorded and the household, i.e. losses during storage and transportation. Losses occurring during the pre-harvest and harvesting stages are excluded (see note on „Production"). Technical losses occurring during the transformation of the primary commodities into processed products are taken into account in the assessment of respective extraction/conversion rates. Post-harvest losses in most countries are substantial owing to the fact that most of the grain production is retained on the farm so as to provide sufficient quantities to last from one harvest to the next. Farm storage facilities in many countries tend to be primitive and inadequately protected from the natural competitors of man for food. Losses become even more serious in countries where agricultural products reach consumers in urban areas after passing through several marketing stages. In fact, one of the major causes of food losses in some countries is the lack of adequate marketing systems and organization. Much food remains unsold because of the imbalances of supply and demand. This is particularly true of perishable foods, such as fresh fruit and vegetables. Post-harvest losses of fruit and vegetables of between 25 and 40 percent occur in many countries, mainly as a result of untimely harvesting and improper packing and/or transport. The waste of both edible and inedible parts of the commodity occurring in the household, e.g. in the kitchen, also is excluded."[2]

OTHER USES: „In order not to distort the picture of the national food pattern, quantities of the commodity in question, consumed mainly by tourists, are included here (see also „12. Per Caput Supply") as well as the amounts of the commodity in question used during the reference period for the manufacture for non-food purposes (e.g. oil for soap). Also statistical discrepancies are included here. They are defined as an inequality between supply and utilization statistics. The food balance sheets are compiled using statistics from various sources. Where no official data are available, other sources of information may be used. Many of the supply and utilization elements compiled from available information will not balance. Bringing together data from different sources would almost always result in an imbalance. Beyond the problem of data sources, imbalances usually fall into one of the following three situations: those occurring mainly in developed countries where there is no shortage of of-

---

1 http://www.fao.org/economic/ess/fbs/en/, retrieved 12.08.2014
2 ibid.

ficial statistics but the information is not internally consistent; cases in which the data are consistent but incomplete; and situations where data are both inconsistent and incomplete."[1]

FOOD: „This comprises the amounts of the commodity in question and of any commodities derived therefrom not further pursued in the food balance sheet that are available for human consumption during the reference period. The element food of maize, for example, comprises the amount of maize, maize meal and any other products derived therefrom, like cornflakes, available for human consumption. The food element for vegetables comprises the amount of fresh vegetables, canned vegetables, and any other products derived therefrom. But the element food of milk relates to the amounts of milk available for human consumption as milk during the reference period, but not as butter, cheese or any other milk product provided for separately in the food balance sheet. It is important to note that the quantities of food available for human consumption, as estimated in the food balance sheet, reflect only the quantities reaching the consumer. The amount of food actually consumed may be lower than the quantity shown in the food balance sheet depending on the degree of losses of edible food and nutrients in the household, e.g. during storage, in preparation and cooking (which affect vitamins and minerals to a greater extent than they do calories, protein and fat), as plate-waste, or quantities fed to domestic animals and pets, or thrown away."[2]

PER CAPUT SUPPLY: „Under this heading estimates are provided of per caput food supplies available for human consumption during the reference period in terms of quantity, caloric value, and protein and fat content. Per caput food supplies in terms of quantity are given both in kilograms per year and grams per day, calorie supplies are expressed in kilo-calories (calories) per day, while supplies of protein and fat are provided in grams per day. It is proposed to retain the traditional unit of calories for the time being until such time as the proposed „kilojoule" gains wider acceptance and understanding (1 calorie = 4.19 kilojoules)."[3]

FOOD LOSSES AND FOOD WASTES

FOOD LOSSES: „refer to the decrease in edible food mass throughout the part of the supply chain that specifically leads to edible food for human consumption. Food losses take place at production, postharvest and processing stages in the food supply chain (Parfitt et al., 2010). Food losses occurring at the end

---

[1] http://www.fao.org/economic/ess/fbs/en/, retrieved 12.08.2014
  ibid.
[3] ibid.

of the food chain (retail and final consumption) are rather called "food waste", which relates to retailers' and consumers' behavior. (Parfitt et al., 2010)."[1]

FOOD WASTE OR LOSS: „is measured only for products that are directed to human consumption, excluding feed and parts of products which are not edible. Per definition, food losses or waste are the masses of food lost or wasted in the part of food chains leading to "edible products going to human consumption". Therefore food that was originally meant to human consumption but which fortuity gets out the human food chain is considered as food loss or waste even if it is then directed to a non-food use (feed, bioenergy…). This approach distinguishes "planned" non-food uses to "unplanned" non-food uses"[2]

---

[1] http://www.fao.org/fileadmin/user_upload/suistainability/pdf/Global_Food_Losses_and_Food_Waste.pdf, retrieved 12.08.2015

[2] Global Food losses and food waste FAO p.2

X. Appendix

Additional Simulation Results   **380**

Impressum   **391**

APPENDIX

CLIMATE DATA OF VIENNA - LIGHT AND TEMPERATURE NEEDS OF L. esculentum

DAYLENGTH				month	day	week	AVAILABLE SOLAR RADIATION				
							direct/diffuse	direct/diffuse	PAR	DLI available	PPFD
available	available	needed	Δ								Ø/s
h	sec/d daylength	sec/d daylength	Diff. Daylength				Wh/m²/w	Ø Wh/m²/d	Wh/m²/d	mol/m²	μmol/m²/s
8.93	32,160	57,600	-25,440	JAN	01.-07.	1	5,300	757.14	378.57	6.23	193.66
9.26	33,345	57,600	-24,255	JAN	07.-14.	2	5,605	800.71	400.36	6.59	197.53
9.59	34,530	57,600	-23,070	JAN	14.-21.	3	5,624	803.43	401.71	6.61	191.40
9.92	35,715	57,600	-21,885	JAN	21.-28.	4	5,690	812.86	406.43	6.69	187.22
10.25	36,900	57,600	-20,700	JAN FEB	28.-04.	5	10,924	1,560.57	780.29	12.84	347.89
10.53	37,905	57,600	-19,695	FEB	04.-11.	6	9,890	1,412.86	706.43	11.62	306.61
10.81	38,910	57,600	-18,690	FEB	11.-18.	7	7,232	1,033.14	516.57	8.50	218.42
11.09	39,915	57,600	-17,685	FEB	18.-25.	8	15,530	2,218.57	1,109.29	18.25	457.22
11.51	41,423	57,600	-16,178	MAR FEB	25.-04.	9	13,717	1,959.57	979.79	16.12	389.15
11.92	42,910	57,600	-14,690	MAR	04.-11.	10	24,497	3,499.57	1,749.79	28.79	670.88
12.19	43,900	57,600	-13,700	MAR	11.-18.	11	17,383	2,483.29	1,241.64	20.43	465.32
12.51	45,032	57,600	-12,568	MAR	18.-25.	12	15,841	2,263.00	1,131.50	18.62	413.38
12.80	46,065	57,600	-11,535	MAR APR	25.-02.	13	21,538	3,076.86	1,538.43	25.31	549.45
13.23	47,633	57,600	-9,968	APR	02.-09.	14	25,513	3,644.71	1,822.36	29.98	629.43
13.67	49,200	57,600	-8,400	APR	09.-16.	15	19,265	2,752.14	1,376.07	22.64	460.14
14.04	50,535	57,600	-7,065	APR	16.-23.	16	25,243	3,606.14	1,803.07	29.66	587.00
14.41	51,870	57,600	-5,730	APR	23.-30.	17	39,132	5,590.29	2,795.14	45.99	886.56
14.78	53,205	57,600	-4,395	MAY	01.-08.	18	35,239	5,034.14	2,517.07	41.41	778.33
15.15	54,540	57,600	-3,060	MAY	08.-15.	19	32,907	4,701.00	2,350.50	38.67	709.03
15.25	54,898	57,600	-2,703	MAY	15.-22.	20	41,777	5,968.14	2,984.07	49.09	894.28
15.55	55,995	57,600	-1,605	MAY	22.-29.	21	22,905	3,272.14	1,636.07	26.92	480.70
15.72	56,590	57,600	-1,010	MAY JUN	29.-05.	22	45,754	6,536.29	3,268.14	53.77	950.12
15.81	56,900	57,600	-700	JUN	05.-12.	23	29,255	4,179.29	2,089.64	34.38	604.20
15.92	57,300	57,600	-300	JUN	12.-19.	24	35,369	5,052.71	2,526.36	41.56	725.37
15.82	56,955	57,600	-645	JUN	19.-26.	25	37,498	5,356.86	2,678.43	44.07	773.69
15.73	56,610	57,600	-990	JUL JUN	26.-03.	26	47,768	6,824.00	3,412.00	56.13	991.60
15.63	56,265	57,600	-1,335	JUL	03.-10.	27	44,748	6,392.57	3,196.29	52.59	934.60
15.53	55,920	57,600	-1,680	JUL	10.-17.	28	36,006	5,143.71	2,571.86	42.31	756.66
15.21	54,765	57,600	-2,835	JUL	17.-24.	29	41,442	5,920.29	2,960.14	48.70	889.26
14.89	53,610	57,600	-3,990	JUL	24.-31.	30	35,141	5,020.14	2,510.07	41.30	770.30
14.57	52,455	57,600	-5,145	AUG	01.-08.	31	30,878	4,411.14	2,205.57	36.29	691.76
14.25	51,300	57,600	-6,300	AUG	08.-15.	32	45,008	6,429.71	3,214.86	52.89	1,031.01
13.83	49,793	57,600	-7,808	AUG	15.-22.	33	40,222	5,746.00	2,873.00	47.27	949.27
13.41	48,285	57,600	-9,315	AUG	22.-29.	34	22,026	3,146.57	1,573.29	25.88	536.06
12.99	46,778	57,600	-10,823	SEP AUG	29.-05.	35	31,996	4,570.86	2,285.43	37.60	803.80
12.58	45,270	57,600	-12,330	SEP	05.-12.	36	24,913	3,559.00	1,779.50	29.28	646.70
12.19	43,900	57,600	-13,700	SEP	12.-19.	37	21,986	3,140.86	1,570.43	25.84	588.54
11.93	42,930	57,600	-14,670	SEP	19.-26.	38	25,929	3,704.14	1,852.07	30.47	709.77
11.62	41,847	57,600	-15,753	SEP	26.-03.	39	17,636	2,519.43	1,259.71	20.72	495.25
11.32	40,764	57,600	-16,836	SEP OCT	03.-10.	40	21,176	3,025.14	1,512.57	24.88	610.46
10.83	39,000	57,600	-18,600	OCT	10.-17.	41	16,771	2,395.86	1,197.93	19.71	505.34
10.45	37,628	57,600	-19,973	OCT	17.-24.	42	7,716	1,102.29	551.14	9.07	240.98
10.07	36,255	57,600	-21,345	OCT	24.-31.	43	7,822	1,117.43	558.71	9.19	253.54
9.69	34,883	57,600	-22,718	NOV	01.-08.	44	8,940	1,277.14	638.57	10.51	301.18
9.31	33,510	57,600	-24,090	NOV	08.-15.	45	9,282	1,326.00	663.00	10.91	325.51
9.11	32,813	57,600	-24,788	NOV	15.-22.	46	6,247	892.43	446.21	7.34	223.73
8.92	32,115	57,600	-25,485	NOV	22.-29.	47	6,041	863.00	431.50	7.10	221.05
8.73	31,418	57,600	-26,183	NOV DEC	29.-06.	48	4,177	596.71	298.36	4.91	156.24
8.53	30,720	57,600	-26,880	DEC	06.-13.	49	4,770	681.43	340.71	5.61	182.47
8.73	31,423	57,600	-26,178	DEC	13.-20.	50	4,311	615.86	307.93	5.07	161.22
8.92	32,099	57,600	-25,501	DEC	20.-27.	51	4,179	597.00	298.50	4.91	152.99
8.92	32,120	57,600	-25,480	DEC	27.-02.	52	3,561	508.71	254.36	4.18	130.28
638.56	2,298,801.63	2,995,200.00	-696,398.38						3,075.05	1,537.53	25.30	525.51
					ANNUAL RESULTS		1,119,320.00		561,197.53		

Tab.26.   Vienna - Daylength, available solar radiation, light- and temperature need for **L. esculentum**

APPENDIX

LIGHT NEED FOR PHOTOSYNTHESIS AND AVAILABILITY							TEMPERATURE		
DLI needed	DLI needed	PAR needed	PAR needed	PPFD	PPFD	Temp.	Temp.	Temp.
	Δ		Δ	∅/s	Δ			
mol/m²	mol/m²	Wh/m²/d	Wh/m²/d	µmol/m²/s	µmol/m²/s	media low	media	media high
8.64	-2.41	528.00	-149.43	150	43.66	-9.70	-1.86	14.80
9.04	-2.46	528.00	-127.64	157	40.53	-8.10	-0.90	5.60
9.73	-3.13	552.64	-150.93	169	22.40	-7.20	-3.13	1.40
13.42	-6.73	594.88	-188.45	233	-45.78	-4.00	3.58	13.60
16.93	-4.10	820.16	-39.87	294	53.89	-5.30	-4.16	10.80
17.28	-5.66	1,034.88	-328.45	300	6.61	-5.60	-0.92	11.20
17.11	-8.61	1,056.00	-539.43	297	-78.58	-5.93	3.49	9.40
17.05	1.20	1,045.44	63.85	296	161.22	-6.20	1.37	9.90
16.99	-0.87	1,041.92	-62.13	295	94.15	-6.55	3.47	17.70
16.93	11.85	1,038.40	711.39	294	376.88	-5.40	6.19	9.20
16.93	3.49	1,034.88	206.76	294	171.32	-3.63	6.38	17.30
16.82	1.80	1,034.88	96.62	292	121.38	-2.60	1.70	21.00
16.65	8.66	1,027.84	510.59	289	260.45	-0.70	6.51	28.00
16.47	13.51	1,017.28	805.08	286	343.43	0.00	7.09	27.80
16.13	6.51	1,006.72	369.35	280	180.14	2.00	6.64	27.10
15.84	13.82	985.60	817.47	275	312.00	3.60	11.46	29.00
15.61	30.38	968.00	1,827.14	271	615.56	2.90	17.11	26.70
15.32	26.09	953.92	1,563.15	266	512.33	7.20	12.83	25.90
15.03	23.64	936.32	1,414.18	261	448.03	6.50	13.25	26.10
14.75	34.35	918.72	2,065.35	256	638.28	7.05	19.58	36.90
14.46	12.46	901.12	734.95	251	229.70	6.55	14.94	31.20
14.17	39.60	883.52	2,384.62	246	704.12	6.90	15.25	33.40
13.88	20.50	865.92	1,223.72	241	363.20	7.55	14.94	34.00
13.59	27.97	848.32	1,678.04	236	489.37	9.10	15.71	33.40
13.31	30.76	830.72	1,847.71	231	542.69	9.20	18.43	31.00
13.02	43.12	813.12	2,598.88	226	765.60	11.30	21.72	36.80
12.73	39.86	795.52	2,400.77	221	713.60	10.50	22.32	39.20
12.44	29.87	777.92	1,793.94	216	540.66	11.50	20.84	36.00
12.15	36.55	760.32	2,199.82	211	678.26	10.85	17.42	37.40
11.87	29.43	742.72	1,767.35	206	564.30	11.50	20.68	37.40
11.58	24.71	725.12	1,480.45	201	490.76	11.20	18.89	38.50
11.29	41.60	707.52	2,507.34	196	835.01	10.60	22.14	31.00
11.00	36.26	689.92	2,183.08	191	758.27	7.70	23.30	35.00
10.71	15.17	672.32	900.97	186	350.06	6.60	16.47	30.00
10.43	27.17	654.72	1,630.71	181	622.80	4.20	16.96	27.00
10.14	19.14	637.12	1,142.38	176	470.70	5.60	14.74	26.00
9.85	15.99	619.52	950.91	171	417.54	5.50	17.18	25.50
9.56	20.91	601.92	1,250.15	166	543.77	3.30	16.58	24.00
9.27	11.45	584.32	675.39	161	334.25	3.50	13.98	23.20
8.99	15.90	566.72	945.85	156	454.46	1.00	11.15	22.50
8.70	11.01	549.12	648.81	151	354.34	2.00	13.72	21.90
8.41	0.66	531.52	19.62	146	94.98	-0.35	12.76	21.10
8.12	1.07	513.92	44.79	141	112.54	-1.20	8.49	10.00
7.83	2.67	496.32	142.25	136	165.18	-1.70	1.00	19.70
7.55	3.36	478.72	184.28	131	194.51	-2.00	4.35	18.20
7.26	0.08	461.12	-14.91	126	97.73	-2.45	4.41	16.50
6.97	0.13	443.52	-12.02	121	100.05	-3.00	5.03	14.70
6.68	-1.77	425.92	-127.56	116	40.24	-3.20	4.20	13.30
6.39	-0.79	408.32	-67.61	111	71.47	-4.00	-0.08	13.70
6.11	-1.04	390.72	-82.79	106	55.22	-6.45	0.64	14.05
5.82	-0.91	373.12	-74.62	101	51.99	-7.00	1.39	13.90
5.76	-1.58	355.52	-101.16	100	30.28	-9.70	-2.61	14.80
11.98	13.32	735.21	802.32	207.90	317.61	1.41	9.94	22.96
		268,350.25	292,847.28					

ADDITIONAL SIMULATION RESULTS

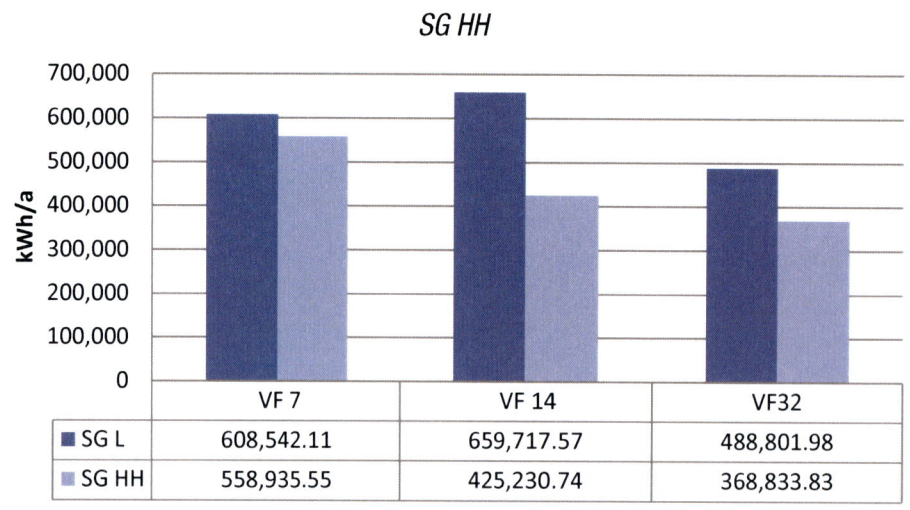

SG HH

	VF 7	VF 14	VF32
SG L	608,542.11	659,717.57	488,801.98
SG HH	558,935.55	425,230.74	368,833.83

Tab.27. All VF - Total Heating Demand for SG and HH-Schedule (kWh/a)

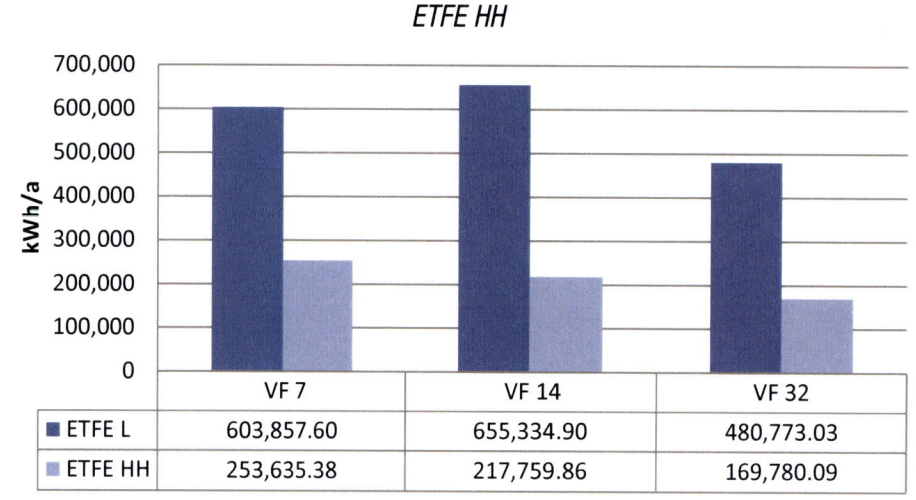

ETFE HH

	VF 7	VF 14	VF 32
ETFE L	603,857.60	655,334.90	480,773.03
ETFE HH	253,635.38	217,759.86	169,780.09

Tab.28. All VF - Total Heating Demand for ETFE and HH-Schedule (kWh/a)

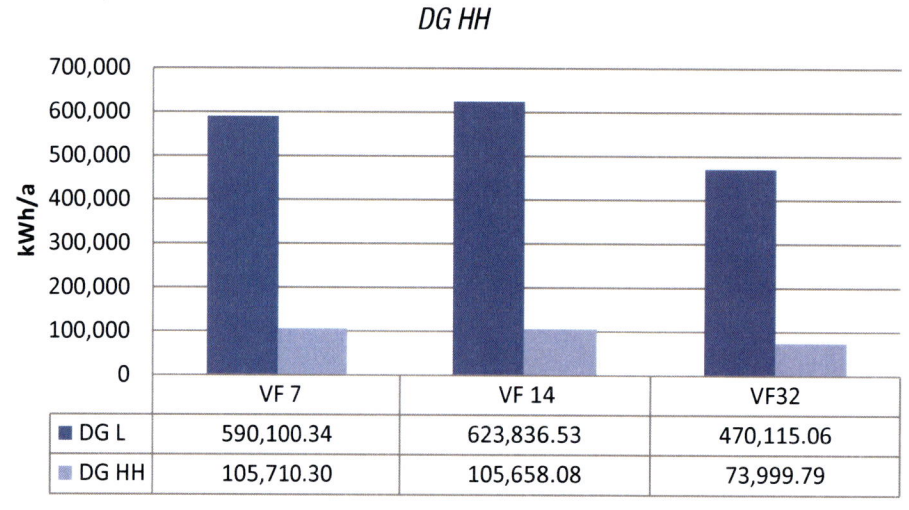

DG HH

	VF 7	VF 14	VF32
DG L	590,100.34	623,836.53	470,115.06
DG HH	105,710.30	105,658.08	73,999.79

Tab.29. All VF - Total Heating Demand for DG and HH-Schedule (kWh/a)

APPENDIX

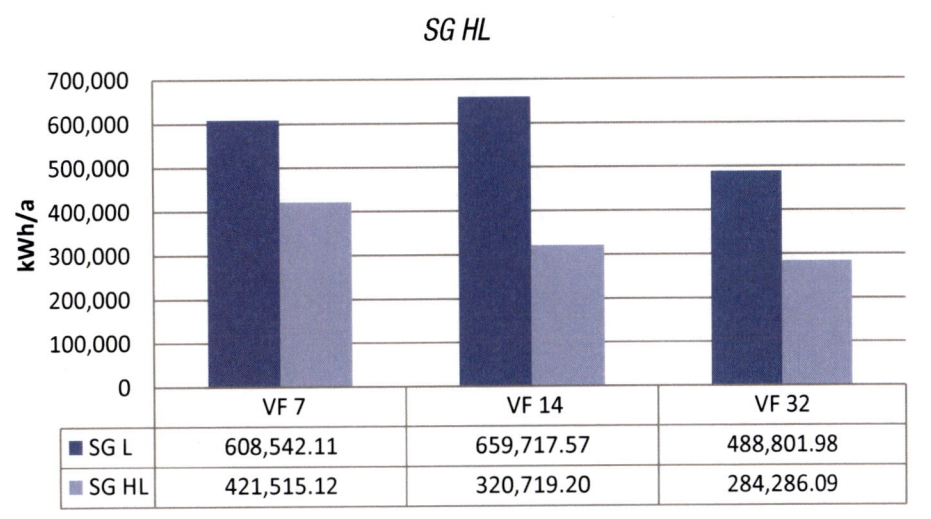

SG HL

	VF 7	VF 14	VF 32
SG L	608,542.11	659,717.57	488,801.98
SG HL	421,515.12	320,719.20	284,286.09

Tab.30.   All VF - Total Heating Demand for SG and HL-Schedule (kWh/a)

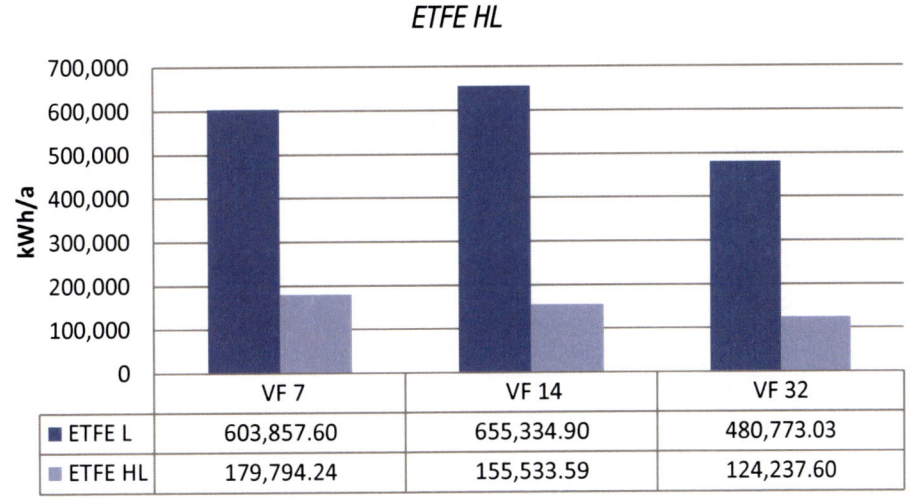

ETFE HL

	VF 7	VF 14	VF 32
ETFE L	603,857.60	655,334.90	480,773.03
ETFE HL	179,794.24	155,533.59	124,237.60

Tab.31.   All VF - Total Heating Demand for ETFE and HL-Schedule (kWh/a)

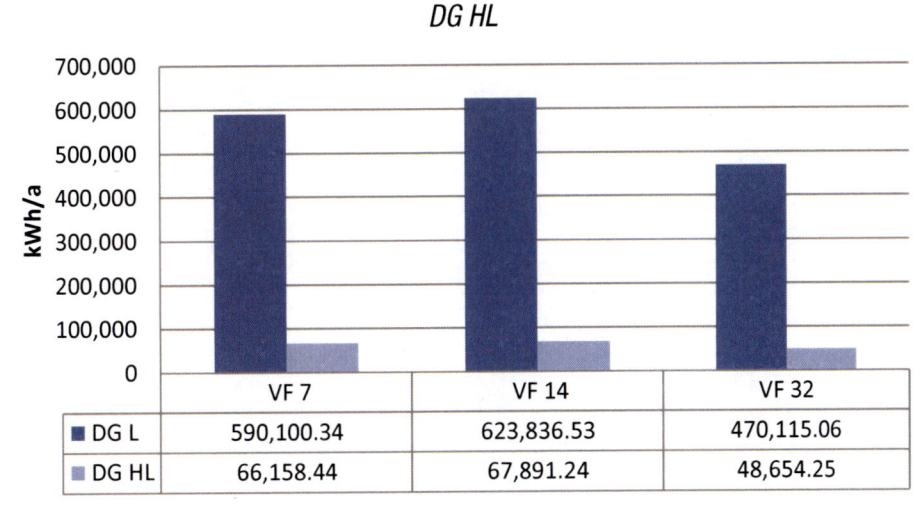

DG HL

	VF 7	VF 14	VF 32
DG L	590,100.34	623,836.53	470,115.06
DG HL	66,158.44	67,891.24	48,654.25

Tab.32.   All VF - Total Heating Demand for DG and HL-Schedule (kWh/a)

APPENDIX

ZONE	VF 7					
	SG		ETFE		DG	
	HH	HL	HH	HL	HH	HL
16	87,548.31	70,220.67	42,423.67	33,128.08	19,155.98	14,196.09
15	26,304.31	19,435.66	12,101.95	8,311.14	5,050.25	2,980.13
14	26,835.60	19,860.14	12,250.49	8,427.64	5,088.25	3,009.08
13	27,245.10	20,187.00	12,364.22	8,515.93	5,116.18	3,029.52
12	27,600.84	20,472.36	12,472.82	8,601.31	5,150.08	3,055.01
11	27,921.33	20,729.75	12,575.35	8,681.82	5,184.61	3,080.87
10	28,208.73	20,960.96	12,668.25	8,754.91	5,216.48	3,104.87
9	28,470.95	21,172.53	12,754.46	8,822.83	5,246.61	3,127.61
8	28,712.68	21,367.77	12,834.73	8,886.25	5,274.96	3,149.11
7	28,938.97	21,550.80	12,910.70	8,946.44	5,302.18	3,169.79
6	29,159.80	21,729.72	12,987.23	9,007.06	5,330.82	3,191.51
5	29,364.47	21,896.05	13,058.05	9,063.41	5,357.17	3,211.55
4	29,564.37	22,058.66	13,128.64	9,119.51	5,384.08	3,231.97
3	29,756.92	22,215.46	13,197.31	9,174.11	5,410.53	3,252.09
2	29,943.02	22,367.11	13,264.35	9,227.47	5,436.58	3,271.90
1	30,424.84	22,794.75	13,607.51	9,523.88	5,676.54	3,467.23
0	42,935.31	32,495.73	19,035.65	13,602.45	7,329.00	4,630.11
kWh/a	558,935.55	421,515.12	253,635.38	179,794.24	105,710.30	66,158.44
kWh/m²/a	126.85	95.66	57.56	40.80	23.99	15.01

Tab.33. VF7 - Total Heating Demand

ZONE	VF 32					
	SG		ETFE		DG	
	HH	HL	HH	HL	HH	HL
2	232,447.74	183,177.86	101,983.16	77,154.11	40,709.33	28,057.66
1C	1,187.11	317.54	1,301.19	355.60	1,233.36	331.67
1N	14,894.40	11,031.99	7,430.64	5,187.39	3,759.56	2,358.32
1E	12,263.16	9,031.57	6,216.66	4,322.38	3,115.85	1,941.13
1S	11,353.97	8,113.08	5,408.47	3,561.02	2,577.18	1,479.02
1W	12,566.24	9,313.14	6,480.59	4,543.68	3,363.78	2,137.51
0C	433.90	35.75	427.93	37.43	480.63	46.13
0N	25,642.36	19,642.93	12,720.39	9,266.82	6,183.87	4,187.69
0E	19,797.81	15,016.18	9,683.06	7,000.15	4,412.84	2,907.28
0S	17,843.22	13,110.25	8,000.61	5,462.39	3,363.37	2,023.43
0W	20,403.92	15,495.80	10,127.39	7,346.63	4,800.02	3,184.41
kWh/a	368,833.83	284,286.09	169,780.09	124,237.60	73,999.79	48,654.25
kWh/m²/a	106.72	82.26	49.13	35.95	21.41	14.08

Tab.34. VF32 - Total Heating Demand

VF 7			
ZONE	SG	ETFE	DG
16	25,209.66	25,122.89	24,596.39
15	36,691.57	36,401.79	35,563.60
14	36,691.57	36,401.79	35,563.60
13	36,691.57	36,401.79	35,563.60
12	36,691.57	36,401.79	35,563.60
11	36,691.57	36,401.79	35,563.60
10	36,691.57	36,401.79	35,563.60
9	36,691.57	36,401.79	35,563.60
8	36,691.57	36,401.79	35,563.60
7	36,691.57	36,401.79	35,563.60
6	36,691.57	36,401.79	35,563.60
5	36,691.57	36,401.79	35,563.60
4	36,691.57	36,401.79	35,563.60
3	36,691.57	36,401.79	35,563.60
2	36,691.57	36,401.79	35,563.60
1	36,691.57	36,401.79	35,563.60
0	32,958.90	32,707.86	32,049.95
kWh/a	608,542.11	603,857.60	590,100.34
kWh/m²/a	138.10	137.04	133.92

Tab.35. VF7 - Total Lighting Demand

VF 32			
ZONE	SG	ETFE	DG
2	112,447.30	112,050.43	109,579.39
1C	115,891.92	110,567.74	113,482.66
1N	26,766.76	26,459.97	25,683.69
1E	21,790.52	21,054.73	19,708.65
1S	21,884.41	21,709.11	20,383.24
1W	21,085.45	20,903.37	19,565.62
0C	91,105.87	90,540.74	86,369.14
0N	21,309.13	21,309.13	20,559.55
0E	18,319.57	18,087.47	17,571.50
0S	20,150.00	20,104.66	19,717.78
0W	18,051.05	17,985.68	17,493.84
kWh/a	488,801.98	480,773.03	470,115.06
kWh/m²/a	141.44	139.11	136.03

Tab.36. VF32 - Total Lighting Demand

ZONE	VF 14					
	SG		ETFE		DG	
	HH	HL	HH	HL	HH	HL
7.00	128,294.57	102,018.61	58,164.72	44,673.60	24,361.51	17,358.45
6C	1,252.66	624.98	1,256.96	627.12	1,254.07	623.17
6N	15,222.42	11,346.46	7,771.22	5,500.13	3,859.45	2,452.57
6E	6,379.09	4,871.80	3,633.29	2,667.44	1,795.22	1,223.28
6S	11,319.83	8,113.05	5,693.55	3,778.32	2,468.41	1,393.06
8W	6,485.83	4,967.42	3,630.30	2,674.21	1,911.17	1,316.88
5C	1,122.94	527.27	1,126.86	529.03	1,124.50	525.69
5N	14,955.21	11,098.03	7,569.33	5,313.07	3,694.76	2,311.20
5E	6,267.23	4,771.06	3,550.27	2,591.16	1,733.70	1,169.46
5S	10,959.47	7,809.68	5,427.83	3,563.04	2,292.93	1,261.09
5W	6,364.76	4,858.93	3,541.49	2,594.34	1,842.70	1,257.43
4C	1,116.40	522.66	1,120.30	524.41	1,117.97	521.10
4N	14,832.53	10,999.71	7,530.53	5,283.13	3,685.33	2,304.36
4E	6,205.20	4,720.00	3,527.20	2,572.46	1,725.27	1,162.98
4S	10,832.55	7,709.66	5,383.62	3,529.54	2,277.72	1,250.47
4W	6,305.77	4,810.55	3,520.23	2,577.17	1,834.87	1,251.28
3C	1,109.08	517.54	1,112.96	519.26	1,110.66	515.99
3N	14,697.56	10,892.37	7,490.84	5,253.00	3,678.53	2,300.02
3E	6,138.20	4,665.04	3,504.31	2,554.04	1,717.99	1,157.46
3S	10,686.24	7,594.59	5,333.58	3,491.71	2,260.91	1,238.76
3W	6,241.11	4,757.69	3,498.24	2,559.49	1,827.70	1,245.68
2C	1,100.67	511.59	1,104.53	513.28	1,102.27	510.08
2N	14,548.86	10,776.19	7,452.13	5,224.59	3,677.94	2,300.84
2E	6,056.40	4,598.07	3,476.20	2,531.55	1,709.08	1,150.74
2S	10,512.36	7,458.21	5,275.28	3,447.83	2,242.08	1,225.70
2W	6,165.88	4,696.42	3,473.64	2,539.85	1,820.52	1,240.11
1C	1,091.02	504.67	1,094.83	506.33	1,092.62	503.20
1N	14,369.62	10,636.45	7,413.26	5,196.19	3,685.38	2,307.98
1E	5,956.00	4,515.71	3,443.28	2,505.19	1,699.83	1,143.83
1S	10,299.93	7,292.43	5,209.58	3,398.69	2,224.49	1,213.63
1W	6,076.17	4,623.34	3,446.87	2,518.34	1,814.66	1,235.45
0C	1,502.57	676.50	1,281.73	577.16	1,501.35	672.15
0N	24,321.67	18,631.65	12,555.99	9,209.84	6,045.62	4,113.62
0E	9,800.20	7,630.78	5,650.22	4,313.48	3,001.03	2,165.58
0S	16,593.01	12,122.67	7,819.83	5,339.07	3,269.94	1,948.49
0W	10,047.73	7,847.42	5,674.86	4,336.53	3,195.90	2,319.46
kWh/a	425,230.74	320,719.20	217,759.86	155,533.59	105,658.08	67,891.24
kWh/m²/a	102.53	77.33	52.51	37.50	25.48	16.37

Tab.37. VF14 - Total Heating Demand

APPENDIX

VF 14			
ZONE	SG	ETFE	DG
7.00	49,308.71	49,308.71	49,149.83
6C	21,956.82	21,737.25	21,407.90
6N	29,962.04	29,662.42	29,362.80
6E	7,317.47	7,270.38	6,425.49
6S	23,978.50	23,960.36	21,639.59
6W	7,262.77	7,197.68	6,361.24
5C	21,956.82	21,737.25	21,407.90
5N	29,962.04	29,662.42	29,362.80
5E	7,317.47	7,270.38	6,425.49
5S	23,978.50	23,960.36	21,639.59
5W	7,262.77	7,197.68	6,361.24
4C	21,956.82	21,737.25	21,407.90
4N	29,962.04	29,662.42	29,362.80
4E	7,317.47	7,270.38	6,425.49
4S	23,978.50	23,960.36	21,639.59
4W	7,262.77	7,197.68	6,361.24
3C	21,956.82	21,737.25	21,407.90
3N	29,962.04	29,662.42	29,362.80
3E	7,317.47	7,270.38	6,425.49
3S	23,978.50	23,960.36	21,639.59
3W	7,262.77	7,197.68	6,361.24
2C	21,956.82	21,737.25	21,407.90
2N	29,962.04	29,662.42	29,362.80
2E	7,317.47	7,270.38	6,425.49
2S	23,978.50	23,960.36	21,639.59
2W	7,262.77	7,197.68	6,361.24
1C	21,956.82	21,737.25	21,407.90
1N	29,962.04	29,662.42	29,362.80
1E	7,317.47	7,270.38	6,425.49
1S	23,978.50	23,960.36	21,639.59
1W	7,262.77	7,197.68	6,361.24
0C	16,028.48	15,868.19	15,627.77
0N	21,872.29	21,653.57	21,434.84
0E	6,110.09	6,070.77	5,365.28
0S	17,504.31	17,491.06	15,796.90
0W	6,028.10	5,974.07	5,279.83
kWh/a	659,717.57	655,334.90	623,836.53
kWh/m²/a	159.08	158.02	150.42

Tab.38. VF14 - Total Lighting Demand

APPENDIX

VF 7

Zone	m²	SG LIGHTING	SG HL	SG HH	ETFE LIGHTING	ETFE HL	ETFE HH	DG LIGHTING	DG HH
16	259.20	97.26	270.91	337.76	96.92	127.81	163.67	94.89	73.90
15	259.20	141.56	74.98	101.48	140.44	32.06	46.69	137.21	19.48
0	259.20	127.16	125.37	165.65	126.19	52.48	73.44	123.65	28.28

	SG			ETFE			DG	
Σ LIGHT			608,547.11			603,857.64		590,100.34
Σ HH			556,939.55			259,635.53		105,710.50
Σ HL			421,555.12			179,744.21		66,198.48
LIGHT/m²			138.10			137.04		133.92
HH/m²			126.85			57.56		23.96
HL/m²			95.66			40.80		15.01

VF 14

Zone	m²	SG LIGHTING	SG HH	ETFE LIGHTING	ETFE HH	DG LIGHTING	DG HH
7	518.40	95.12	247.48	95.12	112.20	94.81	46.99
6C	114.40	191.93	10.95	190.01	10.99	187.13	10.96
6N	155.00	193.30	98.21	191.37	50.14	189.44	24.90
6E	47.00	155.69	135.73	154.69	77.30	136.71	38.20
6S	155.00	154.70	73.03	154.58	36.73	139.61	15.93
6W	47.00	154.53	138.00	153.14	77.24	135.35	40.66
0C	114.40	140.11	13.13	138.71	11.20	136.61	13.12
0N	155.00	141.11	156.91	139.70	81.01	138.29	39.00
0E	47.00	130.00	208.51	129.17	120.22	114.15	63.85
0S	155.00	112.93	107.05	112.85	50.45	101.92	21.10
0W	47.00	128.26	213.78	127.11	120.74	112.34	68.00

	SG		ETFE		DG	
Σ LIGHT		659,717.57		555,334.90		613,836.54
Σ HH		425,230.74		217,759.88		105,658.08
Σ HL		320,719.20		155,533.59		67,891.24
LIGHT/m²		159.08		158.02		150.42
HH/m²		102.53		52.51		25.48
HL/m²		77.33		37.50		16.37

VF 32

Zone	m²	SG LIGHTING	SG HH	ETFE LIGHTING	ETFE HH	DG LIGHTING	DG HH
2	1,152.00	97.61	201.78	97.27	88.53	95.12	35.34
1C	572.00	202.61	5.86	193.30	2.27	198.40	2.16
1N	155.00	172.69	86.25	170.71	47.94	165.70	24.26
1E	135.00	161.41	75.97	155.96	46.05	145.99	23.08
1S	155.00	141.19	80.42	140.06	34.89	131.50	16.63
1W	135.00	156.19	80.46	154.84	48.00	144.93	24.92
0C	572.00	159.28	0.76	158.29	0.75	151.00	0.84
0N	155.00	137.48	165.43	137.48	82.07	132.64	39.90
0E	135.00	135.70	146.65	133.98	71.73	130.16	32.69
0S	155.00	130.00	115.12	129.71	51.62	127.21	21.70
0W	135.00	133.71	151.14	133.23	75.02	129.58	35.56

	SG		ETFE		DG	
Σ LIGHT		483,801.98		480,773.03		470,115.08
Σ HH		368,833.83		169,780.09		73,999.73
Σ HL		284,286.09		124,237.60		48,654.73
LIGHT/m²		141.44		139.11		136.03
HH/m²		106.72		49.13		21.41
HL/m²		82.26		35.95		14.08

Tab.39. All VF - End Energy Demand for representative Zones

VF 7

Zone	m²	SG LIGHTING	SG HL	ETFE LIGHTING	ETFE HL	DG LIGHTING	DG HL
16	259.20	97.26	270.91	96.92	127.81	94.89	54.77
15	259.20	141.56	74.98	140.44	32.06	137.21	11.50
0	259.20	127.16	125.37	126.19	52.48	123.65	17.86

VF 14

Zone	m²	SG LIGHTING	SG HL	ETFE LIGHTING	ETFE HL	DG LIGHTING	DG HL
7	518.40	95.12	196.80	95.12	86.18	94.81	33.48
6	518.40	174.53	57.72	173.28	29.41	164.35	13.52
0	518.40	130.29	90.49	129.36	45.86	122.50	21.64

VF 32

Zone	m²	SG LIGHTING	SG HL	ETFE LIGHTING	ETFE HL	DG LIGHTING	DG HL
2	1,152.00	97.61	159.01	97.27	66.97	95.12	24.36
1	1,152.00	180.05	30.65	174.21	15.60	172.59	7.16
0	1,152.00	146.65	54.95	145.86	25.27	140.37	10.72

Tab.40. All VF - End Energy Demand for representative Levels

APPENDIX

VF 7

Zone	m²	SG			ETFE			DG		
		LIGHTING	HL		LIGHTING	HL		LIGHTING	HL	
16	259.20	97.26	270.91		96.92	127.81		94.89	54.77	
15	259.20	141.56	74.98		140.44	32.06		137.21	11.50	

Zone	m²	SG LIGHTING	HL	ETFE LIGHTING	HL	DG LIGHTING	HL
0	259.20	127.16	125.37	126.19	52.48	123.65	17.86

	SG	ETFE	DG
Σ LIGHT	608,542.11	603,857.60	590,100.34
Σ HH	558,985.55	553,695.38	505,710.38
Σ HL	421,515.12	179,794.24	66,158.44
LIGHT/m²	138.10	137.04	133.92
HH/m²	126.85	125.75	102.53
HL/m²	95.66	40.80	15.01

VF 14

Zone	m²	SG		ETFE		DG	
		LIGHTING	HL	LIGHTING	HL	LIGHTING	HL
7	518.40	95.12	196.80	95.12	86.18	94.81	33.48
6C	114.40	191.93	5.46	190.01	5.48	187.13	5.45
6N	155.00	193.30	73.20	191.37	35.48	189.44	15.82
6E	47.00	155.69	103.66	154.69	56.75	136.71	26.03
6S	155.00	154.70	52.34	154.58	24.38	139.61	8.99
6W	47.00	154.53	105.69	153.14	56.90	135.35	28.02
0C	114.40	140.11	5.91	138.71	5.05	136.61	5.88
0N	155.00	141.11	120.20	139.70	59.42	138.29	26.54
0E	47.00	130.00	162.36	129.17	91.78	114.15	46.08
0S	155.00	112.93	78.21	112.85	34.45	101.92	12.57
0W	47.00	128.26	166.97	127.11	92.27	112.34	49.35

	SG	ETFE	DG
Σ LIGHT	659,717.57	655,334.90	623,836.53
Σ HH	425,330.74	217,759.86	105,658.08
Σ HL	330,719.22	155,532.59	67,891.24
LIGHT/m²	159.08	158.02	150.42
HH/m²	102.53	52.51	25.48
HL/m²	77.33	37.50	16.37

VF 32

Zone	m²	SG		ETFE		DG	
		LIGHTING	HL	LIGHTING	HL	LIGHTING	HL
2	1,152.00	97.61	159.01	97.27	66.97	95.12	24.36
1C	572.00	202.61	1.57	193.30	0.62	198.40	0.58
1N	155.00	172.69	63.88	170.71	33.47	165.70	15.21
1E	135.00	161.41	55.95	155.96	32.02	145.99	14.38
1S	155.00	141.19	57.46	140.06	22.97	131.50	9.54
1W	135.00	156.19	59.63	154.84	33.66	144.93	15.83
0C	572.00	159.28	0.06	158.29	0.07	151.00	0.08
0N	155.00	137.48	126.73	137.48	59.79	132.64	27.02
0E	135.00	135.70	111.23	133.98	51.85	130.16	21.54
0S	155.00	130.00	84.58	129.71	35.24	127.21	13.05
0W	135.00	133.71	114.78	133.23	54.42	129.58	23.59

	SG	ETFE	DG
Σ LIGHT	488,801.98	480,773.03	470,115.06
Σ HH	368,833.83	169,780.09	73,999.79
Σ HL	284,286.09	124,237.60	48,654.25
LIGHT/m²	141.44	139.11	136.03
HH/m²	106.72	49.13	21.41
HL/m²	82.26	35.95	14.08

Tab.41. All VF - End Energy Demand for representative Zones

VF 7

Zone	m²	SG		ETFE		DG	
		LIGHTING	HH	LIGHTING	HH	LIGHTING	HH
16	259.20	97.26	337.76	96.92	163.67	94.89	73.90
15	259.20	141.56	101.48	140.44	46.69	137.21	19.48
0	259.20	127.16	165.65	126.19	73.44	123.65	28.28

VF 14

Zone	m²	SG		ETFE		DG	
		LIGHTING	HH	LIGHTING	HH	LIGHTING	HH
7	518.40	95.12	247.48	95.12	112.20	94.81	46.99
6	518.40	174.53	78.43	173.28	42.41	164.35	21.78
0	518.40	130.29	120.11	129.36	63.62	122.50	32.82

VF 32

Zone	m²	SG		ETFE		DG	
		LIGHTING	HH	LIGHTING	HH	LIGHTING	HH
2	1,152.00	97.61	201.78	97.27	88.53	95.12	35.34
1	1,152.00	180.05	43.67	174.21	23.30	172.59	12.20
0	1,152.00	146.65	73.02	145.86	35.56	140.37	16.70

Tab.42. All VF - End Energy Demand for representative Levels

IMPRESSUM

Daniel Podmirseg
Windmühlgasse 9/23
1060 Wien

daniel@podmirseg.com

www.paratufello.com

All graphics and tables created by
Daniel Podmirseg

Lectured by y'plus,Graz

Photos:
All photos retreived from the official website mentioned on each page of the
Vertical Farms selected for this work.
Tomato on p. 5: „Temptation". Retrieved: http://www.galaxytomatoes.co.nz/tomato-varieties

up!

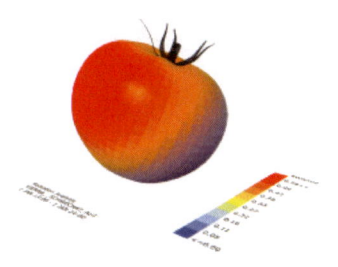

**Contribution of Vertical Farms to increase the overall
Energy Efficiency of Cities**

"How inappropriate to call this planet ‚Earth', when it is clearly 'Ocean'."

Arthur C. Clarke

DISSERTATION

up!

CONTRIBUTION OF VERTICAL FARMS TO INCREASE THE OVERALL
ENERGY EFFICIENCY OF CITIES

Academic advisor:
Univ.-Prof. B.Sc.(Hons). CEng MCIBSE Brian Cody
Institute for Buildings and Energy
Graz University of Technology

© 2015 - Copyright of all graphics: Daniel Podmirseg